Frontispiece (see p. 252): Llama singing with the emperor (Poma de Ayala 1936, p. 318).

THE WALKING LARDER

TITLES OF RELATED INTEREST

THE WALKING LARDER

Patterns of domestication, pastoralism, and predation

Edited by Juliet Clutton–Brock
Department of Zoology, British Museum (Natural History)

London
UNWIN HYMAN
Boston   Sydney   Wellington

Published by the Academic Division of
Unwin Hyman Ltd
15/17 Broadwick Street, London W1V 1FP, UK

Unwin Hyman Inc.
8 Winchester Place, Winchester, Mass. 01890, USA

Allen & Unwin (Australia) Ltd
8 Napier Street, North Sydney, NSW 2060, Australia

Allen & Unwin (New Zealand) Ltd in association with the
Port Nicholson Press Ltd
60 Cambridge Terrace, Wellington, New Zealand

First published in 1989

British Library Cataloguing in Publication Data

The walking larder: patterns of domestication, pastoralism, and
predation.
   1. Animals. Relationships with humans
   I. Clutton–Brock, Juliet   II. Series
   591.6
   ISBN 0–04–445013–3

Library of Congress Cataloging in Publication Data

The walking larder: patterns of domestication, pastoralism, and
predation/edited by Juliet Clutton–Brock.
     p.   cm.
Chiefly proceedings of the World Archaeological Congress held in
Sept. 1986 in Southampton, England.
Bibliography: p.
Includes index.
ISBN 0–04–445013–3 (alk. paper)
   1. Domestication–Congresses.   2. Human ecology–Congresses.
3. Animal remains (Archaeology)–Congresses.   I. Clutton–Brock,
Juliet.
II. World Archaeological Congress (1986: Southampton, Hampshire)
SF41.W35 1988   88–10127
636'.009–dc19; ff06 03–

Set in 10 on 11 point Bembo by Columns of Reading
Printed in Great Britain
at the University Press, Cambridge

List of contributors

Pekka Aikio, The Research Institute of Northern Finland, The University of Oulu, Finland.

Chow Ben-Shun, The Institute of Archaeology, Beijing, China.

Sandor Bökönyi, Archaeological Institute, Budapest, Hungary.

Sytze Bottema, Biologisch–Archaeologisch Instituut, Groningen, Netherlands.

Gordon Brotherston, Department of Literature, University of Essex, Colchester, UK.

David L. Browman, Department of Anthropology, Washington University, St. Louis, Missouri, USA.

Juliet Clutton–Brock, Department of Zoology, British Museum (Natural History), London, UK.

Sarah M. Colley, Department of Prehistory, The Research School of Pacific Studies, ANU, Canberra, ACT, Australia.

Richard G. Cooke, Smithsonian Tropical Research Institute, Miami, Florida, USA.

Iain Davidson, Department of Archaeology and Palaeoanthropology, University of New England, Armidale, NSW, Australia.

M. K. Dhavalikar, Department of Archaeology, Deccan College, Pune, India.

Pierre Ducos, Laboratoire d'Archéozoologie, Saint André de Cruzières, France.

Anne Eastham, Independent Researcher, Bolton, Lancashire, UK.

John G. Galaty, Department of Anthropology, McGill University, Montreal, Canada.

Valerius Geist, Faculty of Environmental Design, University of Calgary, Canada.

Colin P. Groves, Department of Prehistory and Anthropology, ANU, Canberra, ACT, Australia.

Frank Hole, Department of Anthropology, Yale University, New Haven, Connecticut, USA.

Alex Hooper, Independent Researcher, Truro, Cornwall, UK.

Rhys Jones, Department of Prehistory, The Research School of Pacific Studies, ANU, Canberra, ACT, Australia.

Tom McGreevy, Independent Researcher, Toronto, Canada.

Richard H. Meadow, Department of Zooarchaeology, Peabody Museum, Harvard University, Cambridge, Massachusetts, USA.

Mario A. Rabey, Faculdad de Humanidades y Ciencias Sociales, Universidad Nacional de Jujuy, San Salvador de Jujuy, Argentina.

Anthony J. Ranere, Department of Anthropology, Temple University, Philadelphia, Pennsylvania, USA.

Peter Robertshaw, British Institute in Eastern Africa, Nairobi, Kenya.

James Serpell, Sub-Department of Animal Behaviour, University of Cambridge, UK.

Valentin Pavlovich Shilov, Institute of Archaeology, Moscow, USSR.

Derek Sloan, Independent Researcher, Barry, S. Glamorgan, UK.

Dirk H. R. Spennemann, Department of Prehistory, The Research School of Pacific Studies, ANU, Canberra, ACT, Australia.

Yutaka Tani, Research Institute for Humanistic Studies, Kyoto University, Japan.

P. K. Thomas, Department of Archaeology, Deccan College, Pune, India.

Hans–Peter Uerpmann, Institut für Urgeschichte, Tübingen, West Germany.

Elizabeth S. Wing, Department of Zooarcheology, Florida State Museum, Gainesville, USA.

Juris Zarins, Department of Anthropology, Southwest Missouri State University, Springfield, Missouri, USA.

Foreword

This book is one of a major series of more than 20 volumes resulting from the World Archaeological Congress held in Southampton, England, in September 1986. The series reflects the enormous academic impact of the Congress, which was attended by 850 people from more than 70 countries, and attracted many additional contributions from others who were unable to attend in person.

The *One World Archaeology* series is the result of a determined and highly successful attempt to bring together for the first time not only archaeologists and anthropologists from many different parts of the world, as well as academics from a host of contingent disciplines, but also non-academics from a wide range of cultural backgrounds, who could lend their own expertise to the discussions at the Congress. Many of the latter, accustomed to being treated as the 'subjects' of archaeological and anthropological observation, had never before been admitted as equal participants in the discussion of their own (cultural) past or present, with their own particularly vital contribution to make towards global, cross-cultural understanding.

The Congress therefore really addressed world archaeology in its widest sense. Central to a world archaeological approach is the investigation not only of how people lived in the past but also of how, and why, changes took place resulting in the forms of society and culture which exist today. Contrary to popular belief, and the archaeology of some 20 years ago, world archaeology is much more than the mere recording of specific historical events, embracing as it does the study of social and cultural change in its entirety. All the books in the *One World Archaeology* series are the result of meetings and discussions which took place within a context that encouraged a feeling of self-criticism and humility in the participants about their own interpretations and concepts of the past. Many participants experienced a new self-awareness, as well as a degree of awe about past and present human endeavours, all of which is reflected in this unique series.

The Congress, was organized around major themes. Several of these themes were based on the discussion of full-length papers which had been circulated some months previously to all who had indicated a special interest in them. Other sessions, including some dealing with areas of specialization defined by period or geographical region, were based on oral addresses, or a combination of precirculated papers and lectures. In all cases, the entire sessions were recorded on cassette, and all contributors were presented with the recordings of the discussion of their papers. A major part of the thinking behind the Congress was that a meeting of many hundreds of participants that did not leave behind a published record of its academic discussions would be little more than an exercise in tourism.

Thus, from the very beginning of the detailed planning for the World Archaeological Congress in 1982, the intention was to produce post-Congress books containing a selection only of the contributions, revised in the light of discussions during the sessions themselves as well as during subsequent consultations with the academic editors appointed for each book. From the outset, contributors to the Congress knew that if their papers were selected for publication, they would have only a few months to revise them according to editorial specifications, and that they would become authors in an important academic volume scheduled to appear within a reasonable period following the Southampton meeting.

The publication of the series reflects the intense planning which took place before the Congress. Not only were all contributors aware of the subsequent production schedules, but also session organizers were already planning their books before and during the Congress. The editors were entitled to commission additional chapters for their books when they felt that there were significant gaps in the coverage of a topic during the Congress, or where discussion at the Congress indicated a need for additional contributions.

One of the main themes of the Congress was devoted to 'Cultural Attitudes to Animals, including Birds, Fish and Invertebrates'. The theme was based on discussion of precirculated full-length papers, covering four and a half days, and was under the overall control of Dr Tim Ingold, Senior Lecturer in the Department of Social Anthropology, University of Manchester, and Mark Maltby, Research Fellow in the Faunal Remains Unit of the Department of Archaeology, University of Southampton. The choice of this topic for a major theme arose from a desire to explore, from an interdisciplinary perspective, the many facets of the varying relationships that have developed between humans and animals, as these are reflected by the historical diversity of cultural traditions.

Discussions during the Congress were grouped around four main headings, each of which has led to the publication of a book. The first, organized by Tim Ingold, was concerned with 'What is an Animal?' leading to the book of the same title. The second subtheme, on 'The Appropriation, Domination and Exploitation of Animals', lasted for over a day and a half and was under the control of Juliet Clutton–Brock, editor of this book. A day was devoted to discussion of the 'Semantics of Animal Symbolism' and the co-ordinator, Roy Willis, is also the editor of the resulting book on *Signifying animals: human meaning in the natural world*. Howard Morphy was in charge of the fourth subtheme on 'Learning from Art about the Cultural Relationships between Humans and Animals', and has edited the volume on *Animals into art*.

The overall theme took as its starting point the assumption that there is no *one* human attitude consistently maintained towards a particular species of animal, and that similar human sentiments have been attached to a huge variety of different animals at different times and in different places. It set out to investigate the similarities and differences in practices and beliefs connected with animals, including birds, fish and invertebrates, across both time and space.

Prior to this century, in the West, animal behaviour was usually portrayed and interpreted in terms of a contrast with human behaviour. Darwin was not alone in his frequent adoption of an anthropocentric perspective in formulating questions and in presenting hypotheses and interpretations. It has often been claimed that people of non-Western cultures generally view animals quite differently. Another aim of the Congress theme was to explore such contrasts and to suggest some of the factors underlying both anthropomorphic and anthropocentric perceptions of animals which are currently prevalent at least in Western society.

Ecological, psychological, cultural, and utilitarian considerations are all involved in peoples' attitudes to, and treatment of, other species. These factors were considered not only from a wide, interdisciplinary point of view but also, as befits a world archaeological context, especially in an historical perspective, giving due emphasis to their changes over time.

For example, in the West when those of us who live in towns and cities think of dogs and cats we usually think of them as companions, although dogs are also, in other contexts, considered essential for herding, guarding, and hunting other animals. In ancient Egypt, cats were often shown in artwork as pets, but they were possibly also used to hunt and catch birds. In many present-day cultures across the world people think of quite different animals, such as cattle and pigs, as friends or companions. On the other hand, the hyena is normally considered by the layman today to be wild and untrainable, yet an ancient Egyptian representation appears to show one being handled. Once we move beyond the normal level of trying to ascertain from any excavation simply what animals were eaten or used for transportation, we are bound to look again at the nature of the relationships and interactions between human groups and the animals in their environments. Another aim of this theme, therefore, was to investigate how different people think, and thought, about different classes of animals, to discover the principles of classification involved, and to show how these principles constituted logical systems of belief and action. The presence of so many Congress participants from the so-called Third and Fourth Worlds made it possible to embrace a truly cross-cultural perspective on these issues.

One point of interest lies in the investigation, on a world-wide basis, of the reasons why particular animals have been domesticated by humans – whether for food, such as meat or milk, or for other reasons, such as for ritual purposes.

Contributors to the theme on 'Cultural Attitudes to Animals' adopted a variety of perspectives for looking at the complex ways that past and present humans have interrelated with beings they classify as animals. Some of these perspectives were predominantly economic and ecological, others were symbolic, concerned with the classification of both the physical and the social environment, and still others were primarily philosophical or theological. All these different perspectives are required for a full interpretation of the artworks of the past, which in their representations of humans and animals reveal some of the foci and inspirations of cultural attitudes to animals.

In focusing on the nature of the varying relationships that can develop between humans and animals, one is led inevitably to the question: what actually is an animal or a human? By asking such a question, archaeologists and others are forced to become aware of their own individual and cultural preconceptions, and to pay attention to a set of problems concerning attitudes.

The main themes in *The walking larder* have been detailed in its editorial Introductions (pp. 1–3, 7–9, 115–18, 279–81). My aim in what follows is to examine a few of the points which have struck me personally as being of particular note or fascination.

In this book Juliet Clutton–Brock and her contributors explore the complex interactions which can exist between humans and mammals, birds, fish, and invertebrates, in any society, of whatever period, and in whatever part of the world, within a symbiosis which we often ignorantly assume to be a simple matter of economic contract and necessity. During the Congress these topics were discussed under four main headings, 'Early Husbandry and the Exploitation of Domestic Stock', 'Pastoralism', 'Hunting and Collecting', and 'The Interaction of Human and Animal Behaviour and the Control of Animals'. The unravelling of these complex symbioses from the past depends on a fascinating interplay of evidence from both archaeology and zoology which is often difficult to interpret. It is never easy to determine from subfossil animal bones alone whether or not they have changed in shape or size sufficiently to warrant the deduction that full domestication of the animal species had indeed occurred. Similarly, it is often difficult or impossible to determine from the size of settlement and the type of tools alone whether the economy of a population was dependent on domesticated animals or on hunting.

It is not only the technical and detailed problems of how to recognize animal domestication, and even how precisely the term should be defined and used, that are presented here. The book also brings to life much of the actual experience and meaning of the close relationships human beings have with other species. There is much in this book about humans and their capacity, or lack of capacity, to form relationships and to live in close proximity, and in mutual interdependence, with other living creatures. In this way *The walking larder* is linked to another of the *One World Archaeology* books, *What is an animal? The walking larder* makes clear what a *human* is. It is an organism which not only has an outstanding capacity to make use of other living organisms to provide itself with food, but also creates around itself a conceptual and social framework. The relationship between humans and other living species, as revealed over time by archaeology, seems rarely, if ever, to have been limited to the domain of killing and eating, without involving social identity or beliefs. This does not in any sense deny the economic importance of the domestication process, a process which used, until recently, to be considered so profound that it was termed a 'revolution' in human affairs, and which many still consider an apt description of the peak of the domestication process in the Neolithic (and see *Foraging and farming*, edited by D. R. Harris & G. C. Hillman).

In 1969 Geoffrey Dimbleby and I wrote (p. xx), in a review of the development of domestication studies, that 'domestication was a process extending over several thousand years and that it had its own special characteristics in different areas of the ancient world. Domestication did not, of course, happen only once but has recurred time and time again in different parts of the world and at different times. Domestication as a process still continues.' *The walking larder* explores this diversity of the past as well as documenting a range of social and cultural attitudes which cannot be disentangled from a purely economic domestication process. Perhaps most important of all, this book moves out of the past into the present, and demonstrates that, as in the past, the intimate cultural relationships between humans and other animals continue to be modified, interwoven with, and altered by the encroachment and innovations of different cultures.

P. J. Ucko
Southampton

Reference

Ucko, P. J. & G. W. Dimbleby 1969. Introduction: content and development of studies of domestication. In *The domestication and exploitation of plants and animals*, P. J. Ucko and G. W. Dimbleby (eds), xvii–xxi. London: Duckworth.

Contents

Preface

When Peter Jewell and I were invited by Peter Ucko in 1983 to organize one of the sessions of the Congress within the main theme on 'Cultural Attitudes to Animals' we had to avoid overlap with the themes of the Fourth Conference of the International Council for Archaeozoology held in London in 1982. We also had to avoid duplication with the Fifth Conference which was held in Bordeaux in the week following the Congress in Southampton. We decided to place the emphasis of the session on behavioural studies, and to try to bring together the past and the present in discussions that would be focused on concepts rather than on the results of specialist analyses. In devising a framework for the session we were helped by talks with Eitan Tchernov from the Department of Zoology at the Hebrew University, Jerusalem. Later Professor Tchernov resigned from the Congress over the banning of South African participants. Peter Jewell also resigned, as described so lucidly by Ucko (1987).

There were 50 contributions to the Congress session on 'The Appropriation, Domination, and Exploitation of Animals' and these provided a basis for wide-ranging discussions with three themes predominating: the process and meaning of domestication, studies of pastoralism today and herding practices of the past, and the relationships between predators and their prey.

All but three of the chapters in this book were first presented as contributions to the World Archaeological Congress. James Serpell had intended to come to the Congress but was unable to attend and so sent his chapter in at a later date. Serpell's contribution (Ch. 1) on pet-keeping is linked to *What is an animal?* (edited by T. Ingold) in its discussion on owner–pet relationships and human perception of animals. Many dog-owners would agree that their dog is 'a person without verbal language'. On the other hand, perhaps few archaeologists would agree with Serpell that the domestication of animals for food followed directly from early pet-keeping.

Pierre Ducos did not attend the Congress but was invited to write his chapter on defining domestication in response to that of Sandor Bökönyi, as the two authors have held a long-standing debate on this subject (Chs 2 & 3).

Sytze Bottema's chapter on the modern domestication of waterfowl was given to the book by special request of the editor because the observations it contains are of such particular relevance to discussions on the process of domestication (Ch. 4).

J. Galaty's Chapter 20 was presented to the Congress in the section on 'Indigenous meanings' within the subtheme 'Semantics of Animal Symbolism'.

I am deeply grateful to all the authors of this book who produced their revised manuscripts in a very short time after the World Archaeological Congress and who responded so readily to my troublesome demands for standardization. In particular, I should also like to thank Tim Ingold for his helpful advice over the planning of the session at the Congress and for chairing the section on 'Pastoralism'. Other chairpersons I wish to thank are Mark Maltby, Richard Meadow, and Elizabeth Wing. But, above all, I wish to express my gratitude to Peter Ucko for his enthusiastic, if rigorous, approach to the production of this book, and without whom there would have been no Congress.

My thanks are due to Kim Dennis–Bryan, Department of Zoology, British Museum (Natural History), for checking all the bibliographies, and to Caroline Jones, Department of Archaeology, University of Southampton, for her help with typing and the solving of innumerable small problems.

Juliet Clutton–Brock
London

Reference

Ucko, P. 1987. *Academic freedom and apartheid: the story of the World Archaeological Congress.* London: Duckworth.

Introduction

JULIET CLUTTON–BROCK

History is what you choose to believe. To my father archaeology was the study of antiquities, pottery, and flint implements, and I was told at school that I could not be an archaeologist because I had not learned ancient Greek. Forty years ago animals had little to do with archaeology, except in artistic representation, and the relationships between humans and animals, if thought about at all, were classed as anthropology or biology. To the non-biologist, wild animals were ferocious beasts to be shot on sight, nature was red in tooth and claw, and natural history was pursued with a butterfly net.

Archaeologists and biologists of my generation have mostly spent their lives attempting to dispel these beliefs. To me, archaeology is the history of humans in their environment, and the history of their progressive domination of a world that appears to be rapidly diminishing in size as we travel higher and faster around it. I look at the past through the debris of human settlement and the present through studies of human and animal behaviour, and I see that the master predator is indeed an animal with unique consciousness and capabilities, but an animal all the same.

Over the past 40 years great progress has been made in the understanding of human history and its interlocking with the animal world. Archaeology is no longer only the study of sherds and flints, and a whole new field of work, archaeozoology (or zooarchaeology), has become an essential part of the subject. This began with the simple identification of food remains on archaeological sites but it now encompasses almost all aspects of the physical life of humans and animals in the past, and reaches back to the evolution of the early hominids.

The chapters in this book cover the last 40 000 years but they are not intended to represent a sequence from the past to the present, nor are they to be read as a series of research topics in archaeozoology. The three sections (on domestication, pastoralism, and predation) may appear to be in reverse order to many readers who see the development of civilization as a progression from hunting to early agriculture to livestock husbandry; but the chapters have been arranged intentionally in this order to free them from this concept of progression. The aim of *The walking larder* is to present evidence for the manifold relationships that occur between humans and animals, and to demonstrate that such relationships continue. The hunting of some animals and the keeping of others as valued companions was as much a part of human nature 10 000 years ago as it is today. Just as the domestic dog has the same behavioural patterns as the wolf, so the modern human probably differs little in his or her genetically inherited behaviour from the earliest *Homo sapiens*. It is only the development of culture and the ensuing pressures of social systems that change.

This book is called *The walking larder* because the provision of food is by far the most important function of animals in all human societies, and while they are alive the animals are almost always on the move. Like wolves, humans roam over huge home ranges; they must have protein to eat but they are compulsive travellers who are always searching to widen their territories and find new resources. Wild animals are followed and hunted; domestic animals are driven along with their human owners as a store of meat on the hoof, indeed as *livestock*. There are, of course, sedentary populations of humans and animals, as are discussed in this book. Sloan (Ch. 27) describes the exploitation of shellfish in northern Europe during the Mesolithic, and suggests that the abundance of food available from this source enabled settlements to survive in one place for very long periods. A similar argument is put forward by Spennemann (Ch. 28) for prehistoric settlement on Tonga in the Pacific, where molluscs were the main source of protein. However, sooner or later such food supplies will dwindle and new resources have to be found.

Apart from the provision of food, the most important function of human–animal relationships is mutual companionship. The wolf was the first animal to be domesticated and it was probably not for its meat, but either as object of affection, as a helper in the hunt, or as a useful scavenger of human debris – most likely it was for all three reasons.

Besides the obvious ways in which humans differ from animals, they are unique in their ability to communicate on an interspecific level. This has meant that humans and animals can have much closer relationships than are normally found amongst different groups of vertebrates. A human family can enfold a dog and a cat into the home and the dog can have some sort of rapport with the cat through their mutual owner, but a wild dog would very rarely, if ever, be friendly with a wild cat. This ability to communicate is the basis for the phenomenon of taming and domestication which has been a nucleus of human cultures for at least the past 10 000 years. There can be no civilization without beasts of burden (even if these are mechanized as in many countries at the present day).

For as long as people have travelled around the world they have taken their dogs, livestock, pets, and parasites with them. This has resulted in continuous change in the assemblages of animals and plants in every land, as well as changes in the environment, to an extent that is only just beginning to be realized. One of the aims of this book is to highlight these changes and to emphasize their ancient beginnings. People sailing to Australia took their dogs with them at least 3000 years ago, this being the earliest radiocarbon date for the remains of dingo. Today, living dingoes are the descendants of these ancient dogs, which soon established themselves as feral populations that have had a determining effect on the marsupial fauna. Neolithic people moving around the Mediterranean region, 7000 years ago, may have exterminated the last of the endemic fauna of pigmy elephant and hippo on the islands. The sheep and goats they took with them became part of the stock from which our present breeds are descended, except for a few that became feral in the mountains where they have lived ever since. It used to be thought that the wild sheep and goats of Sardinia, Corsica, Crete, and Cyprus were relics of the ancient endemic wild fauna, but it is now realized (from the lack of fossil evidence) that these are feral populations of Neolithic origin. This topic is reviewed by Groves in Chapter 5.

The rabbit and the fallow deer were both introduced into Europe north of the Mediterranean by the Romans, and, of course, the horse was taken to the Americas by the Spanish. Cattle began to move south through Africa at least 5000 years ago. These are just a few examples of the movements of large mammals – innumerable others could be mentioned on every continent, but perhaps the greatest change that has been brought about for the sake of the walking larder has been the result of deforestation and burning of the land to provide grazing for livestock and fuel for cooking. The cutting of thorn scrub and brushwood for making corrals and enclosures for livestock has also contributed greatly to the destruction of vegetation in arid regions.

By studying, often in painstaking detail, the remains of animals and plants that have been found in association with human settlement on archaeological sites, the changes wrought in the world over the past 40 000 years can be traced in remarkable detail. It is a slow task but a very important one, and there may not be much time left, for the knowledge gained must be used to show politicians and economists that modern technology cannot cure all ills. To dig deeper and deeper wells in the Sahel will not halt the spread of desertification, indeed it will increase it by causing more land to be overgrazed and more land to be trampled around the wells by the ever-increasing numbers of people and cattle. It would be better to listen to the nomadic camel-herders whose ancestors have lived in equilibrium with the desert for thousands of years.

DOMESTICATION

Introduction to domestication

JULIET CLUTTON–BROCK

Human beings are the most highly social and gregarious of all the primates, and it may be postulated that their complicated patterns of social behaviour evolved from the necessity to provide food for a community or family group of early hominids that included helpless infants and aged relatives. The communal hunting of large ungulates for meat could have been part of the basis for the evolution of much human behaviour during the latter part of the Pleistocene. To progress from the hunting of wild animals to the herding of tame ones may seem a small step but it is, in fact, a very large one, because what separates hunting from herding is the concept of ownership of the animals.

The problem of what *is* domestication is of continuing concern to biologists and archaeologists. During the last century it was a subject that preoccupied Darwin, and the conclusions he came to about the origin of species by natural selection were in great part the result of his knowledge about the breeding of domestic animals. During this century the history of domestication has been much studied, notably by Hilzheimer, Zeuner, Herre, and Bökönyi, and yet the process of domestication is still little understood and we are still arguing about its definition.

In Chapter 2 of this volume Bökönyi presents his views on the meaning of domestication, and, because he takes up in debate the definition of Pierre Ducos from an article written in the 1970s, it was only right that Ducos should be invited to reply, which he does in Chapter 3. Bökönyi states in his chapter that his definition of domestication is very close to Clutton–Brock's, the only difference being that 'she explains it in a more colloquial style'.

My definition of a domesticated animal is 'one that has been bred in captivity for purposes of economic profit to a human community that maintains complete mastery over its breeding, organization of territory, and food supply'. Like Bökönyi and Ducos, I believe that domestication is both a cultural and a biological process and that it can only take place when tamed animals are incorporated into the social structure of the human group and become objects of ownership. The morphological changes that are produced in the animal follow after this initial integration.

The biological process of domestication may be seen as a form of evolution in which a breeding group of animals has been separated from its wild conspecifics by taming. These animals constitute a founder population that is changed over successive generations by both natural and artificial selection, and is in reproductive isolation. In wild populations reproductive isolation will often lead to the evolution of subspecies; in the domestic animal it leads to the development of breeds. Unlike a subspecies, a breed is not restricted by geographical locality but, like a subspecies, it has a uniform appearance that is heritable and distinguishes it from other populations.

Considering the large numbers of mammals that have been exploited by humans, it is, at first glance, surprising that so few species have been fully domesticated. On the other hand, considering the intensity of the relationship between domestic animals and humans, perhaps it is even more surprising how many species it has been achieved with and how diverse are the breeds. The answer lies in the behaviour of the species of animals that undergo domestication and in the inherent diversity of the genetic constitution of wild mammals. It is only those species that have inherited behavioural patterns that correspond to those of humans that can survive the process, as discussed by Bökönyi in Chapter 2. In Chapter 1, Serpell considers the closest of all conscious human–animal relationships, that of pet-keeping, and he summarizes its roles in past and present societies, from Palaeolithic hunters to modern societies.

Within recent years considerable attention has been given to attempts to domesticate new species of animals to increase the supply of meat and other resources, such as antler. Bottema, in Chapter 4, describes his observations on the breeding of waterfowl in captivity. The birds undergo a process of domestication and their morphological changes are described.

Chapters 5, 6, and 7 discuss some of the effects of the introduction of new animals, both wild and domestic, to previously closed ecosystems. Groves, in Chapter 5, describes the feral status of the sheep, goats, pigs, and cats on the Mediterranean islands, and discusses the ancient introductions of many other taxa of small mammals, carnivores, and deer. Davidson, in Chapter 6, discusses the thieving of newly introduced stock by hunter-gatherers, and he suggests that this activity, as carried out by the Australian Aborigines in the last century, could be postulated for the endemic Mesolithic people of the Mediterranean. This is a most interesting concept, and it could be an important means by which early domestic animals spread through the ancient world. There is sound documentary evidence for the stealing of livestock from European explorers in the Americas and southern Africa; Cornwallis Harris, for example, in his *The wild sports of southern Africa* published in 1839, described in a chapter headed 'Plundered by Bushmen Hordes, and left a wreck in the desert' how all the cattle of their expedition were stolen by Bushmen. It is of interest to note that these Bushmen, who had no domestic animals of their own, other than a few dogs, were adroit at driving cattle in any direction that they wished and at great speed.

Chapter 7, by Wing, describes the impact in the 16th century AD of the introduction of domesticates to Florida and Haiti, and this can be seen as a parallel to the introduction of animals to the Mediterranean, thousands of years earlier. One particular comment by Wing is noteworthy for its relevance, for she claims that 'goats became feral in the mountains of Jamaica at an early date', just as they did in the Mediterranean islands, the British Isles, and many other parts of the world.

Chapters 8–12 deal with the evidences for domestication. Meadow

(Ch. 8) discusses the ways in which the process of domestication can be deduced from the archaeological record, and he continues the discussion, as do Uerpmann and Hole, on the definitions of domestication. Meadow makes the important point that with domestication there is a change in focus on the part of the humans from the dead to the living animal and more especially to its progeny.

Uerpmann, in Chapter 9, reviews the terminology used for the period of transition from hunter-gathering to sedentism in western Asia. This chapter, however, has wider implications than the often contentious subject of terminology. Uerpmann contrives to explain the beginnings of domestication through the periods that immediately precede the Neolithic and discusses the interpretation of socio-economic systems based on shellfish, which are described further by Sloan in Chapter 27, and he also examines the economic structure of the so-called Mesolithic of southern France which had domestic sheep, as described further by Davidson in Chapter 6.

Hole, in Chapter 10, makes use of his long familiarity with archaeology in western Asia to review the beginnings of domestication in this core area. He discusses the structure and timing of the stages of the seasonal re-use of 'camps', intensive plant use, the replacement of hunted gazelle by herded sheep and goats, and the development of a full agricultural economy. Hole concludes by discussing the environmental distributions of the wild progenitors of domestic livestock and their herding behaviour which enabled them to be controlled.

With the exception of the last two chapters, all those in this section of The walking larder are about the definitions of domestication, its processes during the prehistoric period and at the present day, and the effects of the introduction of domestic animals on endemic faunas. Lastly, there are two chapters that discuss specific examples of early domestication. In Chapter 11 Chow Ben–Shun presents the evidence for the origins of the domestic horse in China, a subject that is very little known outside China. In Chapter 12 Thomas reviews the first domestication of animals in India and discusses the beginnings of milking and the use of cow dung, two products that have been of great importance in the history of India.

1 Pet-keeping and animal domestication: a reappraisal

JAMES SERPELL

Introduction

Just over a century ago, Charles Darwin's cousin, Francis Galton, proposed a new theory about the process of animal domestication (Galton 1883). Using a dossier of anecdotal material provided by 18th- and 19th-century explorers and naturalists, Galton noted that the habit of capturing and nurturing tame wild animals as pets was widespread among what he regarded as 'primitive' people. He then made the assumption that prehistoric hunting societies indulged in similar practices, and argued from this that the original domestication of animals arose as a natural consequence of mankind's pet-keeping tendencies. 'Savages may be brutal,' he said, 'but they are not on that account devoid of our taste for taming and caressing young animals.' It was this taste for taming and caressing pets, he believed, that led to the birth of animal husbandry and farming.

Although Galton's idea has been supported by more recent authorities (e.g. Sauer 1952, Reed 1954, Zeuner 1963, Scott 1968, Clutton–Brock 1981) – particularly in relation to the domestication of dogs and, to a lesser extent, pigs and poultry – it has lately become less fashionable as a general explanation for the origins of domestication. More to the point, perhaps, its existence has done little to stimulate anthropological research into the meaning or functional significance of pet-keeping in either Western or non-Western societies (Serpell 1986). Although this trend largely reflects an increasing emphasis on the ecological and environmental antecedents of domestication rather than on the processes by which it occurred (Harris 1978), it may also be due in part to long-standing prejudices about pets and people's reasons for keeping them. The present review examines some of these preconceptions, and reconsiders the general validity of Galton's theory in the light of recent research on the nature of the human–pet relationship.

A definition of pets

The Oxford English Dictionary (OED) defines a pet as: 'Any animal that is domesticated or tamed and kept as a favourite, or treated with indulgence and fondness.' In practice, however, the word tends to be used more loosely as a blanket description for animals that are kept for no obvious practical or economic purpose – i.e. pets, as opposed to livestock or working animals. This has led to a certain amount of confusion in the literature since, clearly, there are a variety of different reasons for keeping animals that have no direct economic significance. Animals, for example, are kept for symbolic purposes; they are used to advertise status and prestige; they are employed as living adornments, and even as animated playthings or toys. The word 'pet' has been applied in each case (Tuan 1984). Yet an important difference exists between these animals and 'pets' as defined by the OED. The former tend to be viewed and treated essentially as objects or things, whereas the latter are generally perceived as subjects or quasi-persons – hence the tendency to indulge and fondle them. Recently, the term 'companion animal' has been widely adopted to distinguish true pets from other kinds of non–utilitarian animal (see Katcher & Beck 1983), but the present review will continue to use the original and less cumbersome term according to its dictionary definition.

The extent of the pet-keeping phenomenon

The keeping of dogs, cats, budgerigars, and other species as household pets is so widespread in Western societies that it tends to be taken for granted. Roughly half of the households in Britain contain at least one pet animal, and per capita pet-ownership is considerably higher in some other countries, such as France and the United States (Serpell 1986). More often than not, this level of enthusiasm for non–utilitarian animals is regarded as a peculiarly Western (and essentially middle-class) expression of material affluence and bourgeois sentimentality. The prevalence of seemingly identical behaviour among subsistence hunters and horticulturalists does not, however, support this point of view.

Individual non-utilitarian animals are (or were until recently) treated with 'indulgence and fondness' by a considerable range of different societies around the world. Indeed, the only notable exception seems to be Africa, where pet-keeping is nowadays infrequent, but where it may have been more common in pre-colonial days (Speke 1863, Zeuner 1963). In Australia the Aborigines kept dingoes, wallabies, possums, bandicoots, rats, cassowaries, and even frogs as pets (Zeuner 1963, Meggit 1965), and in Southeast Asia indigenous groups kept dogs, cats, pigs, monkeys, and various birds (Evans 1937, Leach 1964, Harrison 1965, Cipriani 1966). The Polynesians and Micronesians favoured dogs, pigeons, parrots, fruit bats, lizards, and eels (Jesse 1866, Galton 1883, Luomala 1960), and the Indians of North America kept deer, moose, bison, racoons, wolves, dogs, bears, turkeys, hawks, crows, and a variety of other small wild mammals and birds (Galton 1883, Elmendorf & Kroeber 1960). Among the southern Amerindians, particularly those inhabiting the Amazon basin, the taming and keeping of wild animal pets was practically a minor industry. The English naturalist, Bates, recorded a list of 'twenty-two species of quadrupeds' which he found living tame in Indian settlements (Galton 1883), and later observers have specified dogs, cats, deer, tapir, peccaries, monkeys, sloths, opossums, foxes, coatis, margay, ocelot, jaguar, chickens, ducks, cormorants, parrots, and an extraordinary variety of small birds and rodents (Roth 1934, Wilbert 1972, Basso 1973, Fleming 1984, Hugh–Jones pers. comm.). One author seems to sum the situation up when he says that 'few, indeed, are the vertebrate animals which the Indians have not succeeded in taming' (Roth 1934).

In general, Western observers have nearly always evinced surprise at the pet-keeping activities of so-called 'indigenous' peoples. Not only was the scale of the phenomenon impressive, but also the intensity of feeling it seemed to evoke in its adherents. On a trip to South America during the 18th century, the Spanish explorers, Juan and Ulloa, were evidently astonished at the Amerindians' affection for pet birds. Such animals were kept about the house, and the Indians never ate them:

> . . .and even conceive such a fondness for them, that they will not sell them, much less kill them with their own hands. So that if a stranger who is obliged to pass the night in one of their cottages, offers ever so much for a fowl, they refuse to part with it, and he finds himself under the necessity of killing the fowl himself. At this his landlady shrieks, dissolves into tears, and wrings her hands, as if it had been an only son. (Juan & Ulloa 1760.)

The resemblance to 'an only son' was not, apparently, an exaggeration. In a recent (and rare) anthropological account of pet-keeping among the Brazilian Kalapalo Indians, Basso (1973) describes the owner–pet relationship as particularly interesting because its distinctive features: 'are also those which define the filiative relationship, or that between human parents and their children. Children and pets alike are ideally supposed to be fed, reared and kept protected within the confines of the house. Often pets are kept secluded like human adolescents "to make them more beautiful".' Although they may belong to species which are otherwise classified as edible, Kalapalo pets are never killed or eaten, and when they die they are often buried close to the house or hammock of the former owner. This manner of burial is one ordinarily reserved for infants that die at birth or before being named.

In modern Europe or North America, a woman would probably be charged with indecency if she attempted to suckle a puppy or kitten at her breast. Yet in many hunting and horticultural societies the suckling of young mammals is considered perfectly normal and natural. Sir John Richardson, one of Galton's informants, not only observed that the 'red races' of North America were fond of pets, such as bison calves, wolves, and other species, but also that it was not unusual for them 'to bring up young bears, the women giving them milk from their own breasts' (Galton 1883). Early visitors to Hawaii were similarly impressed by the inhabitants' fondness for dogs, which 'in spite of their stupidity were in high favour with the women who could not have nursed them with a more ridiculous affection if they had really been ladies of fashion in Europe' (Luomala 1960). Another noted that 'every woman has a pet animal; and mothers who are nursing their own offspring will suckle a puppy at the same time in a rivalry by no means in favour of the strength and number of their own progeny' (Jesse 1866). Evans (1937) made similar observations among the Semang Negritos of Malaysia, where young pigs and monkeys were commonly suckled by the women, and where he reported seeing a woman with 'a child at one breast and a monkey at the other.' He also noted that animals reared in this way were not killed.

Even in societies with relatively strict utilitarian attitudes to domestic animals, exceptions are sometimes made. Among the Inuit and the Indians of the Canadian arctic, for example, dogs are ordinarily regarded as working animals, and treated in a detached and often brutal way. Nevertheless, childless individuals and couples will occasionally adopt one particular puppy into the household and rear it as an indulged pet (Savishinsky 1983). Thereafter, these animals are not expected to engage in any kind of useful work.

Although women seem to be the main animal-tamers and pet-keepers, the practice is certainly not confined to women. According to Luomala (1960), Polynesians of all ages, sexes, and social ranks 'fondled, pampered and talked to their pets, named them, and grieved when death or other circumstances separated them.' Grief over the death of a pet dog was often expressed through tears, poetical eulogies, and ceremonial burial. Amerindian males sometimes travelled around accompanied by their pets. The 18th-century explorer Hearne encountered one particular Indian who had two tame moose as pets. When the man made a trip by canoe 'the moose always followed him along the bank of the river; and at night, or on any other occasions that the Indians landed, the young moose generally came and fondled on them [sic], as the most domestic animal would have done' (Galton 1883). Australian Aborigine males displayed comparable attachments. According to the Swedish explorer, Lumholtz (1884), the Aborigines reared dingo pups:

> . . .with greater care than they bestow on their own children. The dingo is an important member of the family; it sleeps in the huts and gets plenty to eat, not only of meat but also of fruit. Its master never strikes, but merely threatens it. He caresses it like a child, eats the fleas off it, and then kisses it on the snout.

It needs to be emphasized that fondness for pets in all societies is largely independent of the animals' contribution to the local or family economy. Lumholtz, for example, attributed the Aborigines' affection for dingoes to the fact that these animals were useful as hunting aids. More recently, Cipriani (1966) and Harrison (1965) used precisely the same argument to explain the excessive devotion to dogs exhibited by the Andamanese and the Dyaks of North Borneo. But, clearly, mere economic utility provides no guarantee of affection. The B'Mbuti pygmies of Zaïre, for instance, rely heavily on dogs as hunting partners, yet they have a reputation for treating them with extreme brutality (Singer 1968). Conversely, other cultures make little practical use of dogs, but nevertheless regard them with overt affection. Linton's (1936) observations on the North American

Comanche provide a good example. The Comanche economy was based largely on buffalo-hunting and raiding neighbouring tribes, both of which depended on horses and horsemanship. Yet the Comanche possessed a detached and strictly utilitarian attitude to horses. Comanche dogs, on the other hand, were of no practical or economic value whatsoever, and were kept purely as pets. But the average Comanche warrior regarded the loss of a dog as far more devastating than the loss of several horses, and most of them would spend hours discussing with fondness all the dogs they had ever owned.

As already indicated, many of the species kept as pets among indigenous societies were also regularly hunted and killed for food. But once adopted as pets, these same species were generally exempt from slaughter. Referring to the Guiana Indians, Roth (1934) is adamant in stating that the 'native will never eat the bird or animal he has himself tamed any more than the ordinary European will think of making a meal of his pet canary or tame rabbit.' Similar inhibitions also existed in cultures where the species involved was raised commercially as an item of food. In Hawaii, for instance, dogs were reared primarily as livestock, but pet dogs were rarely slaughtered or eaten, and never without vociferous complaints from the owner (Luomala 1960). Even when it was pointed out to them by Europeans, many of these peoples failed to acknowledge the potential economic value of their pets. When Sir John Richardson tried to buy pet bear cubs from the Indians, he noticed that 'in purchasing them there is always the unwillingness of the women and children to overcome, rather than any dispute about price' (Galton 1883). Similarly, Fleming (1984) found that the Caraja people of Brazil refused to sell some of their parrots regardless of how much he offered to pay for them. When he suggested that they train their pet cormorants to catch fish by fastening rings around their necks, they treated the whole idea as a joke: 'In conception, rather than in execution, this project amused them very much; it is clear that they thought of the birds always as guests, never as servants.'

In other words, attitudes or former attitudes to pets among subsistence hunter-gatherers and horticulturalists are not substantially different from those characteristic of Western societies. Pets are raised, suckled if necessary, and cherished like children. They are protected, named, and cared for during life and, after death, they are often mourned. Pet animals may, in addition, serve practical functions but they are not indulged for this reason. The mere idea of killing and eating them is typically greeted with horror.

The functions of pet-keeping

Anthropologists have devoted surprisingly little attention to the possible functions of pet-keeping, although attitudes to pets and other domestic animals have featured prominently in 'structuralist' discussions about the origins of dietary and sexual taboos. Some have argued, for example, that people avoid killing and eating their pets because the animal has been included in the social world of humans, and its consumption is therefore tantamount to cannibalism (Lévi–Strauss 1966, Sahlins 1976). Others have proposed variations on the same theme. Leach (1964) suggests that people tend to prohibit or taboo things which are difficult to classify, and that pet-eating is taboo because pets occupy an uncertain, ambiguous territory between humans and non-humans. The symbolic association between the act of eating a pet and the act of sexual intercourse between close relatives has also attracted attention. According to Tambiah (1969), people don't eat their pets because it would be metaphorically equivalent to committing incest. These authors make no attempt to explain why societies keep pets in the first place. They are solely concerned with people's reluctance to slaughter and devour such animals.

The best-known anthropological explanation for the function of pet-keeping in hunting societies stresses its educational value (Laughlin 1968). In many hunting cultures, hunters have been observed to catch or collect small live animals – insects, frogs, lizards, nestling birds, and rodents – which they later turn over to their children. These 'pets' serve as temporary playthings and, like most toys, tend to be badly treated and short-lived. Often they end up the objects of target practice or mutilation. According to the theory, children receiving such gifts, and having the opportunity to play with them, acquire valuable experience of animals and animal behaviour; experience which will help them to become more efficient and successful hunters in later life. The idea makes sense as far as it goes, but it confuses true pets, i.e. animals treated with indulgence and fondness, with pets as educational toys. The latter are ephemeral objects which may or may not serve as useful childhood instruction. The former are cherished companions and the subjects of strong emotional bonds (Serpell 1986). Indeed, from an educational standpoint, one could equally argue that treating potential prey species with indulgence and affection is actually counter-productive.

The widespread human practice of suckling young mammals has also recently been subjected to functional interpretation. According to Fildes (1986), rich women in Europe used to suckle puppies in order to relieve painful distension of the breasts both before, and in the early stages of, breast-feeding. Since this is evidently a common problem, it could also be used to account for the suckling of animals in the many other cultures in which this activity occurs. Fildes takes the argument even further by suggesting that the breast-feeding of animals could also be used to promote and extend lactation, prevent conception, and develop 'good' nipples in the latter stages of pregnancy. In view of current interest in the links between lactation and human fertility (Short 1984), these fascinating ideas would repay more detailed investigation but, again, they cannot and were not intended to be used to explain the functions of pet-keeping. For one thing, if an animal is being employed as a lactational aid, there is no functional reason why the owner should nurture or cherish it beyond weaning. Yet in most pet-keeping societies the animal is cared for assiduously until it dies of natural causes. Secondly, in many societies, such as the Kalapalo of Brazil, the favourite pets are birds, which cannot be suckled (Basso 1973).

Anthropological reluctance to speculate about the functions of pet-keeping may stem from misunderstandings about the nature of this activity, or its impact on pre-agricultural societies. According to the old-fashioned, ethnocentric view of cultural development, human societies evolved progressively upward towards an increasingly complex and advanced state of civilization. Hunter-gatherers and horticulturalists were seen as occupying the lowest rungs of this developmental ladder, and their lives were assumed to be correspondingly arduous and uncomfortable (Harris 1978). Conceived of in these terms, hunting economies allowed no room for non-productive activities such as pet-keeping and, for this reason, the practice was best ignored, explained away as aberrant, or accommodated within some form of contrived utilitarian hypothesis. Fortunately, such ideas can now be substantially revised.

Within the past 20 years, perceptions of hunting and gathering have changed dramatically. Research on modern hunter-gatherers (e.g. Lee 1969), and the work of palaeoarchaeologists and pathologists (Cohen & Armelagos 1984), has revealed that subsistence hunters often enjoy more leisure, and are generally healthier and better nourished than many agricultural populations. Indeed, according to this view, the Neolithic switch from hunting to farming was not a cultural advance but rather the inevitable outcome of the need to intensify food production in the face of population pressure and diminishing resources (Boserup 1965, Cohen 1977, Harris 1978). In other words, hunters and horticulturalists may be relatively affluent in economic terms, and they are probably less restricted about keeping pets than intensive agriculturalists.

Perceptions of pets have also altered in recent years. Modern theories about the functions of pet-keeping rest primarily on the assumption that humans have social as well as material needs, and that social interactions with pets are able to fulfil at least some of these affiliative requirements (see Serpell 1986). Loneliness and social deprivation are now known to produce deleterious effects on human mental and physical health (Schachter 1959, Bowlby 1980, Perlman & Peplau 1981), and evidence has also accumulated that involvement in positive social relationships can help to buffer people from the damaging influence of negative life-events and stress (Lynch 1977, Duck 1983). Analogous psychophysiological processes have been demonstrated in other social species, and they also appear to operate in relationships between species (Nerem et al. 1980, Gross & Siegel 1982). It has recently been shown that social interactions with pets can produce measurable reductions in heart-rate and blood-pressure in artificially stressed human subjects (Katcher 1981, Friedmann et al. 1983). Further research is needed on the potential social and emotional benefits of pet-ownership, but existing results suggest that people use pets mainly to complement and augment their social relationships, and so enhance their own psychological and physical welfare. Pet-keeping is therefore functional in a broad sense, although one cannot easily evaluate its function in economic terms (Serpell 1986). This hypothesis does not, of course, maintain that pet-keeping is universally beneficial, since, like any activity, the net benefits must be weighed against the costs. It does, however, predict that, where adequate time and resources are available, pet-keeping will occur as a beneficial outcome of human social proclivities.

Pet-keeping and domestication

Although there are good theoretical grounds for arguing that Palaeolithic hunters and incipient agriculturalists reared and nurtured wild animals as pets, there are a number of objections to Galton's theory that pet-keeping led to domestication. The fact that hunters and horticulturalists exhibit moral inhibitions about killing and eating their pets poses a problem, but it is one which many societies have found ways of overcoming. Among the Indians of California, for example, rules existed which required people to disown pet deer and return them to the wild once they reached adulthood (Elmendorf & Kroeber 1960). Presumably, these animals then became fair game without arousing serious ethical conflict. Among the Ainu of Japan, bear cubs were suckled and reared by women and brought up as members of the family before, eventually, being sacrificed and consumed. The Ainu came to terms with the moral contradictions inherent in this relationship by means of self-justifying myths. According to Ainu legend, the bear was a temporary visitor from the spirit world whose ultimate objective in life was to return there. The Ainu believed that they were helping the animal to perform this transmigration by killing it (Campbell 1984). Similarly, the Tungus of Siberia shifted responsibility for the slaughter of pet reindeer by evoking malevolent and capricious supernatural powers who required the animals' death as a token of appeasement (Ingold 1980). As already suggested, the end of the Palaeolithic was probably associated with population growth in certain regions and the depletion of natural food resources. Faced with such pressures, it is likely that some Palaeolithic pet-owners were obliged to convert their pets into livestock by inventing similar methods of overcoming their scruples about exploiting them for food.

Downs (1960) disputed Galton's idea on the grounds that, while pet-keeping is widespread in hunting societies, the phenomenon of domestication is strictly localized in its early stages. Under the circumstances, he argues, pet-keeping and animal domestication should have been coincident in all areas. Again, this objection can be overcome using modern theories about the causes of the Neolithic revolution. The postulated intensification of plant and animal exploitation which occurred at this time was the result of overpopulation and resource depletion in certain areas. Domestication was therefore localized because only some Palaeolithic cultures were subjected to the necessary pressure to change their mode of subsistence. Where these problems did not arise, hunting and gathering remained profitable, and there would have been no incentive to exploit pets more intensively. It should also be emphasized that not all wild animal species are equally amenable to life under domestication (Reed 1954, Clutton–

Brock 1981), and it may have been simpler for post-Neolithic peoples to have adopted the domestic animals of neighbouring groups as livestock or, indeed, as pets, rather than to have domesticated existing tame wild animals. Cultural transmission of this kind would account for the rapid and almost universal spread of species such as the dog and cat.

The failure of many wild animal pets to breed in captivity has also been proposed as a reason why pet-keeping could not have led directly to domestication. Zeuner (1963), for example, accepted that pet-keeping may have 'provided one of the bases on which domestication on an economic scale developed later on,' but he also states that the form of Palaeolithic man's economy 'prevented him from developing this relationship to full domestication.' The statement, however, is based on recent observations of marginal hunter-gatherers, such as the Australian Aborigines, where pets are often too undernourished and restricted to be able to breed. There seems to be little justification for assuming that Palaeolithic pets were similarly constrained and, unfortunately, no adequate information exists on the breeding success or failure of pets among less marginal hunting populations. Indeed, one could turn the argument on its head by asserting that the pets which were eventually domesticated were precisely the ones that managed to breed, despite the rigours of captive existence (Galton 1883). It is worth pointing out, in this respect, that most recently domesticated species – budgerigars, canaries, brown rats, house mice, hamsters, gerbils, etc. – were originally adopted as exotic pets but have since acquired economic uses, for example as laboratory animals, as a result of their ability to breed in captivity.

The apparent adoption of more sedentary lifestyles towards the end of the Palaeolithic may have promoted the captive breeding of pets, since species, such as dogs, pigs, and birds, which rear their young in stationary dens or nests may have been somewhat limited by the formerly nomadic habits of their hunting and gathering owners. The earliest remains of a domestic dog, for instance, have been recovered from a Natufian site in Israel where they were associated with some of the oldest known permanent or semi-permanent villages (Davis & Valla 1978). On the other hand, species with precocious young, such as ungulates, would be unlikely to be adversely affected by nomadism and, of course, there is no obvious reason to assume that Palaeolithic pet-owners would necessarily have abandoned puppies, piglets, or nestling birds whenever they moved.

Conclusions

Recent changes in our perceptions of both hunting and gathering and pet-keeping reinforce, rather than detract from, Galton's (1883) hypothesis. On the basis of present evidence, it is probable that pet-keeping was common, if not universal, among Palaeolithic hunters and incipient agriculturalists. Judging from the wealth of different species employed for this purpose by recent hunter-gatherers, it is also likely that all of our current domestic species, as well as many which were never domesticated, began their association with humans in this essentially non-economic role. Since domestication is invariably associated with some form of captive or controlled breeding, it is possible that the species we now classify as 'domestic' were simply those that bred most readily as pets within the hunter-gatherer milieu.

Pet-keeping, both in the industrial West and other societies, is perhaps best regarded as a leisure activity. But this need not imply that the practice is necessarily without function, any more than it could be said that play or other recreational pursuits serve no functional purpose. The majority of hunter-gatherers and horticulturalists appear to possess adequate time and resources to engage in leisure activities, and the fact that so many choose to invest these resources in pet-keeping suggests that its social and emotional rewards are far from negligible. As in the West, the role that pets seem to occupy in hunting societies is most often analogous to that of infants and young children; a fact which suggests that pets are used primarily as non-reproductive outlets for parental behaviour. One hesitates to speculate too far in this direction, but it is perhaps significant that fecundity is relatively low in hunter-gatherer populations, as it is in the industrial West (Short 1984). The popular belief that pets are simply 'child substitutes' is doubtless an exaggeration (see Serpell 1986) but, at the same time, it is difficult to ignore the almost universal similarities between people's attitudes to pets and their attitudes to children.

The decision to exploit pet animals as sources of food or labour may have been forced upon certain Palaeolithic groups by the necessities of survival in a world of increasing food shortages. In the absence of such economic and ecological pressures, there would have been little incentive to exploit pet animals more intensively or to have embarked on the relatively laborious task of maintaining and breeding them as captive, domestic populations. On its own, Galton's theory cannot be used to explain why animal domestication occurred when and where it did. It does, however, provide a plausible scenario for the development of more intensive systems of animal exploitation when and where such systems were required.

References

Basso, E. B. 1973. *The Kalapalo Indians of central Brazil*. New York: Holt, Rinehart & Winston.

Boserup, E. 1965. *The conditions of agricultural growth*. Chicago: Aldine Press.

Bowlby, J. 1980. *Loss, sadness and depression: attachment and loss*, Vol. 3. London: Hogarth Press.

Campbell, J. 1984. *The way of the animal powers*. London: Times Books.

Cipriani, L. 1966. *The Andaman islanders*. London: Weidenfeld & Nicolson.

Clutton–Brock, J. 1981. *Domesticated animals from early times*. London: British Museum (Natural History) & Heinemann.

Cohen, M. N. 1977. *The food crisis in prehistory*. New Haven: Yale University Press.

Cohen, M. N. & G. J. Armelagos 1984. Paleopathology at the origins of agriculture: editors' summation. In *Paleopathology at the origins of agriculture*, M. N. Cohen & G. J. Armelagos (eds), 585–601. New York: Academic Press.

Davis, S. J. M. & F. R. Valla 1978. Evidence for the domestication of the dog 12,000 years ago in the Natufian of Israel. *Nature* **276**, 608–10.

Downs, J. F. 1960. Domestication: an examination of the changing social relationships between man and animals. *Kroeber Anthropological Society Papers* **22**, 18–67.

Duck, S. 1983. *Friends for life*. Brighton: Harvester Press.

Elmendorf, W. W. & K. L. Kroeber 1960. *The structure of Twana culture with comparative notes on the structure of Yurok culture*. Washington University Research Studies, Monograph 2, 1–28.

Evans, I. H. N. 1937. *The Negritos of Malaysia*. Cambridge: Cambridge University Press.

Fildes, V. 1986. *Breasts, bottles and babies: a history of infant feeding*. Edinburgh: Edinburgh University Press.

Fleming, P. 1984. *Brazilian adventure*. London: Penguin.

Friedmann, E., A. H. Katcher, S. A. Thomas, J. J. Lynch & P. R. Messent 1983. Interaction and blood pressure: influence of animal companions. *Journal of Nervous and Mental Diseases* **171**, 461–5.

Galton, F. 1883. *Inquiry into human faculty and its development*. London: Macmillan.

Gross, W. B. & P. B. Siegel 1982. Socialization as a factor in resistance to infection, feeding efficiency, and response to antigen in chickens. *Journal of Veterinary Research* **43**, 2010–12.

Harris, M. 1978. *Cannibals and kings*. London: Collins.

Harrison, T. 1965. Three 'secret' communication systems among Borneo nomads (and their dogs). *Journal of the Malaysian Branch of the Royal Asiatic Society* **38**, 67–86.

Ingold, T. 1980. *Hunters, pastoralists and ranchers*. Cambridge: Cambridge University Press.

Jesse, G. R. 1866. *Researches into the history of the British dog*. London: Robert Hardwicke.

Juan, G. & A. Ulloa 1760. *Voyage to South America*, Vol. 1. London.

Katcher, A. H. 1981. Interactions between people and their pets: form and function. In *Interrelations between people and pets*, B. Fogle (ed.), 41–67. Illinois: Charles C. Thomas.

Katcher, A. H. & A. M. Beck (eds) 1983. *New perspectives on our lives with companion animals*. Philadelphia: Pennsylvania University Press.

Laughlin, W. S. 1968. Hunting: an integrating biobehavior system and its evolutionary importance. In *Man the hunter*, R. B. Lee & I. DeVore (eds), 3–12. Chicago: Aldine Press.

Leach, E. 1964. Anthropological aspects of language: animal categories and verbal abuse. In *New directions in the study of language*, E. H. Lenneberg (ed.), 170–9. Cambridge, Mass.: MIT Press.

Lee, R. B. 1969. !Kung Bushmen subsistence: an input–output analysis. In *Environment and cultural behavior*, A. Vayda (ed.), 47–79. Garden City, New York: Natural History Press.

Lévi–Strauss, C. 1966. *The savage mind*. Chicago: Chicago University Press.

Linton, R. 1936. *The study of man: an introduction*. New York: Appleton–Century–Crofts.

Lumholtz, C. 1884. *Among cannibals*. London: John Murray.

Luomala, K. 1960. The native dog in the Polynesian system of values. In *Culture in history*, S. Diamond (ed.), 190–240. New York: Columbia University Press.

Lynch, J. J. 1977. *The broken heart: the medical consequences of loneliness.* New York: Basic Books.

Meggit, M. J. 1965. The association between Australian Aborigines and dingoes. In *Man, culture and animals*, A. Leeds & A. Vayda (eds), 7–26. Washington DC: American Association for the Advancement of Science.

Nerem, R. M., M. J. Levesque & J. F. Cornhill 1980. Social environment as a factor in diet-induced atherosclerosis. *Science* **208**, 1475–6.

Perlman, D. & L. A. Peplau 1981. Toward a social psychology of loneliness. In *Personal relationships in disorder: personal relationships*, Vol. 3, S. Duck & R. Gilmour (eds), 31–56. New York: Academic Press.

Reed, C. A. 1954. Animal domestication in the prehistoric Near East. *Science* **130**, 1629–39.

Roth, W. E. 1934. *An introductory study of the arts, crafts and customs of the Guiana Indians.* 38th Annual Report of the Bureau of American Ethnology, 25–745.

Sahlins, M. 1976. *Culture and practical reason.* Chicago: Chicago University Press.

Sauer, C. O. 1952. *Agricultural origins and dispersals.* Cambridge, Mass.: MIT Press.

Savishinsky, J. 1983. Pet ideas: the domestication of animals, human behaviour and human emotions. In *New perspectives on our lives with companion animals*, A. H. Katcher & A. M. Beck (eds), 112–31. Philadelphia: Pennsylvania University Press.

Schachter, S. 1959. *The psychology of affiliation.* London: Tavistock.

Scott, J. P. 1968. Evolution and domestication of the dog. *Evolutionary Biology* **2**, 243–75.

Serpell, J. A. 1986. *In the company of animals.* Oxford: Basil Blackwell.

Short, R. V. 1984. Breast-feeding. *Scientific American* **250**, 35–41.

Singer, M. 1968. Pygmies and their dogs: a note on culturally constituted defence mechanisms. *Ethos* **6**, 270–9.

Speke, J. H. 1863. *Journal of the discovery of the source of the Nile.* London: W. Blackwood.

Tambiah, S. J. 1969. Animals are good to think and good to prohibit. *Ethnology* **8**, 452–3.

Tuan, Yi–Fu. 1984. *Dominance and affection: the making of pets.* New Haven: Yale University Press.

Wilbert, J. 1972. *Survivors of Eldorado: four Indian cultures of South America.* New York: Praeger.

Zeuner, F. E. 1963. *A history of domesticated animals.* London: Hutchinson.

2 Definitions of animal domestication

SANDOR BÖKÖNYI

Attempts to define the term 'domestication' began more than 100 years ago when research on the nature and origin of domestic animals began. The stages in the search for a satisfactory definition can be traced rather easily, for they have always been connected with the development of the natural and social sciences. However, this chapter is not concerned with history, but with my own definitions of domestication. My first published definition in 1969 (p. 219) was later extended in 1985 (p. 571) and was based on earlier discussions on domestication published in German by Keller (1902), Klatt (1927), Röhrs (1961–62), and others on the one hand and on my own experience on the other. My definition is as follows: 'The essence of domestication is the capture and taming by man of animals of a species with particular behavioural characteristics, their removal from their natural living area and breeding community, and their maintenance under controlled breeding conditions for mutual benefits.'

Since the 1970s, several authors have discussed this definition although Ducos is the only one who rejected it, on the basis that it contained '*a priori* propositions on the causes, mechanism and consequences of domestication' (Ducos 1978, p. 54). Nevertheless, it seems to me that Ducos himself has confused the process of domestication (which I am defining) with the owning of fully domesticated animals. Furthermore, he has failed to cover the keeping of newly domesticated livestock which are not isolated from their wild forms and are not under human control, say in corrals. In fact, domestic cattle, pigs, sheep, and goats are kept free in this way in western Africa today, as Ducos claims, but only because their wild forms do not live there and because they are at a high level of domestication. Ducos's own definition of domestication will be discussed later.

Another author, Hecker, is dissatisfied with the term 'domestication' as a whole and suggests the use of the term 'cultural control' instead of it. However, when the main components of this cultural control are summarized (Hecker 1982, p. 219) it emerges that they are more or less identical with the main points of my definition. In his definition, cultural control does not, however, necessarily mean domestication; wild animals can be culturally controlled without being domesticated. In this respect Clutton–Brock's definition of domestication is very close to mine (Clutton–Brock 1981, see p. 7 this volume), the only difference is that she explains it in a more colloquial style because of the popular nature of her book.

Let us approach the term 'domestication' again without prejudice or preconception. Surely, there is agreement that domestication is a highly developed man–animal relationship that emerged in a rather late phase of mankind's history. It certainly had Palaeolithic and Neolithic antecedents in the form of the isolated taming and keeping of dogs and pigs, two animal species which could live on the remnants and debris of human food and did not need large quantities of vegetable fodder (Turnbull & Reed 1974, p. 100, Bökönyi 1978, 38ff., Nobis 1979, p. 610, 1984, p. 74, Altuna & Mariezkurrena 1985, 11ff.). In this sense the first attempts at domestication date as far back as the Magdalenian, c. 16 000 BP (Altuna & Mariezkurrena 1985, p. 111). Nevertheless, since dog and pig ate the same foodstuffs as prehistoric man, they became competitors for food and this may have hindered their large-scale domestication. A complete and profitable animal husbandry, which essentially changed man's economy, in other words which made the switchover to food production possible, could not develop from these isolated attempts at domestication. Large-scale domestication and a more or less complete Neolithic domestic fauna started hand-in-hand with the beginning of cereal production, which provided the large amounts of rough fodder necessary for the caprovines, the leading species of the earliest animal husbandry at the advent of the Neolithic.

The basis of the first large-scale domestication – and for that matter the later ones as well, because the process did not cease to exist after the acquisition of the first wave of domestic animals – was specialized hunting. This certainly was a kind of cultural control, though not every specialized hunting economy necessarily led to domestication, as clearly demonstrated by the predominant gazelle-hunting of early Jericho (Clutton–Brock 1971, Legge 1972, p. 123) or the similar onager- and gazelle-hunting at Umm Dabaghiyah (Bökönyi 1973, p. 61). In fact, these cases also demonstrate the complicated nature of domestication and its connection with the behavioural characteristics of animal species. Hediger stated more than 40 years ago (in a neglected work – 1942, p. 160) that wild species which keep a large inter-individual distance are not suitable for domestication because they keep a large distance from man too. Gazelles (and onagers) surely belong to the solitary type (Clutton–Brock 1978, p. 50). However, the hunting of these animals was useful for man because in this way he could increase his biological knowledge, and he undoubtedly used it in Umm Dabaghiyah, keeping all five Neolithic domestic species, besides the hunting of herds of wild onagers and gazelles.

But returning to the extremely complex nature of domestication one has to turn to Ducos's formula for the definition of domestication. He admits that in his view, 'domestication must be defined with reference to human society'. Consequently, he suggests the following definition: 'Domestication can be said to exist when living animals are integrated as objects into the socio-economic organization of the human group, in the sense that, while living, those animals are objects for ownership, inheritance, exchange, trade, etc., as are the other objects (or persons) with which human groups have something to do.' (Ducos 1978, p. 54).

This definition is only partially correct, for it is over-simplified and one-sided. Domestication is the beginning of a symbiosis that needs at least two partners, and it is simplistic to view it from the side of one of the partners alone. It is indeed true that domestication is a special kind of symbiosis (see also Röhrs 1961–62, p. 8, Herre & Röhrs 1973, p. 9) in the sense that one of the partners, man, influences the other by isolating, taming, controlling, breeding, and taking animals into new habitats, etc., but the animal itself also plays an essential part in this process. As one can see from the examples of gazelle and onager cited earlier, only animals with particular behavioural characteristics can be domesticated, and this essential point is missing from Ducos's definition. It is important to realize that gazelle and onager are not the only examples, wall-paintings and reliefs suggest that the ancient Egyptians tried to domesticate whole series of animals, from hyenas to antelopes, without much success. These species certainly have some behavioural barrier that blocks their domestication. This fact cannot be omitted in a definition because the behavioural patterns of animals are of crucial importance, and through their study a lot of questions can be elucidated. Similarly, a satisfactory definition must express the fact that through domestication animals also influence man and society as well, though their influence is not as strong as that of man on them.

According to Ducos (1978, p. 54) 'living conditions are among the consequences of domestication, not the mark of it'. In my opinion this contradicts Ducos's original definition, because if domestication means integration of living animals as objects into the socio-economic group, one necessarily has to change the living conditions of the animals by isolating them, corralling them, etc. At the same time, items such as inheritance, exchange, trade, etc. are consequences, not components, of domestication.

In practice, the discussion about domestication, its antecedents and consequences, is a rather academic one, just as is the definition of cultural control, herding, and animal husbandry, because these factors overlap each other and clear-cut borderlines cannot be determined. It is the same with the morphology of the bones of animals, and most probably it was the same with the appearance and way of life of the living animals themselves in the early prehistoric period.

Although it is statistically possible to determine the presence of already domesticated animals in a given assemblage of faunal remains when the sample is very large, in the early phase of domestication, developed from specialized hunting, one cannot determine in every case whether a particular bone represents a domestic or a wild animal. In other words, there are no exact boundaries between specialized hunting and animal domestication, and between wild and domestic forms, except in the highly developed phases of domestication (animal husbandry) and hunting without local domestication. An example of local domestication providing a large number of indigenous domestic individuals, which makes possible the definitive distinction between wild and domestic cattle populations, can be seen in the late Neolithic sites of Hungary (Bökönyi 1974, p. 112ff). On the other hand, in Switzerland, for example at the Neolithic site of Seeberg, Burgäschisee–Süd where it has been claimed there was no evidence for local domestication, it was much easier to distinguish the wild *Bos primigenius* bones from the bones of domestic cattle (Boessneck *et al.* 1963, p. 160ff.).

Thus domestication can be seen as a gradual and dynamic, though not always irreversible, process. For example, the 'wild' rabbits of Europe east of Spain and north of southern France are in fact feral animals that escaped from captivity in ancient times and quickly returned to the wild state. Since domestication is a complex interaction between man and animal, its consequences are influenced by society, economy, ideology, environment, way of life, etc. Any successful definition of domestication must reflect all these possible aspects of the process. The result of domestication is the domesticated animal that first culturally and later morphologically differs from its wild form.

Another question centres on the morphological changes caused by domestication. Darwin was the first to deal with these in detail (Darwin 1868). In respect of these changes there are two main questions: first, which are the main types of such changes? Secondly, how quickly do they appear after domestication? Three selected changes will be considered here: (a) size decrease, (b) crowded teeth, and (c) the hornless skull.

Regarding size decrease, it is undoubtedly true that under certain natural conditions, e.g. isolated populations on islands, the size of the animals can decrease. With animal remains from prehistoric sites, however, decrease of size can only indicate domestication. As for crowded teeth, which are common in pigs and dogs, rare in cattle, and which I have seen in only one single horse skull, this is a reasonably sure proof of domestication. In wild ruminants, such as horses and pigs, it never occurs, but there are wolf skulls in which crowded teeth are found (Musil 1974, p. 49), although only in modern wolves, especially those from zoos. In fossil wolf skulls crowded teeth have only been seen in specimens later in date than the Gravettian (Upper Palaeolithic), a fact that suggests early attempts at wolf domestication. Hornless skulls are never found in wild cattle or goats, but they do appear as a rare occurrence in populations of wild sheep. However, it is not clear whether these are truly wild sheep or whether they have interbred with feral or domestic sheep at some stage in the history of the population. In general, hornlessness is a reliable indication of domestication when found amongst animal bones from an archaeological site.

Morphological changes do not appear quickly after domestication. Recent experiments show that measurable changes need about 30 generations after domestication before they appear. Nevertheless, some recent authors suppose that such changes do not need more than a couple of generations. One must not forget, however, that the relevant counting is in animal generations and not in human generations, so that only 2–3 years is required for one generation in small species and 4–5 years for large mammals. It is possible that changes can appear quickly in animals kept in zoological gardens or in modern experimental stations, but they were much slower in prehistoric times. If this were not so, there would not be so many transitional individuals apparent from the metrical analysis of animal bones, for example from sites with local domestication of cattle.

Another consequence of domestication is the development of animal husbandry, which itself has two phases. The first is primitive animal-keeping without conscious, but with unintentional, breeding selection, the existence of which can be seen in any evidence for castration or the killing of a high proportion of young males. The second is developed animal breeding with conscious selection, the main aim of which is the increase of productivity and the existence of which is seen in the occurrence of different breeds in a given population.

There is a difference between the term 'animal husbandry' as formulated by Higgs & Jarman (1972, p. 8) and as used here. For Higgs & Jarman, animal husbandry 'stresses the important human element in the man–animal relationships and includes in a single category pastoralists, herders, herd followers and the like, where some form of intentional conservation was practised', whereas, for me, animal husbandry is a developed category which follows domestication, although it cannot always be separated from it.

Finally, it should be re-emphasized that animal domestication is a very complex man–animal relationship, all aspects of which have not yet been elucidated. This is particularly true for behavioural aspects, not only of the animal but also of the human.

References

Altuna, J. & K. Mariezkurrena 1985. Bases de subsistencia de los pobladores de Erralla: macromamiferos. In *Casadores magdalenienses en la cueva de Erralla (Cestona, Pais Vasco)*, J. Altuna, A. Baldeon & K. Mariezkurrena (eds), 87–117. San Sebastian: Munibe 37.

Boessneck, J., J. P. Jequier & H. R. Stampfli 1963. Seeberg, Burgäschisee–Süd; Die Tierrests. *Acta Bernensia* **II**(3), 5–215.

Bökönyi, S. 1969. Archaeological problems and methods of recognizing animal domestication. In *The domestication and exploitation of plants and animals*, P. J. Ucko & G. Dimbleby (eds), 219–29. London: Duckworth.

Bökönyi, S. 1973. The fauna of Umm Dabaghiyah: a preliminary report. *Iraq* **35**, 9–11.

Bökönyi, S. 1974. *History of domestic mammals in central and eastern Europe*. Budapest: Akadémia Kiadö.

Bökönyi, S. 1978. The vertebrate fauna of Vlasac. In *Vlasac: a mesolithic settlement in the Iron Gate*, D. Srejovic & Z. Letica (eds), 35–65. Beograd: Serbian Academy of Science and Arts Monograph DXII.

Bökönyi, S. 1985. Problèmes archéozoologiques. In *La protohistoire de l'Europe*, J. Lichardus & M. Lichardus–Itten (eds), 571–81. Paris: Nouvelle Clio.

Clutton–Brock, J. 1971. The primary food animals of the Jericho Tell from the proto-Neolithic to the Byzantine period. *Levant* **3**, 41–55.

Clutton–Brock, J. 1978. Bones for the zoologist. In *Approaches to faunal analysis in the Middle East*, R. H. Meadow & M. A. Zeder (eds), 49–51. Harvard University: Peabody Museum Bulletin 2.

Clutton–Brock, J. 1981. *Domesticated animals from early times*. London: Heinemann & British Museum (Natural History).

Darwin, C. 1868. *The variation of animals and plants under domestication*. London: John Murray.

Ducos, P. 1978. 'Domestication' defined and methodological approaches to its recognition in faunal assemblages. In *Approaches to faunal analysis in the Middle East*, R. H. Meadow & M. A. Zeder (eds), 53–6. Harvard University: Peabody Museum Bulletin 2.

Hecker, H. M. 1982. Domestication revisited: its implications for faunal analysis. *Journal of Field Archaeology* **9**, 217–36.

Hediger, H. 1942. *Wildtiere in Gefangenschaft*. Basel: Benno Schwabe und Verlag.

Herre, W. & M. Röhrs 1973. *Haustiere – zoologisch gesehen*. Stuttgart: Gustav Fischer.

Higgs, E. S. & M. R. Jarman 1972. The origins of animal and plant husbandry. In *Papers in economic prehistory*, E. S. Higgs (ed.), 3–13. Cambridge: Cambridge University Press.

Keller, C. 1902. *Die Abstammung der ältesten Haustiere*. Zürich: B.G. Teubner.

Klatt, B. 1927. *Entstehung der Haustiere*. Berlin: Paul Parey.

Legge, A. J. 1972. Prehistoric exploitation of the gazelle in Palestine. In *Papers in economic prehistory*, E. S. Higgs (ed.), 119–24. Cambridge: Cambridge University Press.

Musil, R. 1974. Tiergesellschaft der Kniegrotte. In *Die Kniegrotte*, R. Feustel (ed.), 30–95. Weimar: Museum für Ur–und Frühgeschichte.

Nobis, G. 1979. Das älteste Haushund lebte vor 14,000 Jahren. *Umschau* **610.**

Nobis, G. 1984. Die Haustiere im Neolithikum Zentraleuropas. In *Die Anfänge des Neolithikums vom Orient bis Nordeuropa, 9, Der Beginn der Ilaustierhaltung in der 'Alten Welt'*, H. Schwabedissen (ed.), 73–105. Köln: Böhlau Verlag.

Röhrs, M. 1961–62. Biologische anschauungen über Begriff und Wesen der Domestikation. *Zeitschrift für Tierzüchtung und Züchtungsbiologie* **76**, 7–23.

Turnbull, P. F. & C. A. Reed 1974. The fauna from the terminal Pleistocene of Palegawra Cave, a Zarzian occupation site in northeastern Iraq. *Fieldiana anthropology* **63**(3), 81–146.

3 Defining domestication: a clarification

PIERRE DUCOS

(translated by Marie Matthews)

During a symposium in 1975 in Dallas, involving a small number of participants studying the fauna of the Levant, I formulated a definition of domestication (1978a).[1] It was not meant to be radically new but was the result of a critical study of several previously formulated definitions, specifically what may be termed the 'classic definition' best formulated by Bökönyi (1969, and ch. 2, this volume).

In returning to the subject after a period of 12 years, I want first to ask what is the purpose of defining domestication? If it is purely a semantic exercise (even in several different languages) then it is a purely academic discussion. In fact, neither the Bökönyi definition (called 'definition B' in this chapter), nor mine ('definition D') are limited to semantics or are purely academic. Both are meant to be practical.

In my opinion, definition B describes *a moment* in the evolution of man–animal relationships and places a milestone in its history. Definition D, however, strives to justify a methodological approach, being, as it is, in the domain of logic.

Both definitions have been quoted in full in the previous chapter, and they represent two different approaches, resulting in what seem to be two contrasting positions. These positions are located, however, in two quite different 'fields' and cannot be opposed to one another. Between these two points of view there is a difference in concept and, for the definition itself, of content.

The difference in concept is epitomized by the beginning of the phrasing 'The essence of domestication . . .' (B) and 'Domestication can be said to exist' (D). Thus, definition D deals with a linguistic problem ('can be said'), within the field of logic, while definition B attempts, as concisely as possible, to isolate the 'essence', the core of a phenomenon, an objective fact, and is thus situated in the field of observation.

It was stated earlier that definition B appears to delimit *a moment* in the history of man–animal relationships. For B, domestication is a specific situation: the capture of a species (presuming, I suppose, a significant number of individuals of both sexes). Once this capture occurs it produces several consequences.

Thus, domestication is necessarily the first stage in a process. It is not itself a process; domestication is the beginning of symbiosis, the result of domestication is the domesticated animal, and animal husbandry implies a higher category (in man–animal relationships) which follows domestication. But on page 25 this volume, Bökönyi writes, 'domestication can be seen as a gradual and dynamic, though not always irreversible, process.' However, the example that he refers to (the wild rabbits in Spain and France), clearly demonstrates that we are still dealing with the period of 'capture'. Also, when he states that animal husbandry is a highly specialized phase of domestication (thus contradicting the last of his quoted statements), he does so only to explain the difficulty in differentiating, within a bone assemblage, domesticated animals from wild ones. Here again: domestication = capture. Thus, I do not think I am wrong when stating that for Bökönyi domestication (concretely: 'capture for profit') is the initial and necessary stage from which will evolve, on the one hand, domestic species and breeds and, on the other, highly modern and evolved forms of man–animal relationships, such as animal husbandry. This point is not one that I would necessarily reject and, in fact, definition D does not exclude it.

In Bökönyi's view, definition D is necessarily incomplete (p. 24 this volume), but it is on purpose that definition D excluded all human behaviour other than that mentioned. It refused to see in the animals' living conditions, particularly that of 'capture', anything other than the consequences of this essential relationship with the animal world: 'living conditions are among the consequences of domestication, not the mark of it.'

I have always believed that the idea (popular in the 1950s) that domestic animals are osteologically distinguishable from wild ones in all periods, is scientifically unsound. This assumption concealed other possible situations, where the man–animal relationship was no longer a hunting relationship but had not, as yet, filtered down to the skeleton and modified the bones. (This was also the position of Higgs and his school.) It is for this reason that it is imperative to define a cut-off point within the 'continuum' of the man–animal relationship.

It was consciously for this reason that I wrote: 'domestication can be said to exist . . .'. In French I would have formulated it as: '*Il y a domestication lorsque et seulement lorsque . . .*'. I was thus looking for a phraseology to express the relationship character of the state of domestication. In this case, domestication does not exclude '*proto-élevage*' situations or the capture of isolated individuals without creating domesticated populations or species that are anatomically different from their corresponding wild counterparts.

For this definition to be useful, it has to permit the creation of a particular methodology. Since the object of search is located in something preceding the act of domestication, it is human action that we must examine.

Incidentally, I do not believe that 'domestication is the beginning of a symbiosis that needs at least two partners' (Bökönyi, p. 24 this volume). These are not partners. I believe (although others may argue against this) that domestication is not a natural state – it exists because humans (and not the animal) wished it.

Having said all this, there are surely other, and possibly better, definitions than definition D. I would like to stress, however, that a method based on the definition that I gave (Ducos 1978a, pp. 54, 59) helped to identify a situation of '*proto-élevage*' in the Middle East, mostly in the pre-pottery Neolithic: the ox of Mureybet (Ducos 1978b) and later at Catal Hüyük (Ducos 1988), the Abu Gosh goat (Ducos 1978b), and possibly the Beida goat (Hecker 1975).

Finally, I would like to point out that definition D does indeed identify a state of domestication in the Egyptian representations mentioned by Bökönyi on p. 24 this volume (although he believes that it does not). By using the method that is based on this definition on large osteological samples, a state of domestication can also be inferred even in the absence of artistic representations.

Archaeozoology is an observational discipline, where experimentation is not often used and where the most convincing idioms are reputed to be closer to the truth. We are therefore tempted to state that one of the definitions is more correct than the other. The truth is that neither expresses a reality, but rather that both reflect the approaches chosen to study the complex phenomenon of human–animal relationships. Bökönyi's 'classic definition' probably applies most successfully to domestication from the Neolithic and later periods. My 1975 definition seems, however, to open up useful perspectives for more ancient periods during which humans appear to have deployed novel methods of exploiting nature.

Note

1 I wrote the original text in what I thought was English but the final version was 'translated' into real English by Richard Meadow and Melinda Zeder, and was approved by me.

References

Bökönyi, S. 1969. Archaeological problems and methods of recognizing animal domestication. In *The domestication and exploitation of plants and animals*, P. J. Ucko & G. W. Dimbleby (eds), 219–29. London: Duckworth.

Bökönyi, S. 1989. Definitions of animal domestication. In *The walking larder*, J. Clutton-Brock (ed.), ch. 2, London: Unwin Hyman.

Ducos, P. 1978a. 'Domestication' defined and methodological approaches to its recognition in faunal assemblages. In *Approaches to faunal analysis in the Middle East*, R. H. Meadow & M.A. Zeder (eds), 53–6. Harvard University: Peabody Museum Bulletin 2.

Ducos, P. 1978b. Le faune d'Abou Gosh. Proto-élevage de la chèvre en Palestine au Néolithique pré-céramique. In *Abou Gosh et Beisamun, deux gisements du VIIe millénaire avant l'ère chrétienne en Israël*, M. Lechevallier, Mémoires et travaux du Centre de Recherches préhistoriques français de Jérusalem, No 2.

Ducos, P. 1988. *Archéozoologie quantitative. Les valeurs numériques immédiates à Catal Hüyük*. Les Cahiers du Quaternaire, 1988, Bordeaux, Ed. du CNRS.

Hecker, H. 1975. *The faunal analysis of the primary food animals from pre-pottery neolithic Beidha* (Jordan). Unpublished PhD thesis, Faculty of Political Science, Columbia University.

4 Some observations on modern domestication processes

SYTZE BOTTEMA

Introduction

In this contribution I present information based upon experience with animal breeding which may be useful for the archaeozoologist. My experience in this field has often concerned the propagation of species not kept for economic reasons, but rather for the sake of interest in their behaviour. These experiments were not especially connected with archaeological purposes, but some examples could be selected that are relevant to archaeozoological research.

My observations are meant to provide the kind of information that is less often referred to in the discussion of domestication processes in connection with prehistory, and that could therefore help to broaden the discussion. This approach is not new, and outlines can be found in Herre & Röhrs (1973). They stress that the term domestication is difficult to define. Here, the term domestication is used in the sense referred to by Oldfield (in MacFarland 1981), i.e. that the process of domestication starts as soon as the wild animal has become habituated to humans. It is well known that wild animals retained in captivity for more than a few months are sometimes difficult to re-establish in the wild (Hediger, in MacFarland 1981).

Some observations on morphological changes occurring during the domestication process

At which moment does breeding wild animals in captivity cause visible changes? Everybody seems to accept domestication when distinct changes in the exterior appearance of animals have developed that make them differ from their wild ancestors. It is easily accepted that such changes have a genetic basis. An example shows that one cannot generalize this phenomenon. For instance, the udder in dairy cattle is developed to such a size that it may easily pass for a domestication characteristic. However, cows from a dairy breed develop only a small udder, densely covered with hairs, when left to suckle their own calves. Although the shape of the udder is hereditary, excessive size is induced by the milking regime, whether milking is done by hand or by machine.

I will discuss in more detail some observations on various species of waterfowl. Morphological changes in members of this group can be very different. It is well known that after a few generations of domestication greylag geese (*Anser anser*) become fatter and heavier, losing the power of flight (Delacour 1964). Besides, after some time, early maturing and loss of the permanent and monogamous pair-bond occurs. Colour variations such as white, piebald, and buff appear, and feet turn orange whereas they were originally pink. The fact that greylag geese become heavier after a few generations is not a genetic change, but a result of feeding. Next to this process, a selection on weight took place, resulting in various extraordinarily heavy breeds. It should be stressed that the greylag and its relatives of the *Anser* genus exhibit territorial behaviour during the breeding season. This contradicts the statement (see, for instance, Garrard 1984) that territorial animals are usually unsuitable for domestication. Characteristics such as monogamy and late breeding age soon disappear and more social behaviour develops. It is likely that the mutual tolerance among members of one family facilitates the keeping of geese. Offspring of a pair of wild geese in captivity will only stay with their parents until the next breeding season. Yet the parents are much more tolerant of these birds than of strangers, especially when the breeding season is over. Such family ties can be witnessed in many species of animals, where they exist between mothers and their female offspring, connecting several generations of female animals.

In white-fronted geese (*Anser albifrons*) the picture is quite different. This species has been kept and bred by generations of Frisian goose-catchers who use an intricate system of a net and a group of decoys. These decoys are not simply tame white-fronts. The catcher relies on the strong family tie of the geese, and to make use of this bond he needs a family of two parents with their offspring of that year. It depends upon the system whether part of the family is kept tied up on short strings between two nets or, in the case of one net, upwind of the net (Lebret *et al.* 1976). The other members of the family, especially the gander, are kept in the hide, at some distance from the net. From this hide the catcher pulls the rope that turns over the net. When a flock of wild white-fronts comes within a critical distance, one or more of the birds in the hide are thrown into the air. They will immediately fly to their relatives in the field behind the net, while both groups constantly give their flight contact call. In this way they may attract the other geese towards the net. It is necessary to have well-trained decoy birds, which are kept on the wing but which are so tame that they can easily be gathered after the net has been pulled. Good white-fronts that breed every year are highly valued and have been bred by generations of goose-catchers.

There is only one change visible in the exterior appearance of such decoy white-fronts: some of them have lost the black markings on the lower breast and belly (Figs 4.1 & 4.2). Compared with the greylag they have not developed white specimens or heavier birds that have lost the power of flight.

In the case of the greylag or domesticated forms of mallard (*Anas*

Figure 4.1 Decoy-breeder, Mr P. Wieland, with typical adult white-front gander, lacking the black marks on the belly as are found in the wild white-fronts.

Figure 4.2 Adult white-fronted goose shot at Nylamer (province of Friesland, the Netherlands) showing the black markings typical for this species.

plathyrhynchos), selection has taken place in favour of heavier animals valued for meat, for instance the Toulouse breed, these being three times as heavy as wild geese. In some Anatidae, however, captivity may result in smaller birds, which have not been selected intentionally. This is so with the shelduck (*Tadorna tadorna*) or male pintails (*Anas acuta*). Food conditions in captivity may cause such a size decrease.

When looking at another goose species, the bean goose (*Anser fabalis*), which is also kept and bred as a decoy, the observer could easily be misled into concluding that an increase in size has occurred in the captives. Bean geese are used as decoys for goose-catching in the central and northern parts of the Netherlands. Geese react specifically to some extent, and it is said that one cannot catch white-fronts with bean geese, but the reverse seems to work. If one compares the bean geese used as decoys with the wild ones present in large flocks in the new polders (reclaimed land) of Flevoland, or the ones caught for instance in the Eempolder in the Netherlands, the larger size of the decoys is obvious. Still, this is not a domestication trait. When the present occurrence of bean geese in the Netherlands is studied, it turns out that the majority of these geese wintering here at the moment belong to the subspecies *Anser fabalis rossicus*, the tundra bean goose. According to van den Berg (1983), over 100 000 tundra bean geese wintered in the new polders and other areas in the middle and south of the Netherlands in 1982.

On the Dutch diluvial soils and peat bogs much lower numbers, mainly small flocks, are recorded that belong to *Anser fabalis fabalis*. This subspecies is called the taiga bean goose, originating from Scandinavia, whereas the tundra form comes from Novaya Zemlya and the Yamal, Gyda, and Taymyr peninsulas (Delacour 1954). The taiga form reaches the Netherlands mainly in severe winters. The goose-catchers know this much bigger subspecies as the '*geelbek*' or 'yellow-bill' (Fig. 4.3), as it often has more orange-yellow colour on the bill than the subspecies *A. fabalis rossicus* (Fig. 4.4) which shows more black. The breeding stock used by the catchers consists of this larger subspecies. Moreover, *A. fabalis fabalis* is said to be tamed more easily than *A. fabalis rossicus* (van den Berg, written communication). The distribution and the number of subspecies, if they are still extant, must be studied to have any value in comparative studies (see also Herre & Röhrs 1973).

The keeping and breeding of geese in captivity: how to start domestic stock

When studying the decoy system, the question arises of how the goose-catchers ever developed a 'stock' of tame decoy birds. When asked, the catchers will always stress the fact that for establishing a stock of birds that will produce goslings in captivity only first-year birds can be used. This has been proven in those cases where the wild-caught adults of various *Anser* or *Branta* species had to be kept under optimal conditions for more than 10–15 years to get any breeding success. For instance, Delacour

Figure 4.3 Head of a taiga bean goose (*Anser fabalis fabalis*) (photo G. Müskens). The slender bill shows much orange-yellow. Decoys used by goose-catchers belong to this subspecies.

Figure 4.4 Head of a tundra bean goose (*Anser fabalis rossicus*) (photo G. Müskens). The bill shows little orange, the mandibles are heavier. Although much more abundant than the taiga form, it is not bred as a decoy.

(1964) writes 'As far as we know, the only wild-trapped red-breasted goose ever to have laid in captivity was one at Woburn, Bedfordshire, after fourteen years, and all the present stock (hundreds) in western Europe and North America are descended from her.' Yet the statement of the goose-catchers is not completely true and it needs further explanation.

The propagation of members of the goose family (as in many other bird species) in captivity depends solely upon the females, in the case of birds caught in the wild. Thus, for any breeding success young female geese have to be caught, geese that are about 6–9 months old at the time of catching. When they are brought together with a wild-caught adult male (three or four years old), they may produce offspring at the time when the female has matured. In general, this will take 3–4 years in the *Anser* group. If the reverse combination is made, with a juvenile gander being brought together with an adult wild-caught female, no success will be obtained.

Some Anatidae are renowned for very difficult breeding, although they have been caught in large numbers and have proved to be hardy in captivity. Breeding success of, for instance, the Brent goose (*Branta bernicla bernicla*) and the Baikal teal (*Anas formosa*) mostly dates from the last five years, although these species had been kept for at least the past 100 years. Mexican peasants traditionally collect eggs of the northern red-billed whistling duck (*Dendrocygna autumnalis discolor*) which are hatched under chickens. Some peasants own up to 30 ducks. Keeping red-billed whistling ducks is an old tradition, but the Mexicans state that no duck ever produced offspring in captivity. The reason is still unknown (F. Feekes pers. comm.).

The difference in breeding success can be explained by the difference between sexes, i.e. the anisogamy, the production of different gametes. To guarantee any success for the investments in gametes made by the parents, the environmental conditions are much more critical, in terms of selection pressure, for the female than for the male. The male can produce millions of gametes whereas the female can only produce a clutch of 4–7 eggs. If the weight of an egg is compared with that of a sperm cell, the difference of invested energy is obvious (Krebs & Davies 1982). Thus, the conditions have to be properly fulfilled, otherwise the female will not waste the energy. For wild-caught adult females, conditions in captivity are not properly fulfilled as such geese will have been previously exposed to and imprinted on other conditions in their natural habitats (see below).

Yet the experienced goose-catchers do not claim their statement without reason. The advantage of bringing first-year birds together to establish decoy families is based upon the existence of a rather rigid pair-bond. It is often stressed that geese form a pair-bond for life and that if one of the pair dies, the other remains solitary for the rest of its life. That geese form a firm pair-bond is true, but there is no convincing evidence for the statement that they remain solitary after losing the partner. In captivity new pair-bonds can be formed quite easily and sometimes a pair splits up and one of the partners takes up with another bird. Conditions in captivity are not the same as in the wild, but solitary adults are not mentioned for the distribution range in the breeding season of *Anser fabalis* in Russia (H. Hallander pers. comm.).

If a young female is put together with an adult wild gander, various problems arise. The young female will have lost her parents and the gander, caught on migration, will probably have become separated from his partner and possibly also his own young. The juvenile goose would have broken up her parental relation altogether some months later, but the gander may not be charmed at all by the young female that is not sexually mature. If the gander should finally court her, she may not react adequately because she is too young. Thus, the goose-catchers are right to advise that the best way to start off a decoy group is to take juveniles of both sexes, which still have to form a pair-bond.

I will explain some of the factors that play a role if a wild goose is suddenly locked up under conditions that differ completely from the wild. One can imagine that this is a situation that occurred in prehistoric times with various kinds of animals.

On 10 October 1983 the migration of *Anser brachyrhynchos*, the pink-footed goose, was clearly at its peak over the northern Netherlands. Thousands of pink-footed geese could be seen moving to the west towards the relatively small wintering grounds between Workum and Gaast in southwest Friesland. This concerns about half the population that breeds in Spitsbergen. They arrive in the Netherlands much earlier than the other *Anser* species because of the Arctic conditions in their breeding grounds. They pass Norway, Jutland, and the coast of Schleswig-Holstein to reach their destination, the wintering grounds. On 12 October, a boy brought me a first-year bird (in fact about four months old), that had hit a high-tension cable in the vicinity. According to the peak in migration the bird must have hit the cable on 10 October. The shoulder of the goose was severely fractured and the complete wing had to be amputated as the bird constantly stumbled over it. Within four weeks the bird approached a human being within a distance of two metres to receive its food.

How can such a change in behaviour be explained, if adult birds caught or winged by shooting may stay shy for years? For an explanation we should be informed about the behaviour of geese and the history of this particular juvenile. The pink-footed geese, like other geese, have a rather fixed migration pattern. Invariably they will fly the same routes between their breeding and wintering grounds and they will only move from their wintering grounds if they are forced to do so by, for instance, long-lasting snow cover. For the fixation of the route the strong family tie is very important. The young birds migrating in the year that they are born are guided by their parents and thus they learn the route. Without their parents they are forced to join other geese of their own kind. Young birds that do not have this opportunity will be lost. In this way, we must explain the aberrant nesting of barnacle geese (*Branta leucopsis*) outside their normal range, on Iceland and Gotland. On Gotland it could be ascertained that escaped ringed birds from the Museum of Skansen near Stockholm settled on the island (H. Hallander verb. comm.). Birds that were released from captivity could not establish a migration tradition and chose breeding grounds thousands of kilometres outside their normal range. Bean geese, caught in Switzerland on their spring migration, which escaped after two years of captivity, were recorded back from Archangel and Kaluga (USSR). As these birds had made the trip at least once they were able to return (Bauer & Glutz von Blotzheim 1968).

The particular pink-footed goose mentioned above had undoubtedly been accompanied by its parents, covering large distances in a relatively short time, constantly absorbing new impressions. On Spitsbergen, where it was hatched and raised, it would hardly have had an opportunity to meet a human being. During the rapid journey south such contacts were also few. Its reaction towards humans was purely dictated by its parents and the (sub) adult members of the group it travelled with. If a person approached, the experienced adults would take to the air well out of shotgun range. Nevertheless, our juvenile pink-footed goose will not have been too much impressed by humans, and the behaviour of its parents was far more important than the behaviour of the potential enemy. When the bird hit the high-tension cable it was completely at a loss. It was hardly experienced and, what is very important, it was still in a state of absorbing new impressions that differed very much from the situation on the Arctic tundra. In this highly sensitive state it was learning that humans are dangerous while cattle and sheep are not, that grass and winter crops are edible, and at the same time it had to absorb large parts of the map of northern Europe. In judging new situations it would consciously pay attention to the behaviour of its parents and other geese in the flock. The difference in behaviour between the freshly caught youngster and the pink-footed geese bred in captivity, present in the same fenced meadow, was striking. For instance, a common gull (*Larus canus*) soaring over the meadow made no impression on the 5–7-year-old pink-footed geese also present. However, the newcomer panicked completely, running for vegetal cover where it crouched and remained motionless for some time. As a chick it had obviously learnt from its parents, on Spitsbergen, to fear marauding glaucus gulls (*Larus hyperboreus*), which have a flight pattern comparable with the common gull. During the first weeks the pink-footed goose refused any food and would only graze at some distance from the threatening human being. It would, however, immediately try to contact the geese that were already in the fenced meadow, since belonging to a group is a first prerequisite for survival. In doing so, it encountered the problem of entering a group of geese only partly belonging to its own species and with a fixed 'pecking order'. Captive geese remaining on their territories (all year round because they cannot migrate) are more aggressive towards newcomers than those in flocks where all the members enter the same new situation, mainly feeding and sleeping at the same time. The pink-footed goose was constantly chased away and was thus forced to be more or less solitary, whereas the urge for the juvenile to join a group was very great. After two weeks it learned that the person who entered the meadow twice a day was not doing any harm and that food, new but edible material, was brought that always had to be inspected immediately, even if it was not eaten. It made social contact with a 4-year-old goose of its own kind, which was also solitary. The bond was formed, as is usual in many Anatidae, by synchronizing daily behavioural rites such as grazing, sleeping, and preening. This example shows the behaviour of a freshly captured bird in relation to a new situation created by man.

An analysis of some colour variations occurring in various subspecies of mallard: an example of isolation

One of the first changes visible in ducks kept in captivity is the appearance of white plumage. Many species, such as mallard (*Anas plathyrynchos*), wood duck (*Aix sponsa*), mandarin duck (*Aix galericulata*), Bahama pintail (*Anas bahamensis*), Egyptian goose (*Alopochen aegyptiacus*), occur in white varieties. Birds first kept for ornamental reasons or for shooting, such as the bobwhite quail (*Colinus virginianus*), are now available in various colours, and recently even 'broiler' types have been developed for economic purposes. If a bird, or a group of birds, differ in colour from the wild-coloured relatives, for instance by being white, the factor that causes white may imply other changes too. Thus, in Japanese quail (*Coturnix coturnix japonica*) kept under the same conditions, white colour seems to be linked with tamer behaviour and weaker condition compared with wild colour. Mink breeders know that mink with the gene for 'colmira' colour are always sleepy; animals of 'pearl' colour tend to hold their head in a peculiar way (Crow 1979).

Colour varieties that appeared in the course of domestication were clearly selected either for reasons of subsistence or for the sake of exclusiveness. Such selection may have side effects. A deviating colour such as white may be linked with other recessive factors, or the gene for such a colour may be pleiotropic. A pleiotropic gene influences more than one trait (Goodenough 1978).

What caused the white specimens in various members of the Anatidae? The origin of colour variations in domestic waterfowl is often ascribed to mutation (Delacour 1964). If large numbers of a species are bred in captivity one can expect mutations to appear. This depends upon the mutation frequency of the genes, generally assumed to be in the order of 1 in 10^5–10^6 gametes. In this connection, it is striking that mallards obtained from an area in Sweden where feral ducks can be excluded nevertheless showed white specimens in the third generation (I. Bossema pers. comm.). The same is true for a small breeding group of eiders (*Somateria mollissima*) in captivity in England that produced a deviating colour (buff) in the third generation. When European pintail (*Anas acuta*), normally caught in duck decoys in the Netherlands, were no longer for sale because of conservation measures, a breeding stock was soon developed to meet the demand. Within a few years 'blonde' specimens appeared. This contradicts the statement that it takes 30 generations to breed deviating specimens (Bökönyi, cited in Meadow 1984).

Such rapid appearance of deviating colours in many species cannot be explained by mutation during domestication, but it may be due to recessive factors in the wild population. The colour of wild duck species is generally dominant over other colours. The wild-colour pattern is caused by many genes responsible for the various components or for the distribution of the colours. If a mutant factor is present in a duck in heterozygous form, it will not show up in the appearance of the bird, because of the dominance of the wild-colour factors. In practice, the chances of a duck meeting a partner with the same recessive factor are limited: offspring in which combinations of the factor have occurred, e.g. white in homozygous form, will therefore be very rare. Besides, there is strong selective pressure against these white mutants, as predators can see them from a great distance. For the same reason a dominant white mutant will have little chance of surviving. On the other hand, a recessive factor, if present in heterozygous form, is not visible, cannot be eliminated by selection, and thus survives to produce a colour variant only if the owner meets a partner with the same genetic combination. The trait white, a clear negative property in the wild, can be positively valued in captivity by man. As this is a recessive trait, it will be very easy to develop a pure breeding stock of white ducks.

One is thus tempted to conclude that colour variations in captivity are more likely to be the result of recessive factors already present in the wild population than to mutations. However, the following example shows that the explanation of the mechanism behind colour variation may be more difficult. I refer to a situation in the wild which shows clear parallels with the domestication process of a small group, a situation that may have been normal in prehistory. On the island of Laysan in the Pacific Ocean, a small population of endemic Laysan duck (*Anas laysanensis*) occurs. This small duck, the size of a teal, is thought to have originated from a group of straggling mallards that happened to reach the lonely island. As the only suitable biotope for duck on Laysan (which is *c.* 5 km long) is a lagoon, the population was always limited in numbers because of space. In the 19th century numbers were *c.* 600 at most. Genetically this is a limited number with a high inbreeding rate. As can be seen in many island species of animals, there was a decrease in size compared with its ancestor, the mallard. In addition, it lost its breeding plumage, male and female both showing brown feathers all the year round, and the female, especially, shows partial albinism. Due to the activities of Japanese plumage hunters, the Laysan ducks were reduced to ten individuals in 1909 (Delacour 1954, Kolbe 1972). Conservation measures allowed the Laysan duck to increase again, but it is clear that this species went through a bottleneck, reducing genetic variation. Obviously, no lethal factors were present in the small population, otherwise it would have disappeared due to inbreeding. In 1963 the population had reached its 19th-century level again and some birds were caught for the purpose of breeding in captivity. In fact this meant a second 'genetic drift'. The Laysan ducks were doing very well in captivity and many have been reared since. After about 15 years other colours appeared in captivity. The limited number of birds indicates that mutants are unlikely to occur. On the other hand, it is difficult to accept that the variation in colours developed from one recessive factor. The rate of inbreeding in the ten individuals left in 1909 was so high that one could expect colour variations to show up soon after that year. This holds also for the offspring of the few ducks caught in the sixties from which the breeding stock in captivity was developed. It is not clear which genetical factors are responsible for the black, blue, and buff specimens that are present at the moment. It is possible that in Laysan ducks a colour variation is caused by several factors which have to combine to have any visible effect.

Some observations on the role of temperature in the rearing of nidifugous birds

Abiotic factors may strongly influence domestication processes. It is understandable that animal species kept under conditions that deviate too much from their ecological amplitude will not survive. A factor that is of importance in initial domestication is temperature. It is, for instance, advantageous to rear chicks of various nidifugous species (those that leave the nest soon after hatching) under higher temperatures than those occurring in at least part of their natural habitat. I once saw a little girl herding a group of very young ducklings (about one-week old) near the prehistoric site of Suberde in Anatolia. The little ducklings had no natural heat source in the form of a natural mother but because of the high temperature they were happily running around in search of food. The girl would chase away possible predators and when a thunderstorm approached she guided the herd inside the house before the rain poured down. Such a situation would be impossible in a temperate climate. One can observe that domestic fowl in farmyards in the Near East always have larger numbers of young than domestic fowl on farms in temperate regions. This is connected with a higher survival rate because of higher temperature and not with greater clutch size.

Some experiments demonstrate that a high temperature may compensate for deficiency in the daily diet. Teal (*Anas crecca*) hatched under a bantam and kept outside are difficult to rear at a time when the April–May temperatures in the northern Netherlands can still be low. A good bantam will be a guarantee of warm shelter for the little ducklings but they still may not survive the first critical weeks of their life. The ducklings will simply stay under the bantam and die of hunger although food is available. At the same time, teal ducklings can be seen chasing insects in peat bogs and ditches in the wild, obviously not harmed by the same relatively cold weather. If small teal are put in an electrically heated artificial breeder, they will survive on the same diet that was not sufficient for those raised outside with a bantam. Where a natural diet seems to compensate for loss of body warmth, a warm temperature does the same under artificial conditions where the diet is suboptimal.

I experienced the same phenomenon in the case of lapwing (*Vanellus vanellus*) chicks. When kept in an artificial breeder at 30° C they would give plaintive calls, refusing to eat. If the temperature was raised to 35°C, they would start to feed. At the same time wild lapwings could be seen in the field raising chicks at much lower temperatures. Here also, the natural optimum diet seems to compensate for low temperatures.

Propagating mallard in captivity is far from easy, compared with greylag or white-fronted geese, a fact stressed by Lepiksaar (cited in Prummel 1983). One can imagine that domestication took place where the mallard was most common. The mallard has a widely varying range of habitat and it occurs in Eurasia, North America, and North Africa. The density of the breeding population of mallards in the Netherlands is among the highest in its distribution range. Yet the domesticated duck was introduced into this country in medieval times only. It is more probable that they were imports than they were domesticated locally. Early domestication centres are suggested for Asia. The ancient Egyptians are reported to have kept and bred ducks a few centuries BC (strangely enough, pintails dominate many of their frescos). The Romans kept ducks in enclosures especially for fattening. They are reported from France in the 6th century AD (Bottema 1980). It is rather easy to raise mallard from eggs collected in the wild. Such birds will nest in captivity but lose their offspring when the ducklings are a few days old.

It is acceptable that temperature may determine successful breeding when one starts with chicks. If the possibility of rearing chicks from the egg is ruled out, one may capture adults, but then specially reinforced housing is needed otherwise they will escape. It is better to start off with juvenile ducks, which are caught when they are completely feathered apart from the primaries. Nevertheless, propagation will be a problem as such ducks will not be successful in raising ducklings.

Thus, an area with a high spring temperature may have been a locus for initial duck domestication from which the spread of adapted, or selected, domesticated ducks could take place into temperate regions.

Food as a mechanism of control in the domestication process

The relation between humans and animals is sometimes explained as symbiosis. Herre & Röhrs (1973) stress that the term symbiosis used with reference to domestication does not always correspond to the biological definition. Tchernov (1984), for instance, uses the term 'one-way symbiosis'. The domestication of the fowl is sometimes claimed to have its roots in some kind of symbiosis that developed on the edge of settlements in the jungle. The red junglefowl (Gallus gallus) is thought to have been attracted by food in the settlements. When the inhabitants started to feed the birds on purpose, they were able to become attached to the village. Wayre (1969), however, reports that red junglefowl in Bhutan only occur far away from any human habitation. Practical experience also pleads against domestication developing from such a proposed form of symbiosis. An example of a relation between humans and wild birds is provided by common eiders (Somateria molissima) nesting close to lodgings because of the protection this affords against egg-preying gulls. The tenant of the house may gather the down, without harming the sitting bird. Indeed, lonely houses on Arctic islands may have a complete colony of eiders in front of them. However, after the eggs have hatched the relationship will come to an end.

There is a less conspicuous though interesting connection, that often goes unnoticed, between the magpie (*Pica pica*) and farmyards in the northern Netherlands. The magpie is robbed of its eggs by the crow (*Corvus corone*). Farmyards in the Netherlands will often have a magpie's nest in a tree close to the farm buildings, whereas the nests of crows are found much further away. The crow is afraid of extending its territory close to the threatening buildings. The magpie takes advantage of this situation (I. Bossema pers. comm.). However, the biological term 'symbiosis' still cannot be applied here.

Experience with ducks leads one to think that the reverse is much easier. Domesticated ducks kept on farms easily revert to a feral state and so a process of 'dedomestication' sets in. In times when duck-keeping for the production of eggs was still economically viable in the Netherlands, measures such as regular feeding in combination with locking up at night were necessary, otherwise the ducks soon disappeared. For this reason duck-keeping outside suitable biotopes, such as ditches and marshes, is much easier, because in an unsuitable biotope they have to rely upon food supplied by the owner. Farmers in the Frisian wetlands keep their ducks under feral conditions. The wild population mixes with the domesticated ducks to such an extent that hardly any 'unspoiled' mallards can be found. This is in contrast to the situation on diluvial sandy soils where wild mallard are restricted to small streams and pingo-ruins (small lakes of periglacial origin). The few domestic ducks kept on farms have to rely on food offered there and hardly go astray. Mixing of the two groups has hardly been noticed. The same is true for mute swans (*Cygnus olor*) or domestic geese.

Humans have a means of firm control over domesticated animals, in the form of food. The role of food is pointed out by Oldfield (in MacFarland 1981). Chickens in a farmyard will show at least one sign of 'dedomestication' by selecting nest sites carefully hidden in barns or bushes; by giving them food regularly one can keep them bound to the yard.

Acknowledgements

I am very much indebted to my wife Nicolien and my eldest daughter Fionna for their assistance on our farm, to my youngest daughter Wytske for driving the tractor, to Mr L. van den Bergh for his information on bean geese, to Mrs G. Entjes–Nieborg for preparing the manuscript, and to Mrs S. M. van Gelder–Ottway for correcting the English.

References

Bauer, K. M. & U. N. Glutz von Blotzheim 1968. *Handbuch der Vögel Mitteleuropas.* Vol. II: *Anseriformes (1).* Frankfurt: Akademische Verlagsgesellschaft.

Berg, L. van den 1983. De Rietgans. *Vogels* **19**, 240–4.

Bottema, S. 1980. Eenden. In *Zeldzame Huisdierrassen*, A. T. Clason (ed.), 191–204. Zutphen: Thieme.

Crow, J. F. 1979. *Overzicht van de Genetica.* Groningen: Wolters–Noordhoff.

Delacour, J. 1954. *The waterfowl of the world.* Vol. I. London: Country Life.

Delacour, J. 1964. *The waterfowl of the world.* Vol. IV. London: Country Life.

Garrard, A. 1984. The selection of south-west Asian animal domesticates. In *Animals and Archaeology.* Vol. 3: *Early herders and their flocks*, J. Clutton–Brock & C. Grigson (eds), 117–33. Oxford: BAR International Series 202.

Goodenough, U. 1978. *Genetics.* London: Holt Rinehart & Winston.

Herre, W. & M. Röhrs 1973. *Haustiere – zoologisch gesehen.* Stuttgart: Gustav Fischer.

Kolbe, H. 1972. *Die Entenvögel der Welt.* Neudamm: Neumann.

Krebs, J. R. & N. B. Davies 1982. *An introduction to behavioural ecology.* Oxford: Blackwell.

Lebret, T., T. Mulder, J. Philippona & A. Timmerman 1976. *Wilde ganzen in Nederland.* Zutphen: Thieme.

MacFarland, D. (ed.) 1981. *The Oxford companion to animal behaviour.* Oxford: Oxford University Press.

Meadow, R. H. 1984. Animal domestication in the Middle East: a view from the eastern margin. In *Animals and archaeology.* Vol. 3: *Early herders and their flocks*, J. Clutton–Brock & C. Grigson (eds), 309–39. Oxford: BAR International Series 202.

Prummel, W. 1983. Excavations at Dorestad Vol 2. Early medieval Dorestad, an archaeozoological study. *Nederlandse Oudheden, Kromme Rijn Project.*

Tchernov, E. 1984. Commensal animals and human sedentism in the Middle East. In *Animals and archaeology.* Vol 3: *Early herders and their flocks*, J. Clutton–Brock & C. Grigson (eds), 91–116. Oxford: BAR International series 202.

Wayre, P. 1969. *A guide to the pheasants of the world.* London: Country Life.

5 *Feral mammals of the Mediterranean islands: documents of early domestication*

COLIN P. GROVES

Introduction

The islands of the Mediterranean are inhabited by a variety of wild mammals conspecific with widespread domesticates: sheep, goats, pigs, and cats. In the past these forms were considered to be genuinely wild, vicariant subspecies of species from which the respective domesticates had sprung. In recent years, however, it has become clear that the Pleistocene faunas of many, at least, of the Mediterranean islands were highly differentiated, with endemic species or genera, and with no traces of *Ovis, Capra, Sus,* or *Felis* (Schwartz 1973, Sondaar 1977, Dermitzakis & Sondaar 1979, Azzaroli 1981, 1982). The implication is clear: the sheep, etc. have been introduced by human agency at some time during the Holocene, and so are not naturally occurring subspecies.

It is, therefore, of great interest to know whether these species were brought to the islands as wild individuals, and released for some reason (to act as a food source, perhaps), or as domesticates. The fact that they were previously mistaken for naturally occurring wild forms is an acknowledgement of how very close they are to truly wild representatives, so that if they were brought in as domesticates they were clearly in a stage when domestication had not proceeded very far, giving them an intrinsic interest as a kind of living museum of the initial stages of the domestication process. If, on the other hand, they were brought in as wild stock, there is still considerable interest attached to them as examples of rapid *in situ* evolution, since they are not claimed to be precisely identical to any continental wild form.

As well as (potential) domesticates, a variety of wild species have been brought into these same islands: shrews, hares, mice, dormice, spiny mice, foxes, weasels, martens, badgers, and deer. Again, it is necessary to stress that there is not a trace of these species in Pleistocene deposits, nor, all things considered, is there much likelihood of their having introduced themselves by jumping on to floating logs or other 'sweepstake routes'. Why anyone should want to introduce weasels to Corsica is not immediately apparent; the introduction of badgers to Crete is perhaps less mysterious, given the wide-ranging interests of the eclectic Minoans. But introduced they were, and the sheep (etc.) must be considered in the same context.

The best prospect for determining the feral versus wild status of a wild-living form is relative cranial capacity. Herre & Röhrs (1973) show that the relative brain size (compared to body size) of a domestic form is invariably less than that of its truly wild conspecifics, sometimes massively so; Hemmer (1983) disputes the extreme degree of reduction, but agrees that it has occurred. The data of Kruska & Röhrs (1974), in particular, show that brain size does not increase again in feral forms, over at least 100 generations. Admittedly, this is different from the thousands of years over which the sheep and other animals have inhabited the Mediterranean islands; but it is at least worth investigating the proposition that brain size may have remained small over that period of time.

Material and methods

Skulls of both wild and domestic representatives of *Ovis, Capra, Sus*, and *Felis* were measured. Cranial capacities (to represent brain size) were measured by pouring birdseed into the braincase (after sealing up the optic foramina with Plasticine), shaking it at intervals to pack it down, and then decanting it into a measuring cylinder to measure the volume. Several linear measurements were taken with calipers, partly to determine which would act as the most efficient size standard, and partly to act as accessory means of discrimination.

Results

Wild sheep

Wild-living sheep, known as mouflon, live today on Corsica, Sardinia, and Cyprus (Figs 5.1 & 5.2). The name *Ovis musimon* Schreber has generally been applied to the Corsico–Sardinian mouflon, but Uerpmann (1981) has shown that the correct citation of this name is Pallas, 1811, and that the type locality is Transcaspia; hence the name *musimon* is not available for the mouflon.

The mouflon of Cyprus has been known as *Ovis ophion* Blyth, 1840, and with the reallocation of the name *musimon* this appears to be the earliest name for any mouflon. Pfeffer (1967) found Cypriot and Corsico–Sardinian mouflon to be identical, apart from the former being slightly smaller; Valdez (1982), however, points out that they can still be distinguished on average, because the horns of the rams in Cyprus are supracervical, while those on Corsica and Sardinia are usually homonymous. But in colour and colour pattern, and in all features of the ewes, they are the same.

Pees & Hemmer (1980) found that, among living wild sheep, the Argali (*Ovis ammon*) has a relatively higher cranial capacity than the Urial (*Ovis*

Figure 5.1 *Ovis musimon*: mouflon or wild sheep from Corsica/Sardinia, in winter coat, London Zoo.

Figure 5.2 *Ovis ophion*: mouflon or wild sheep from Cyprus, in summer coat, Hai–Bar Carmel, Israel.

orientalis vignei); the latter in turn has a slightly higher capacity than the Corsico–Sardinian mouflon, and domestic sheep are still further reduced. Some Soay sheep, a long-established feral form from the Scottish Isles, had cranial capacities equivalent to domestic sheep: increasing confidence that brain size remains small over very long periods of time.

Using Pees & Hemmer's data and basic diagram, I added my own data to produce Figure 5.3. As in Pees & Hemmer's findings, *Ovis ammon* has a very high cranial capacity, followed by *O. orientalis* (other subspecies, including the Turkish *O. o. gmelini*, being now added to *O. o. vignei* which Pees & Hemmer had used), followed by Corsico–Sardinian mouflon, followed by domestic and Soay sheep. There are big overlaps between mouflon and *O. orientalis*, on the one hand, and mouflon and domestic sheep on the other; such that a few domestic sheep capacities fall into the mouflon polygon and even overlap *O. orientalis*. From these data, it would appear that the brain in the Corsico–Sardinian mouflon is somewhat reduced from that of wild sheep (Urial), but not much. Unless there has indeed been a reversal of the initial reduction, the explanation that most immediately impresses itself is that the mouflon is a feral relic of a species that had not long been under domestication.

A single skull of a Cypriot mouflon was studied. The specimen is unusually large (in the upper part of the size range for the Corsico–Sardinian mouflon), but its cranial capacity is very small. It would be approximately on the domestic/Soay line, extrapolated upwards. Plausibly, the mouflon of Cyprus has a different, more fully domesticated, ancestry from that of Corsica and Sardinia.

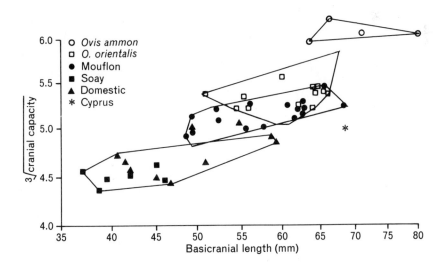

Figure 5.3 Double logarithmic plot showing relative cranial capacity in wild, feral, and domestic sheep.

Wild goats

Wild goats occur on the Aegean islands of Crete, Antimilo (Erimomilos) in the Cyclades, and Yioura (Jura) in the Sporades. Schultze–Westrum (1963) considered the goats of Crete and Antimilo to be truly wild, calling them *Capra aegagrus cretensis* (Fig. 5.4) and *C. ae. pictus* respectively, but those of Yioura to be feral.

I measured the cranial capacities of a large number of wild and domestic goats (Fig. 5.5). Wild (Bezoar) goats from Turkey and the Caucasus (*Capra aegagrus aegagrus*) have the highest cranial capacities; those of *C. ae. blythi* from Iran and Pakistan are somewhat lower on average, a finding which will not be commented upon here. Domestic goats have much lower capacities, the more highly modified breeds (Alpine, Cashmere, Angora) having less reduced levels than the primitive Nilotic goats, suggesting an increase under domestication, after the initial decrease, for some breeds. Some feral goats from Juan Fernandez island off the Chilean coast, a population of some 400 years' standing (Rudge 1984), have capacities of the same relative size as the 'higher' domestic breeds.

The Aegean goats scatter neatly between the wild and the domestic goats, overlapping marginally with both. There is no difference between those from the three islands. That there are differences, especially in horn shape, between them may mean that they originate from different breeds, but their origin fairly clearly is from domestic goats: as in the case of the

Figure 5.4 *Capra aegagrus cretensis*: Cretan wild goat. Hai–Bar Carmel, Israel.

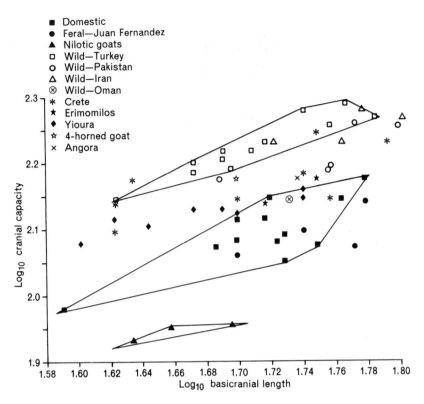

Figure 5.5 Double logarithmic plot showing relative cranial capacity in wild, feral, and domestic goats.

mouflon, their ancestors were 'only just' domestic, implying that they derive from quite an early era in the history of goat domestication.

Pigs

In a general revision of *Sus*, I (Groves 1981) called the Corsico–Sardinian wild pig *Sus scrofa meridionalis* (Fig. 5.6), and assigned the southern Spanish wild pig to the same subspecies. A report on cranial capacity studies was made in a later publication (Groves 1983): cranial capacities are larger in Eurasian wild pigs than in those from Southeast Asia, with domestic pigs falling below both. Some known feral forms fell into the general domestic range, while others of equivocal status could be identified by this method as feral.

In Figure 5.7, some Corsican and Sardinian skulls have been added to the picture. They are, as far as brain size goes, simply small specimens of wild, Eurasian *Sus scrofa*. Their identity in every other respect with wild pigs from southern Spain suggests that they are, indeed, wild pigs which have been introduced from there. This conclusion contrasts markedly with

Figure 5.6 *Sus scrofa meridionalis*: wild pig from Sardinia, in summer coat, Tierpark Hellabrun, Munich.

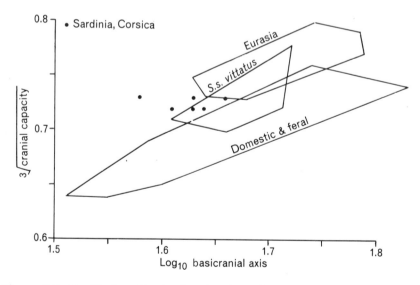

Figure 5.7 Double logarithmic plot showing relative cranial capacity in wild, domestic, and feral pigs. Among wild pigs, those from Eurasia (Europe, Siberia, Japan) and those from Malaysia/Indonesia (*Sus scrofa vittatus*) are depicted separately.

that for the mouflon, and implies that we cannot simply envisage an enterprising group of people who migrated from the Levant to Corsica and Sardinia, bringing their (albeit primitive) domestic stock with them; different species, of different status, were brought for different purposes (and perhaps by different peoples!).

Cats

Whereas wild sheep, pigs, and goats are found on some islands but not on others, wild cats appear to occur on them all; at least, they are well enough known from Sardinia, Corsica, Crete, and even the Balearics to have received subspecific names on each of those islands. These putative subspecies are as follows:

Sardinia: *Felis silvestris sarda* Lataste, 1885.
Corsica: *Felis silvestris reyi* Lavauden, 1929.
Crete: *Felis silvestris agrius* Bate, 1905, and *F. s. cretensis* Haltenorth, 1953.
Mallorca (Balearic Is.): *Felis silvestris jordansi* Schwarz, 1930.

Haltenorth (1953) doubts that *F. s. reyi* is really different from *F. s. sarda*. The same author considered that the type specimen of *F. s. agrius*, from Crete, was actually a feral cat, but that another specimen, collected at the same time, was a genuine wild cat, so he rejected the name *F. s. agrius* and founded a new subspecies, *F. s. cretensis*, on the second specimen.

Wild-living cats are not known from Cyprus. Recently Davis (1987) has figured the mandible of a cat from the Neolithic (*c.* 6000 BC) of Khirokitia on Cyprus, which he suggests was most probably domestic. Whether wild or domestic, this find indicates a special human-cat relationship at a much earlier period than had hitherto been thought probable.

What is remarkable about this picture, and has never as far as I know been specially commented upon, is that wild cats from North Africa have also been referred to as *F. s. sarda*. There are thought (as, for example, by Haltenorth 1953) to be two North African subspecies, *F. s. sarda* and *F. s. lybica*, in more mesic and more xeric habitats, respectively. I have studied a number of specimens assigned to *F. s. sarda* from both the Maghreb and Sardinia, and it is quite true, there is no difference: they have the same buffy-grey colour with a dark spinal stripe, and varyingly (but always rather poorly) expressed striping; head-and-body length averages 58 cm in Sardinia and 59 cm in North Africa (smaller than most European wild cats, larger than most from western Asia); tail length averages 57 per cent of head-and-body length in both places, which is shorter than Arabian races but longer than most other wild cats; zygomatic breadth averages 78 per cent of condylobasal length in both, which is the same as the western Asian and other African wild cats, but more than those from Europe; and the length of the upper carnassial averages 11.3 mm in North Africa and 11.1 mm in Sardinia, these figures being less than most European wild cats but greater than those from western Asia and North Africa. There is, in other words, no room for doubt but that the Sardinian and North African wild cats really are the same.

What, then, of the domestic or wild status of the Sardinian wild cat? Hemmer (1976) has illustrated the values of Schauenberg's Index (skull length divided by cranial capacity, an isometric relationship in cats), and I have used this as a basis for Figure 5.8, to which I have added the values of specimens measured by myself, including some from Sardinia. It is first noticeable that, as Hemmer found, *F. s. lybica* has a higher index (i.e. relatively lower cranial capacity) than *F. s. silvestris* from Europe, with North African *F. s. sarda*, as well as sub-Saharan African and the NW Indian/western Asian *F. s. ornata*, falling in between. Domestic cats, as well as feral cats from the Northern Territory of Australia, have higher values for the index (but overlapping, even nearly covering, the *F. s. lybica* range), and Egyptian mummified cats fall into either the *F. s. lybica* or domestic range. The values for Sardinian wild cats are remarkably low: they could not possibly be feral domestic cats, unless there has been total reversion to the wild brain-size, but they fit exactly the range of North African *F. s. sarda*. I conclude that cats were introduced into Sardinia from the Maghreb region in the wild state: introduced they certainly must be, but feral (in the sense of being reverted from a domestic state) they are not.

A single specimen from Corsica, the type of *F. s. reyi*, fits almost exactly into the Sardinian range: its head-and-body length is 58 cm, right on the Sardinian average; its zygomatic breadth is 76 per cent of the condylobasal, only a little below the Sardinian average but well within the range (73–83 per cent); the skin pattern, as illustrated by Haltenorth (1953, Fig.30) is similar to that of *F. s. sarda*; only the tail is apparently shorter (46.6 per cent of head-and-body length: the *F. s. sarda* range is 47–68 per cent). I have not seen the specimen itself, and no cranial capacity data exist, but I believe that *F. s. reyi* is a synonym of *F. s. sarda*, and that Corsican, like Sardinian cats, were introduced wild from North Africa.

No skulls are available from Crete: only skins. These, as illustrated by Haltenorth (1953, Fig.31–3), and as seen from the types of *F. s. agrius* and *F. s. cretensis*, are rather densely and clearly striped, and very obviously different from *F. s. sarda*, or indeed from any true wild cat known to me. In the absence of skulls, and of external measurements of fresh specimens, I can only record my belief that all Cretan wild cats, despite Haltenorth's proposed separation of them into two types, are actually feral.

The Mallorcan wild cat is smaller than *F. s. sarda*, or indeed most other wild cats: in the type of *F. s. jordansi* the head-and-body length is 51 cm (at the bottom end of the *F. s. sarda* range, 49–62 cm); the tail is short, only 44 per cent of the head-and-body length; the skull is narrow (zygomatic breadth only 70 per cent of condylobasal length, below the *F. s. sarda* range); and the carnassial is 11.0 mm long, within the *F. s. sarda* and western Asian ranges but also within (upper end of) the domestic cat range. As illustrated by Haltenorth (1953, Fig.28–9), the skin is strongly marked, like Cretan specimens. Figure 5.8 shows the Schauenberg Index of a series of skulls in the Geneva Museum: the index is above that for

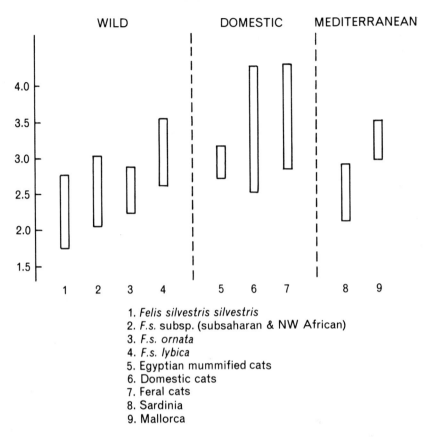

1. *Felis silvestris silvestris*
2. *F.s.* subsp. (subsaharan & NW African)
3. *F.s. ornata*
4. *F.s. lybica*
5. Egyptian mummified cats
6. Domestic cats
7. Feral cats
8. Sardinia
9. Mallorca

Figure 5.8   Relative cranial capacity in wild, feral, and domestic cats. The ranges are for values of Schauenberg's Index (greatest skull length divided by cranial capacity), as described in the text.

Sardinia, with no overlap, and in the range of domestic cats and of *F. s. lybica*. I have no doubt at all that the Mallorcan wild cat is a feral form.

Discussion

The results of the investigations recorded above suggest that mouflon on Corsica and Sardinia are derived from very primitive domestic sheep, those on Cyprus from more fully domesticated sheep; the wild goats of the Aegean islands are derived from very primitive domestic goats; the Corsican and Sardinian wild pigs were introduced, but probably not as domestic animals, from southern Spain; the wild cats of Corsica and Sardinia were introduced, again not as domestic animals, from the

Maghreb, while those of the Balearics and (probably) Crete are feral. Looking at this in another way, the introduced mammals of Sardinia and Corsica include a very primitive domestic sheep from the Middle East, a wild pig from southern Spain, and a cat from North Africa; the Balearics have a feral cat, of uncertain origin; Crete has a very primitive goat from the Middle East (as do two other Aegean islands) and a feral cat of uncertain origin; and Cyprus has a primitive but clearly once domestic sheep.

It was mentioned in the introduction that other wild mammals, of species in which there is (presumably!) no question of domestication, also occur on these Mediterranean islands. It is worth asking what the mainland relationships of these might be, too. As far as I can judge, mainly from descriptions or opinions expressed by commentators in the literature, the island by island relationships appear to be as follows.

Balearics

All subspecies of mammals on the Balearic islands appear to be identical or very close to those on the nearby Spanish mainland. Hemmer *et al.* (1981) record the same thing for a Balearic frog, *Rana perezi*, but find that a toad, *Bufo viridis*, is identical to one on Sardinia and in Israel, and postulate a common Bronze Age origin for them.

Sardinia and Corsica

The two islands have a virtually identical mammalian fauna. The putative indigenous subspecies of *Suncus etruscus*, *Lepus europaeus* (only on Corsica), *Apodemus sylvaticus*, *Glis glis*, *Mustela nivalis*, and *Martes martes* are variously distributed through France, Yugoslavia, and Greece, but all occur in Italy, which on the basis of the (here possibly spurious) parsimony principle may be suggested to be their origin. (Interestingly, *Martes martes latinorum* occurs also on the Balearics: a possible parallel to Hemmer *et al.*'s [1981] toad.) A second group of species has other relationships: *Lepus capensis* (Sardinia only; this being one of the few marked differences between the two islands), *Eliomys quercinus*, and *Vulpes vulpes* of these islands have their closest relatives in southern Spain, like the wild pig; and the distinctive subspecies of red deer, *Cervus elaphus corsicanus*, occurs in both southern Spain and North Africa.

I must admit that there are no direct Middle Eastern affinities in this list. Whoever brought the primitive domestic sheep to these islands seems to have brought nothing else (except toads?). At some other time, trade links involving Sardinia and Corsica, Italy, the Maghreb, southern Spain, and the Balearics succeeded in spreading around a number of other species, all as wild animals. It is tempting to hold the Carthaginians responsible for these transfers (which would, conveniently, provide an ultimate Middle Eastern connection, the Phoenicians), but the evidence will not at present support such a surmise, merely raise it as a hypothesis to be tested.

Aegean islands

Most of the mammals of Crete (the only one of the Greek isles to be tolerably well known faunistically) have detectably mainland Greek affinities: *Erinaceus europaeus, Lepus europaeus, Apodemus mystacinus, Apodemus sylvaticus, Glis glis, Meles meles*. There are, however, a few species whose closest relatives are in Turkey: *Crocidura gueldenstaedti, Acomys cahirinus, Martes foina*. It would, again, be tempting to add 'primitive domestic goats' to this latter list, but at present it would be going beyond the evidence, and I put it forward only as a working hypothesis.

Cyprus

The present-day mammal fauna of Cyprus is little studied. Archaeologically, sheep as well as goat, pig, cat, and fallow deer occur from the Neolithic at least to the early Bronze Age (Davis 1984): apparently occasionally as wild as well as domestic or semi-domestic forms (Schwartz, 1973). The presence of fallow deer, at least, implies a mainland western Asian origin for part of these imports: if the species is correctly reported as *Dama mesopotamica* rather than *D. dama*, then the Levant/Palestine region, not Turkey, is implied.

Conclusions

Some of the 'domesticable' mammals at present living wild on the Mediterranean islands are indeed feral, from very primitive domestic stock; these are the mouflon and the Aegean wild goats, and as such they are extremely valuable documents of the early stages of the domestication process. However, the wild cats and pigs of Corsica and Sardinia show no signs of ever having been domesticated; they would seem to be part of a complex of trade links, involving transport of wild animals, around the ancient western Mediterranean. The wild cats of Crete and the Balearics are almost certainly feral cats, whether particularly primitive or not being at present uncertain.

References

Azzaroli, A. 1981. Cainozoic mammals and the biogeography of the island of Sardinia, western Mediterranean. *Palaeogeography, Palaeoclimatology, Palaeoecology* **36**, 107–11.

Azzaroli, A. 1982. Insularity and its effects on terrestrial vertebrates: evolutionary and biogeographic aspects. In *Palaeontology, essential of historical geology*, E. M. Gallitelli (ed.), 193–213. Modena: S. T. E. M. Mucchi.

Davis, S. J. M. 1984. Khirokitia and its mammal remains: a neolithic Noah's Ark. *Fouilles récents à Khirokitia (Chypre) 1977–1981*, A. Le Brun (ed.), 147–62. Paris: Recherche sur les Civilisations.

Davis, S. J. M. 1987. *The archaeology of animals.* London: Batsford.

Dermitzakis, M. D. & P. Y. Sondaar 1979. The importance of fossil mammals in reconstructing paleogeography with special reference to the Pleistocene Aegean archipelago. *Annales géologiques des Pays helleniques* **29**, 808–40.

Groves, C. P. 1981. *Ancestors for the pigs: taxonomy and phylogeny of the genus Sus.* Canberra: ANU Press.

Groves, C. P. 1983. Pigs east of the Wallace Line. *Journal de la Société des Océanistes* **39**, 105–19.

Haltenorth, T. 1953. *Die Wildkatzen der alten Welt.* Leipzig: Geest & Portig K–G.

Hemmer, H. 1976. Man's strategy of domestication – a synthesis of new research trends. *Expérientia* **32,** 663–6.

Hemmer, H. 1983. *Domestikation: Verarmung der Merkwelt.* Braunschweig/Wiesbaden: Wieweg.

Hemmer, H., B. Kadel & K. Kadel 1981. The Balearic toad (*Bufo viridis balearicus* (Boettger, 1881)), human bronze age culture, and Mediterranean biogeography. *Amphibia–Reptilia* **2**, 217–30.

Herre, W. & M. Röhrs 1973. *Haustiere – zoologisch gesehen.* Stuttgart: Gustav Fischer.

Kruska, D. & M. Röhrs 1974. Comparative–quantitative investigations on brains of feral pigs from the Galapagos Islands and of European domestic pigs. *Zeitschrift für Anatomie und Entwicklungsgeschichte* **144**, 61–73.

Pees, W. & H. Hemmer 1980. Hirngrosse und Aktivitat bei Wildschafen und Hausschafen (Gattung *Ovis*). *Säugetierkundliche Mitteilungen* **28**, 39–45.

Pfeffer, P. 1967. Le mouflon de Corse (*Ovis ammon musimon* Schreber, 1782); position systématique, écologie et éthologie comparées. *Mammalia* **31** (suppl.), 1–262.

Rudge, M. R. 1984. The occurrence and status of populations of feral goats and sheep throughout the world. In *Feral mammals – problems and potential,* P. N. Munton, J. Clutton–Brock & M. R. Rudge (eds), 57–84. Morges: IUCN.

Schultze–Westrum, T. 1963. Die Wildziegen der ägaischen Inseln. *Säugetierkundliche Mitteilungen* **11**, 145–82.

Schwartz, J. H. 1973. The palaeozoology of Cyprus: a preliminary report on recent analysed sites. *World Archaeology* **5**, 215–20.

Sondaar, P. Y. 1977. Insularity and its effect on mammal evolution. In *Major patterns in vertebrate evolution,* M. K. Hecht, P. C. Goody & B. M. Hecht (eds), 671–707. New York: Plenum Publishing.

Uerpmann, H.-P. 1981. *Ovis musimon* Schreber, 1782, oder *Ovis musimon* Pallas, 1811? *Säugetierkundliche Mitteilungen* **29**, 59–60.

Valdez, R. 1982. *The wild sheep of the world.* Mesilla (New Mexico): Wild Sheep and Goat International.

6 Escaped domestic animals and the introduction of agriculture to Spain

IAIN DAVIDSON

Frontiers between fisher-gatherer-hunters and farmers

Contact between fisher-gatherer-hunters and agricultural or pastoral people creates spatial frontiers. Prehistorians may also consider that there is a temporal frontier (see also Dennell 1985, p. 114). Interaction at either of these frontier classes (spatial or temporal) has been considered by Alexander (1976, 1984) and others (Hall *et al.* 1984, Green & Perlman 1985), with special reference to the implications for our understanding of the archaeological record. Archaeologists have turned to this theoretical issue relatively recently, and, therefore, the concerns have been more with recent issues in archaeology related to social organization, than with ecological aspects of the record. In particular, most emphasis has been on the behaviour of people rather than with the ecology of plants and animals at the frontier. This chapter argues that the documented example of the behaviour of animals during the European colonization of Australia suggests that the ecological autonomy of plants and animals was an important part of the interaction between human groups at the frontier.

Dennell (1985) provides a useful introduction to the history of the study of the moving frontier between the first agriculturalists and the non-agricultural indigenous inhabitants of prehistoric Europe. Although there are some difficulties with the account (e.g. claims of lack of evidence for violence at the frontier could be countered by some interpretations of the chronology of the Levantine art of eastern Spain, which depicts opposed groups of humans using bows and arrows), Dennell makes many good points about the archaeology of the European frontiers. This chapter, however, concentrates on an aspect which Dennell only touches on (1985, pp. 130, 132): natural movement across the frontier of the plants and animals used by agriculturalists.

There are, in reality, four categories of people concerned at these frontiers:

A 'Hunters': those who gain their subsistence from plants and animals that are not cultivated or over which breeding control is not exercised.
B Those who have some mixture of subsistence based on cultivation or control and subsistence based on other resources:
    (i) 'cultivating hunters': hunters who cultivate, thus being closer to category A;

(ii) 'hunting agriculturalists': agriculturalists who hunt, thus being closer to category C.

C   Those who gain no *important* subsistence from such non-cultivated or uncontrolled resources ('agriculturalists').

Dennell (1985, p. 114), indeed, goes further and points to the academic frontier which also separates the study of the behaviour of prehistoric fisher-gatherer-hunters from the study of prehistoric agricultural societies. There is, in addition, a curious phenomenon, which has also been noted in the literature on modern fisher-gatherer-hunters (Ingold 1984, p. 5). Scholars tend to classify people out of the category of fisher-gatherer-hunters if they seem to have any attributes of agriculture or pastoralism. The existence of the four categories complicates the interpretation of the archaeological record still further as blurring takes place of the criteria for recognizing which side of the frontier particular archaeological sites may be on.

In general, we are dealing with a literature which tends to describe all those who are *not* in category A (hunters) as having changed in status from fisher-gatherer-hunters to agriculturalists however little the degree of dependence on cultivated and controlled resources. Little attention is paid to understanding the groups which are not in category C (agriculturalists) but cannot be said to be in category A (hunters) because of the lack of importance of the subsistence they gain from cultivation or control. This seems to be the case for archaeological sites as well as in the ethnographic literature.

Definitions must be very carefully applied here. This is because there are important status shifts, as people who routinely obtain their important plant foods from cultivated crops also gather plants from the uncultivated environment as relishes or luxury foods (truffles, or blackberries, for example). By the same token, hunting is now a high status activity in some social groups (grouse shooting, fox hunting) and an important recreation in others (pig or kangaroo shooting, bird shooting in Mediterranean countries). In these circumstances, it is difficult to separate the subsistence and social components of the activity.

The Australian model

When the Australian Aborigines were first seen by agriculturalists from Europe, they were described by those agriculturalists. Few, if any, other fisher-gatherer-hunters were described for posterity by the first group of agriculturalists who saw them. Indeed, Woodburn (1986) has recently suggested that many of the recent groups of fisher-gatherer-hunters have determined their behaviour by their encapsulation by surrounding agricultural societies. In all other cases, there was some local contact between groups where the contact was prehistoric. In the first European colonization of Australia, contact was between groups of non-Aboriginal agriculturalists who wrote of their experiences (see, for example, King

1984) and Aboriginal fisher-gatherer-hunters who did not. In northern Australia, Aborigines were visited by Macassans (MacKnight 1976) or had contact with horticulturalists of Papua (Harris 1979). Apart from these contacts, the Australian Aborigines lived on a continent of fisher-gatherer-hunters. Unlike, say, the San, the Inuit, the Shoshoni, the Tasaday, or the Mbuti, they could have had no experience with a radically different way of life. We can gain some idea from this encounter about how introduced materials and species crossed the frontier between history and prehistory, and between the world of agriculturalists and that of fisher-gatherer-hunters.

Materials crossing the frontier

Reynolds (1982) has documented in detail the extent of penetration of the frontier between Aborigines and non-Aborigines in Australia. Goods doubtless travelled along traditional trading and exchange routes (Mulvaney 1976) between Aboriginal groups which were in contact with non-Aborigines as well as with those which were not. Tamar found a button on a necklace in 1804; Carnegie found an iron tent peg and glass objects; Warburton a butcher's knife and a steel axe. Among the first materials which crossed the frontier were animals.

Escaped domestic animals

The first cattle to land in Australia, five cows and a bull, escaped from the control of the British settlers on 5 June 1788, a mere 19 weeks after landing (King 1984, p. 73). They, or their offspring, were not rediscovered until 1795 when an Aboriginal fisher-gatherer-hunter reported them to a European convict. Clegg (1984) and Lyon & Urry (1979) have shown that these first escaped cattle were depicted by Aborigines in the paintings at Bull cave near Sydney.

Rolls (1984, p. 16) has given an account of this and other escapes from pastoral control. Almost every other useful animal, and some which were not, went wild (Rolls 1969). The spread of rabbits is not only well known to Europeans, but remembered by Aborigines (Reynolds 1982, p. 10). Horses were seen by explorers, such as Leichhardt in 1844 and McKinlay in 1861, way beyond the regions then settled by non-Aborigines (Reynolds 1982, p. 10). Pigs, foxes, goats, dogs, buffalo, camels, and chickens all have feral populations in Australia, but, curiously, sheep do not. One reason for this might be the vulnerability of woolly sheep to fly strike and dingo attack if left unshorn. This would certainly have been a factor with the fat-tailed sheep of the First Fleet (1788) and the *Gorgon* (1791). despite the fact that their coats were primarily of hair and not wool (Garran & White 1985, pp. 7–9). The second introduction of fat-tailed sheep grew 'too fat to breed' (Governor Phillip, quoted by Garran & White 1985, p. 13), a problem which was solved by the import of hairy sheep from India in 1791. Woolly coats need not have been a problem for the escaping sheep in the earliest phases of the invasion. If it was some other ecological intolerance to Australian climates which inhibited the escape of sheep, no such barrier need be supposed in Europe where the species were dispersing near to the same biogeographic province in which they originated.

Discussion about the suitability of the vegetational environment for the introduced species depends on opinion about the extent of environmental modification by non-agricultural people in Europe, on one hand, and in Australia, on the other. Clearly this is one area of interpretation in which understanding is not improved by use of an analogy between recent Australia and prehistoric Europe. On the other hand, we might also argue that the successful survival of escaped domesticates to become feral populations was easier in Australia than it would have been for domestic species dispersing within a single biogeographic province. The situation for interactions between introduced and native fauna is much simpler. Competition with the indigenous fauna would have been greater in Europe, and there would have been more large predators than in Australia. Amongst the predators in both cases were indigenous people.

Other considerations

Australian history also suggests some appropriation of animals from tended herds. Bradley reported the first spearing of a goat on 21 August 1788 (King 1984, p. 89), only 30 weeks after landfall. In a remarkable story (Campbell 1933), one of the massacres of Aborigines near Armidale, in northern New South Wales, was justified by the belief that a group of Aborigines had followed some pastoralists and their sheep down steep wooded country and stolen their whole flock. The Aborigines, fisher-gatherer-hunters with at most a couple of decades of knowledge of the pastoral habits of the non-Aborigines, herded the sheep into the steep gorge of Kunderang Brook. When the sheep and their Aboriginal rustlers were found, massacre was deemed to be the appropriate punishment. Similar stories abound (Reynolds 1982, pp. 156–69). Such theft from pastoralists is, of course, known in other situations of contact between fisher-gatherer-hunters and pastoralists, as Schrire (1980) has documented for the San of southern Africa. Archaeologically, sites of people who stole livestock in this way would look like the result of activities of people in category B(i) (cultivating hunters), although in reality they would not have moved out of our category of hunters.

Jones (1970) documented the rapidity with which Tasmanian Aborigines adopted European dogs. Despite the fact that the Kunderang story documents the ease with which Aborigines could learn the physical manipulation of flocks of sheep, there is no evidence that Aborigines showed any desire to adopt a pastoral way of life. Of course, the history of the European invasion of Australia militated against it.

The early days of European colonization of Australia were times of great hardship for the colonists. Their stock died, were stolen, or escaped. The crops failed to produce sufficient food or withered in the unaccustomed heat. The diet of the Europeans had to be supplemented by the hunting of kangaroos and the catching of fish (see King 1984). In archaeology this might be interpreted as people in category B(ii) (hunting agriculturalists).

In Australia the historical archaeology of such contact would have some relatively simple problems of interpretation because of the differences between the species indigenous to Australia and those which were introduced. In interpreting the archaeology of prehistoric diffusion, as in Europe, the problems would be far more difficult, because of the similarities between species, and because of the possibilities of local indigenous domestications.

The model

It is inconceivable, in a situation of diffusion of new plants and animals into a strange environment, that the introduced plants and animals did not have any opportunities to escape from the control of agriculturalists or pastoralists. Without even appealing to the conditions in the initial non-Aboriginal colonization of Australia, we may follow Lewthwaite (1984, p. 26) in pointing to the escape of pastoral animals from loose herding arrangements. Thus, it is possible to point to an analogy for the escape of animals either in the circumstances of the introduction of domestic stock with groups of invading people, or at any stage in their introduction.

Once we postulate the existence of escaped domestic animals, we then have to assume that these creatures became available as resources for the indigenous populations of people. If we take an over-simplified model of Mesolithic fisher-gatherer-hunters and Neolithic farmers, then we should expect that some of the faunal assemblages of Mesolithic sites should contain bones of sheep or goats which had escaped from the custody of their Neolithic herders and had been hunted as game by the indigenous population.

Either the escape of the domestic animals to an unprotected state, or the theft of beasts from tended herds, should lead to the presence of such animals in Mesolithic archaeological assemblages. Twenty years ago Helbaek (1966) wrote of 'cultivated wild barley' which has since been shown likely to have been roofing material and unrelated to diet (Dennell 1972). Now we must speak of the likelihood of *hunted domestic animals*. People who exploited such animals would be in category A (hunters) and would be difficult to identify as such, because in identifying the escaped domestic species we might be inclined to include them in category B(i) (cultivating hunters). Without these fine divisions there would be a temptation to identify such people as incipient pastoralists.

In this model there should exist sites of groups in category A (hunters), which are prior to those of groups in category C (agriculturalists). Diffusion of agriculture and pastoralism from outside will lead to the existence of sites which are in reality those of groups of hunters, but because the hunted prey are escaped domestic animals they *appear to be* cultivating hunters. Such sites should be contemporary with those of agriculturalists. The Australian model suggests that even the material culture will be an unreliable guide to the true status of these sites because material culture can also cross the frontier easily.

Australian history might suggest that early agriculturalists should also hunt and fish to a significant extent, and hence be hunting agriculturalists. This possibility will be clarified when we understand the process of agricultural spread in such a way as to take into account the possibility of hunted domestic animals.

Two consequences follow. The first is a consideration of the methods which might permit us to recognize such a phenomenon. The second arises because there do not seem to be sites in Spain which show the presence of escaped domesticates which were hunted (or, in the case of plants, gathered). The model would predict that there should be such a body of evidence, and we need to consider the theoretical implications of such a lack.

The frontier in Spain

There are two competing interpretations of the beginning of agriculture and stock-raising in Spain. On the one hand there is the diffusionist model which postulates that all such economic innovations came from the east, and on the other is the model of indigenous development (see Geddes 1980, pp. 56–80 for a summary of the arguments as they apply to southern France, and Lewthwaite 1986 for a comprehensive synthesis of the cultural evidence for the whole western Mediterranean). Even extreme proponents of indigenous development could not account for all of the features which appear for the first time at Coveta de l'Or (Martí *et al.* 1980).

Cattle, boars, wolves, and caprines occurred in the Spanish Pleistocene fauna and, by some bending of the orthodoxy about domestication, could have been the ancestors of domestic animals which appeared in the early Neolithic sites. But there was no wild progenitor of sheep. We know next to nothing about the Pleistocene plant exploitations (see summary in Davidson 1980), but nobody has claimed that there could have been ancestors of wheat or barley in the Peninsula. There is agreement that the Epipalaeolithic and the Neolithic stone industries showed some similarities, but few are happy with the thought that pottery was developed indigenously in the Peninsula. As things stand at the moment, we would probably need to postulate some diffusion into the Peninsula of sheep, wheat and barley, and probably goats and pottery (see Lewthwaite 1986).

It does not matter very much what the conditions of this diffusion were. In particular, we might postulate whether groups of people carried them into territory occupied by indigenous fisher-gatherer-hunters or exchanged them with neighbouring groups whom they instructed in the domestic sciences.

In a small number of cases where determinations are possible, there seems to be a general absence of remains of the introduced animals in the sites with the earliest pottery in the Peninsula: Mallaetes (Davidson 1983); Niño (Davidson 1980, Ch. 10, Fortea *et al.* 1983); Verdelpino (Morales

1977); Botiqueria dels Moros (Altuna 1978); Cocina (Fortea *et al*. n.d.); Nacimiento (Alférez *et al*. 1981); and Valdecuevas (Sarrión 1980). One exception to this is Nerja (Boessneck & von den Driesch 1980, p. 24) where some specimens were found which were described as ovicaprines. It was further stated that they were indistinguishable from ovicaprines which would have been called domestic if they had been found in a Neolithic context. It would seem, from their statement, that Boessneck & von den Driesch (1980) are not willing to apply their morphological criteria without taking into account the cultural context, thus begging the question of the cultural context in which agriculture emerged. The problem here is similar to the problem of the southern French sites such as Chateauneuf and Gramari. Even if we accept both the stratigraphic context of the finds, and the identifications, it is difficult to be clear about the nature of the economy which was responsible for them, as will be argued below.

Two other Spanish sites are crucial here: Matutano and Fosca. Both of these are in Castellón province and show evidence for changes in the fauna through time. These are the sorts of changes which have been identified elsewhere as those associated with the beginnings of domestication (Estévez *et al*. 1983, Estévez pers. comm.). Estévez and his colleagues have argued for the local operation of the process of domestication. The argument depends on the application of the same criteria that have been used in the eastern Mediterranean. These are: the presence of wild and domesticated animals; presence of transitional animals; difference from the composition of a wild population; representation of scenes of capture; survival of aberrant individuals; and different butchery of wild species and species in the process of domestication (Bökönyi 1977).

Let us accept for a moment that hunted domestic animals would have existed. Would our methods be suitable for identifying them? Morphologically and in terms of size the first such animals would have been indistinguishable from their domestic progenitors (cf. Groves, ch.5, this volume), although some changes might be expected after several generations of a more natural selection. Slaughter patterns should follow those of the hunting strategy applied to animals of similar size or behaviour. If the introduction of agriculture involved the movement of species into a similar biogeographical province, one might expect that the previously existing faunal populations would have been well suited to the existing conditions. This suitability might have prevented the rapid successful expansion by the newly escaped domestic species. As a result, populations of the escaped species might have been low and the killed animals so few as to make it difficult to estimate the slaughter patterns from an archaeological assemblage.

Simple interpretation of the fauna suggests the presence of both wild and domestic species, but this tells us nothing about the relationships between the people and the animals. In the case of sheep, there would not necessarily be any transitional animals, although goats might interbreed with the local Spanish ibex (Galindo 1965, pp. 35–6). Galindo suggests that the offspring of such crosses seem to resemble their wild sires more than their domestic dams. Representations of scenes of capture could not enable us to distinguish between the hunting of domestic species and the normal activities of pastoralists.

We are left with two only of the criteria used by Estévez and his colleagues (Bökönyi 1977, Estévez *et al.* 1983) – the survival of aberrant individuals and the patterns of butchery.

Aberrant individuals can be found in large samples of ungulates slaughtered in the wild (Peter Jarman pers. comm. May 1986). They would, however, normally be scarce, for that is why we recognize them as aberrant. The assumption is that under normal conditions of natural selection they would not survive, and that only human protection allows them to reach an age to be slaughtered by their protectors. Jarman's observation suggests that the assumption is a weak one. In addition, they, and their absence, would not necessarily be a satisfactory clue that the relationship was not one of pastoralism. They might simply have escaped human predation, or not been killed at a particular site.

There remains the possibility that hunters would butcher hunted domestic animals in a manner distinct from that used by the contemporary pastoralists, but similar to their practices with wild prey. It is certainly true that we need to pay far more attention to the patterns of butchery, as only Estévez (in Olaria *et al.* 1981) has so far published such details for Mediterranean Spain. Nevertheless, we might appeal to Binford's (1978) study of the economic anatomy of reindeer to suggest that the utility of particular parts of an animal in any given set of environmental conditions is tied relatively strongly to the nature of the animals and less to the cultural practices of the people.

It may be that the identification of hunted domestic animals is beyond our current methods of analysis of the archaeological record.

Of course, Higgs & Jarman (1969) suggested long ago that application of the conventional criteria for the identification of domestication could produce some strange results if selectively applied to the interpretation of partial prehistoric evidence. They, therefore, suggested that a more flexible approach to 'man–animal' relationships (Shawcross 1975 referred to *zoocresis* which would avoid the sexist terminology), needed to be adopted. Consideration of the problems involved in the identification of escaped domestic animals suggests that their caution still deserves respect.

Moreover, the principle that animals will escape from control in many pastoral systems brings into question some of the logic by which scholars have looked for the origins of agriculture in the regions of modern distribution of the supposed wild ancestors of the domesticates (see, for example, the papers by Herre & Röhrs, on animals, and Harlan, on plants, in Reed 1977). It is at least plausible that many of the modern animals and plants escaped from control at some early stage in the history of domestication, to form early *feral* populations in the region of their modern distribution. On this argument, the modern distributions do not represent the prehistoric ranges of the *wild* species concerned, let alone the distribution of the species which were the wild *ancestors* of perhaps both the domesticates and the modern representatives of the non-domestic species.

The agricultural colonization of Spain

How might the expectation that animals escaped beyond the frontier of agricultural advance affect our interpretation of Spanish prehistory?

In some of the eastern Spanish sites (Parpalló, Les Mallaetes, Niño) the beginnings of an indigenous process of domestication or the beginnings of pastoralism cannot readily be identified (Davidson 1980, 1983, Bailey & Davidson 1983). This leads to a consideration of the conditions which operate in a situation where agriculture and pastoralism are introduced from outside.

There is very little evidence for introduced species of plants and animals in the non-pottery post-Pleistocene sites of eastern Spain. This contrasts with the situation in southern France where there are at least claims for sheep, in small numbers, from Chateauneuf (Ducos 1976), Gramari (Poulain 1971) and other Mesolithic sites (Rozoy 1978). There are difficulties in chronology and identification: Ducos, for example, acknowledges that his identifications were made before the publication of the criteria for distinguishing sheep from goat (Boessneck et al. 1964). Nevertheless, the finds are such as one would expect from a situation of escaped animals being hunted by the indigenous non-pastoral human population. Geddes (1985) has reviewed the old evidence and provided detailed descriptions of specimens from more recent excavations at the late Mesolithic sites of Gazel, and Dourgne. Commenting on these and other remains of domestic animals in Mesolithic sites, he attributes them to 'isolated survivals of Mesolithic groups which acquired domestic animals by trade, theft or other social means.' (Geddes 1985, p. 26). Geddes uses the phrase 'incipient animal herding', and suggests 'that indigenous Mesolithic communities could have constituted a principal component in the emergence of settled farming communities in Mediterranean Europe.' (Geddes 1985, pp. 44–5). If this was the case, then why are there so few cases in France, and none in Spain?

One possible explanation is that the early sites with pottery and high proportions of hunted animals are not the sites of the first farmers, but of fisher-gatherer-hunters who obtained pots across the frontier and hunted escaped domestic animals. Some sites, such as Or (Martí et al. 1980) contain such abundant quantities of pottery and of cereal grains that they probably represent the initial occupation by people in category B(ii) (hunting agriculturalists). Other sites, such as Nerja, Parralejo and Dehesilla (see Muñoz 1984) probably fall into category A (hunters).

The site of La Cocina deserves attention. For long regarded as a classic Mesolithic site (Pericot 1945, Fortea 1971), Fortea's recent excavations allow the first interpretation of faunal exploitation by the inhabitants of this cave (Fortea et al., unpublished) through the identifications by Pérez Ripoll. In the layers without pottery there were no remains of species which might have been thought to be domestic, and analysis of the ages at death of the Spanish ibex showed that the high proportions of adult and old individuals contrasted with the high proportions of juveniles at Or (Pérez Ripoll, in Martí et al. 1980). There was no change in this proportion in the upper layers at La Cocina, which contained cardial pottery, although small numbers of bones were identified as *Ovis/Capra*. This looks like a perfect example of a site used by people in category A (hunters), who hunted escaped domestic animals.

Conclusions

The purpose of this chapter has been to draw attention to the continuing difficulty of recognizing the relationships which groups of prehistoric people had with animals. If we confine ourselves to recognizing the status of animals as wild or domestic, then the possibility that people hunted escaped domestic animals would confuse most interpretations. Similar problems might exist with plants, although the properties which allow plants to escape from cultivation to reproduce without human intervention might actually have direct morphological expression.

If we consider a model of diffusion of agriculture into a new environment already populated by fisher-gatherer-hunters, then we would predict that both material culture and animals would cross the frontier between the two ways of life. Some sites are classified as being created by agriculturalists or pastoralists because they contain both pottery and the new species which might be identified as domesticated. In the absence of clear evidence in Spain that the fisher-gatherer-hunters of the Epipalaeolithic or Mesolithic did exploit the escaped domesticates of the early pastoralists, we should consider carefully whether some of the sites which are generally regarded as those of Neolithic agricultural and pastoral groups should be reclassified as the indications of the hunters of escaped domesticates. Full testing of this hypothesis will require more detailed re-analysis of individual sites.

Acknowledgements

This chapter has been improved enormously by the generous and thoughtful comments of J. Clegg, G. E. Connah, J. Driver, R. Fletcher, A. Gilman, L. Godwin, M. Jackes, P. Jarman, D. Lubell, S. Solomon, D. Witter, and J. P. White. I also acknowledge helpful discussions with I. Plug and E. Voigt about the prehistoric frontier in southern Africa. G. E. Connah also suggested that the comparison was worth making – but that would be another chapter. Only I am responsible for the final form of this one.

References

Alexander, J. 1976. The frontier concept in prehistory. In *Hunters, gatherers and first farmers beyond Europe*, J. V. S. Megaw (ed.), 25–40. Leicester: Leicester University Press.
Alexander, J. 1984. Early frontiers in southern Africa. In *Frontiers: southern African archaeology today*, M. Hall, G. Avery, D. M. Avery, M. L. Wilson & A. J. B.

Humphreys (eds), 12–23. Oxford: BAR International Series 207.

Alférez, F., G. Molero, V. Bustos & P. Brea 1981. Apendice II. La fauna de macromamíferos. *Trabajos de Prehistoria* **38**, 139–45.

Altuna, J. 1978. Fauna del yacimiento prehistórico de Botiqueria dels Moros, Mazaleon (Teruel). *Cuadernos de Prehistoria y Arqueologia Castellonense* **5**, 139–42.

Bailey, G. N. & I. Davidson 1983. Site exploitation territories and topography: two case studies from Palaeolithic Spain. *Journal of Archaeological Science* **10** (2), 87–115.

Binford, L. R. 1978. *Nunamiut Ethnoarchaeology*. New York: Academic Press.

Boessneck, J. & A. von den Driesch 1980. Tierknochenfunde aus vier Südspanische Hölen. *Studien über frühe Tierknochenfunde von der Iberischen Halbinsel* **7**, 1–83.

Boessneck, J., H. H. Müller & M. Teichert 1964. Osteologische unterscheidungs-merkmale zwischen Schaf (*Ovis aries* Linné) und Ziege (*Capra hircus* Linné). *Kühn–Archiv* **78**, 1–129.

Bökönyi, S. 1977. *Animal remains from the Kermanshah Valley, Iran*. Oxford: BAR Supplementary Series 34.

Campbell, J. S. 1933. The Kunderang ravines of New England. *Royal Australian Historical Society Journal and Proceedings* **18**, 63–73.

Clegg, J. 1984. Pictures 'of' bulls and boats. In *Under the shade of a coolibah tree: Australian studies in consciousness*, R. A. Hutch & P. G. Fenner (eds), 219–38. Lanham, Maryland: University Press of America.

Davidson, I. 1980. *Late Pleistocene economy in eastern Spain*. Unpublished PhD thesis, Department of Archaeology, University of Cambridge.

Davidson, I. 1983. Site variability and prehistoric economy in Levante. In *Hunter–gatherer economy in prehistory*, G. N. Bailey (ed.), 79–95. Cambridge: Cambridge University Press.

Dennell, R. W. 1972. The interpretation of plant remains. In *Papers in economic prehistory*, E. S. Higgs (ed.), 149–59. Cambridge: Cambridge University Press.

Dennell, R. W. 1985. The hunter-gatherer/agricultural frontier in prehistoric temperate Europe. In *The archaeology of frontiers and boundaries*, S. W. Green & S. M. Perlman (eds), 113–39. New York: Academic Press.

Ducos, P. 1976. Quelques documents sur les débuts de la domestication en France. In *La préhistoire française*, Vol. II, J. Guilaine (ed.), 165–7. Paris: CNRS.

Estévez, J., F. Gusi, C. Olaria, A. Vila & R. Yll 1983. Evolución ambiental y desarrollo de la base subsistencial hasta el 7000 bp en el Levante Ibérico. In *Résumé des Communications. Premières Communautés Paysannes en Méditerranée Occidentale*, p. 27. Montpellier: UISPP.

Fortea, J. 1971. *La Cueva de la Cocina*. Valencia: Trabajos Varios del Servicio de Investigación Prehistórica, Diputación Provincial.

Fortea, J., B. Martí, M. P. Fumanal, M. Dupré & M. Pérez Ripoll n.d. *Epipaleolítico y neolitización en la zona oriental de la Peninsula Ibérica*. (Unpublished.)

Fortea, J., J. M. Fullola, V. Villaverde, I. Davidson, M. Dupré & M. P. Fumanal 1983. Schema paléoclimatique, faunique et chronostratigraphique des industries à bord abattu de la région méditerranéenne espagnole. *Rivista di Scienze Prehistoriche* **38** (1–2), 21–67.

Galindo, F. 1965. La Capra Pyrenaica Hispanica de los Puertos de Beceite (Teruel). *Teruel* **33**, 5–76.

Garran, J. C. & L. White 1985. *Merinos, Myths and Macarthurs. Australian graziers and their sheep, 1788–1900*. Canberra: Australian National University Press.

Geddes, D. S. 1980 (1983). *Patterns of animal exploitation in the late Mesolithic and early Neolithic in the Aude Valley (southern France)*. Ann Arbor: University Microfilms International.

Geddes, D. S. 1985. Mesolithic domestic sheep in west Mediterranean Europe. *Journal of Archaeological Science* **12** (1), 25–48.

Green, S. W. & S. M. Perlman (eds) 1985. *The archaeology of frontiers and boundaries.* New York: Academic Press.

Groves, C. P. 1989. Feral mammals of the Mediterranean islands: documents of early domestication. In *The walking larder*, J. Clutton–Brock (ed.), ch. 5, London: Unwin Hyman.

Hall, M. J., D. M. Avery, M. L. Wilson & A. J. B. Humphreys (eds) 1984. *Frontiers: southern African archaeology today.* Oxford: BAR International Series.

Harlan, J. R. 1977. The origins of cereal agriculture in the Old World. In *Origins of agriculture*, C. A. Reed (ed.), 357–83. The Hague: Mouton.

Harris, D. R. 1979. Foragers and farmers in the western Torres Strait islands: an historical analysis of economic, demographic, and spatial differentiation. In *Social and ecological systems*, P. C. Burnham & R. F. Ellen (eds), 75–109. London: Academic Press.

Helbaek, H. 1966. Pre-pottery neolithic farming at Beidha. *Palestine Exploration Quarterly* **98**, 61.

Herre, W. & M. Röhrs 1977. Zoological considerations on the origins of farming and domestication. In *Origins of agriculture*, C. A. Reed (ed.), 245–79. The Hague: Mouton.

Higgs, E. S. & M. R. Jarman 1969. The origins of agriculture: a reconsideration. *Antiquity* **43**, 31–43.

Ingold, T. 1984. Time, social relationships and the exploitation of animals: anthropological reflections on prehistory. In *Animals and Archaeology*. Vol. 3: *Early herders and their flocks*, J. Clutton–Brock & C. Grigson (eds), 3–12. Oxford: BAR International Series 202.

Jones, R. 1970. Tasmanian Aborigines and dogs. *Mankind* **7** (4), 256–71.

King, J. 1984. *The first settlement. The convict village that founded Australia, 1788–1790.* South Melbourne: Macmillan.

Lewthwaite, J. 1984. The art of corse herding: archaeological insights from recent pastoral practices on west Mediterranean islands. In *Animals and archaeology*. Vol. 3: *Early herders and their flocks*, J. Clutton–Brock & C. Grigson (eds), 25–37. Oxford: BAR International Series 202.

Lewthwaite, J. 1986. From Menton to the Mondego in three steps: application of the availability model to the transition to food production in Occitania, Mediterranean Spain and southern Portugal. *Arqueologia (Porto)* **13**, 95–119.

Lyon, K. & J. Urry 1979. Bull shelter: a cow pastures' conundrum. *A.I.A.S Newsletter* **11**, 39–45.

MacKnight, C. C. 1976. *The voyage to Marege'. Macassan trepangers in northern Australia.* Melbourne: Melbourne University Press.

Martí, B., V. Pascual, M. D. Gallart, P. Lopez García, M. Pérez Ripoll, J. D. Acuña & F. Robles 1980. *Cova de l'Or (Beniarrés, Alicante).* Vol. II. Valencia: Trabajos Varios del Servicio de Investigación Prehistórica, Diputación Provincial.

Morales, A. 1977. Apendice I. Análisis faunístico de Verdelpino (Cuenca). *Trabajos de Prehistoria* **34**, 69–81.

Mulvaney, D. J. 1976. 'The chain of connection': the material evidence. In *Tribes and boundaries in Australia*, N. Peterson (ed.), 72–94. Canberra: Australian Institute of Aboriginal Studies.

Muñoz Amabilia, A. Ma. 1984. La neolitización en España: problemas y lineas de investigación. In *Scripta Praehistorica. Francisco Jordá Oblata*, J. Fortea (ed.), 349–69. Salamanca: Ediciones Universidad de Salamanca.

Olaria, C., F. Gusi, J. Estévez, J. Casabo & M. L. Rovira 1981. El yacimiento magdaleniense superior de Cova Matutano (Villafamés, Castellón). Estudio del sondeo estratigráfico 1979. *Cuadernos de Prehistória y Arqueología Castellonenses* **8**, 21–100.

Pericot, L. 1945. La Cueva de la Cocina (Dos Aguas). *Archivo de Prehistória Levantina* **2**, 39–71.

Poulain, T. 1971. Le camp mésolithique de Gramari à Méthamis (Vaucluse). III. Étude de la faune. *Gallia Préhistoire* **14**, 121–31.

Reed, C. A. 1977. A model for the origin of agriculture in the Near East. In *Origins of agriculture*, C. A. Reed (ed.), 543–67. The Hague; Mouton.

Reynolds, H. 1982. *The other side of the frontier*. Ringwood, Victoria: Penguin.

Rolls, E. C. 1969. *They all ran wild: the story of pests on the land in Australia*. Sydney: Angus and Robertson.

Rolls, E. C. 1984. *A million wild acres*. Ringwood, Victoria: Penguin.

Rozoy, J. G. 1978. *Les Derniers Chasseurs*. Bulletin de la Société Archéologique Champenoise, Charleville.

Sarrión, I. 1980. Valdecuevas. Estación Meso-neolítica en la Sierra de Cazorla (Jaén). *Saguntum* **15**, 23–56.

Schrire, C. 1980. An inquiry into the evolutionary status and apparent identity of San hunter-gatherers. *Human Ecology* **8**, 9–32.

Shawcross, F. W. 1975. Some studies of the influences of prehistoric human predation on marine animal population dynamics. In *Maritime adaptations of the Pacific*, R. W. Casteel & G. I. Quimby (eds), 39–66. The Hague: Mouton.

Woodburn, J. 1986. *African hunter-gatherer social organisation. Is it best understood as a product of encapsulation?* Unpublished paper given at the Fourth International Conference on Hunting and Gathering Societies, London, September.

7 Evidences for the impact of traditional Spanish animal uses in parts of the New World

ELIZABETH S. WING

Columbus' celebrated voyage and subsequent ones brought great changes to the aboriginal way of life and the faunas of the western hemisphere. Even the very first exploratory ventures were accompanied by European domestic animals and the commensal rat, as well as by a pattern of subsistence developed over centuries in the Iberian peninsula. Some of the Spanish traditions of animal use and exploitation had an impact on the aboriginal populations, modifying them in a more profound way than any influence during the previous centuries of stability. Spanish traditions of animal use were likewise modified. Studies of the animal remains from three pairs of prehistoric and historic sites in Florida and Haiti are used here to illustrate the magnitude of the changes and the types of changes which took place. The sites are listed in Table 7.1. These samples show which European domesticates first took hold in the New World environment and the cultural and biological factors determining their success are suggested.

Table 7.1 Names and locations of the prehistoric and historic sites.

Sites reviewed	Location	Dates of occupation	Occupants
Hontoon Island 8V0202	central Florida	AD 928–1470 AD 1540–1758	aboriginal aboriginal with indirect influence of Spanish
Fountain of Youth 8SJ31	St. Augustine Florida	late prehistoric-early historic	aboriginal
Various	St. Augustine Florida	AD 1565–1700	Spanish partially provided with food by aboriginals
En Bas Saline	north coast Haiti	contact	aboriginal and possibly some sailors from Columbus' ship
Puerto Real	north coast Haiti	AD 1502–1578	Spanish

The two faunal assemblages from Hontoon Island differ in many respects (Table 7.2). Most of the same species occur in both components of the site but their relative abundance changes. Aquatic vertebrates predominate during both time periods. Those terrestrial species which were procured during historic times were on the average larger fragments and fragments of larger animals than the prehistoric catches. The most obvious difference between the two components is the decrease in the gastropod molluscs in the historic assemblage. During historic times the long-established practice of gathering snails was discontinued, but mussels were still collected. These changes in procurement of animals are accompanied by substantial changes in uses of plants, which suggest clearing and increased agricultural activities (Newsom 1986). These changes are correlated with the advent of Spanish influence in Florida. Further research on such sites is needed to determine how Spanish influence caused these changes.

Further insight into the effect of Spanish traditions may be seen in a comparison of the faunal assemblages from the prehistoric aboriginal site and a number of 16th-century Spanish sites in St. Augustine (Reitz 1985). As at Hontoon, the remains identified from the aboriginal site are predominantly aquatic, though the species encountered are marine and estuarine rather than the freshwater species. Differences in the fishes used by the Indians and the colonists are noted by Reitz (1985). The 16th-century Spanish faunal samples show a greater dependence on terrestrial fauna, as measured by the numbers of remains identified as terrestrial animals in the faunas (Table 7.3). European domestic animals, however, account for only 3–10 per cent of the vertebrates. The Spanish remains include molluscs, primarily oyster, but these may have been remains of building materials rather than meals.

Table 7.2 Hontoon Island faunal distribution.

	Prehistoric				Historic			
	Count		Weight		Count		Weight	
	No.	%	g	%	No.	%	g	%
Habitat preference of identified species:								
terrestrial vertebrates	33	0.49	29.7	8.46	37	0.52	51.5	17.69
aquatic vertebrates	6740	99.51	321.2	91.54	7067	99.48	239.6	82.31
total vertebrates	6773	100.00	350.9	100.00	7104	100.00	291.1	100.00
Classification of faunal remains:								
vertebrates	6897	49.76	382.5	4.73	7324	99.07	331.6	60.43
invertebrates	6964	50.24	7702.9	95.27	69	0.93	217.1	39.57
total fauna	13861	100.00	8085.4	100.00	7393	100.00	548.7	100.00

Table 7.3 Habitat preferences of species identified from historic and prehistoric sites in St. Augustine (Reitz 1985, Reitz & Scarry 1985).

Habitat preference of identified species	Prehistoric (aboriginal)						Historic (Spanish with some aboriginal influence)							
	8SJ31 Count		34–1 Count		26–1 Count		29–2 Count		36–4 Count		34–2 Count		34–3 Count	
	No.	%	No.	%	No.	%	No.	%	No.	%	No.	%	No.	%
Terrestrial vertebrates	62	2.83	627	21.76	1595	21.99	42	26.75	257	34.87	167	43.49	52	59.77
Aquatic vertebrates	2130	97.17	2254	78.24	5658	78.01	115	73.25	480	65.13	217	56.51	35	40.23
Total vertebrates	2192	100.00	2881	100.00	7253	100.00	157	100.00	737	100.00	384	100.00	87	100.00

The faunal sample from the predominantly aboriginal site of En Bas Saline in Haiti indicates a focus in the economy on fishing and shellfish-gathering (Table 7.4). This sample includes only 5 per cent terrestrial vertebrates, 0.09 per cent of these being animals introduced from Spain. In contrast, the samples from the colonial site of Puerto Real show far greater reliance on terrestrial vertebrates which are predominantly introduced domestic animals (McEwan 1983, Reitz 1986). The aquatic component from locus 33/35 at Puerto Real is primarily composed of pond turtle (*Pseudemys* spp.), which were apparently sought in preference to fishes by the colonists (McEwan 1983). Shellfish remains are scarce at Puerto Real. The faunal remains from locus 39 at Puerto Real are primarily cattle (see Table 7.6). This area of the site is thought to have been a processing area for cattle, established within ten years of Columbus' first voyage to the New World (Reitz 1986).

These three pairs of samples show the same types of differences between the prehistoric and historic sites or components. The historic samples have relatively more, or larger, terrestrial species and less shellfish. European domesticates are represented even in the early historic sites but are not as relatively abundant as wild species, except at locus 39 at Puerto Real. In order to understand which domestic species are included in the first colonial outposts, human cultural characteristics such as refuse disposal practices, social status, and animal husbandry traditions, as well as the biological characteristics of the introduced animals, must be understood (Tables 7.5 & 7.6).

Those animals for which there was a tradition of husbandry, which

Table 7.4 Habitat preferences of species identified from historic and prehistoric sites in Haiti (McEwan 1983, Reitz 1986).

	Prehistoric (aboriginal)		Historic (Spanish)			
	En Bas Saline		Puerto Real locus 33/35		Puerto Real locus 39	
	Count		Count		Count	
	No.	%	No.	%	No.	%
Habitat preference of identified species:						
terrestrial vertebrates	277	5.18	16619	57.55	71098	99.87
aquatic vertebrates	5075	94.82	12256	42.45	89	0.13
total vertebrates	5352	100.00	28875	100.00	71187	100.00
Classification of fauna remains:						
vertebrate	5352	21.75	incomplete data but virtually no			
invertebrate	19256	78.25	invertebrate remains were excavated			
total fauna	24608	100.00	from the Puerto Real site			

Table 7.5 Distribution of domestic species in faunal samples from 16th-century sites in Spanish St. Augustine (Reitz & Scarry 1985).

Species	36–4		34–3		34–2		34–1		29–2		26–1 1979		26–1 1977		Total	
	Count No.	%	Count No.	%	Count No.	%	Count No.	%	Count No.	%	Count No.	%	Count No.	%	Count No.	%
rat	0	0.00	1	14.29	1	6.67	2	2.20	0	0.00	1	2.78	2	0.62	7	1.25
cat	1	1.41	0	0.00	1	6.67	6	6.59	0	0.00	3	8.33	3	0.92	14	2.50
dog	0	0.00	0	0.00	1	6.67	0	0.00	0	0.00	0	0.00	2	0.62	3	0.53
pig	26	36.62	5	71.43	9	60.00	13	14.29	4	25.00	8	22.22	152	46.77	217	38.68
cattle	13	18.31	1	14.29	1	6.67	18	19.78	11	68.75	18	50.00	7	2.15	69	12.30
caprine	0	0.00	0	0.00	0	0.00	0	0.00	0	0.00	0	0.00	2	0.62	2	0.36
chicken	31	43.66	0	0.00	2	13.33	52	57.14	1	6.25	6	16.67	157	48.31	249	44.39
Total	71	100.00	7	100.00	15	100.00	91	100.00	16	100.00	36	100.00	325	100.00	561	100.00

Table 7.6 Distribution of domestic species in faunal samples from 16th-century sites in Spanish Haiti (McEwan 1983, Reitz 1986).

Species	En Bas Saline Count		Puerto Real locus 33/35 Count		Puerto Real locus 39 Count	
	No.	%	No.	%	No.	%
rat	2	40.00	0	0.00	0	0.00
cat	0	0.00	4	0.70	0	0.00
pig	3	60.00	411	72.36	106	10.87
cattle	0	0.00	116	20.42	861	88.31
caprine	0	0.00	1	0.18	5	0.51
horse	0	0.00	8	1.41	0	0.00
chicken	0	0.00	28	4.93	3	0.31
total	5	100.00	568	100.00	975	100.00

could be transported easily, have a high reproductive potential, and were introduced into suitable habitats in the new environment would be expected to be most successful. On this basis one would expect to find the small barnyard animals, such as pigs, chickens, dogs, and cats. Of these likely animals, pigs and chickens are most abundantly represented in the sites reviewed. The hardy introduced range hogs were so successful in Hispaniola that licences to hunt the pigs which had become feral were issued as early as 1508 (Sauer 1966, p. 157). Dogs are known to have accompanied the Spanish. Their scarcity in the sites may reflect a different disposal pattern reserved for these animals. Cats and rats are known to flourish to the detriment of native species in the Caribbean. Their scarcity in the sites may indicate that their remains were disposed of in a different way to the remains of food refuse. Thus, of the six species which would be expected to flourish in early colonization, evidence shows that pigs and chickens prospered above all others. Dogs and cats may have been equally numerous but possible differences in disposal patterns may result in an underestimation of their numbers.

The small herd animals, sheep and goats, were important in traditional Spanish husbandry and would have been easily transported. However, their remains are rare in the sites reviewed. Goats are reported to have adapted well in parts of the Caribbean. They became feral in the mountains of Jamaica at an early date (Sauer 1966, p. 181). They did not adapt as well as other stock to the humid lowland environments. Sheep do not adapt well to new environments, or adjust to new diets as well as other livestock (Williamson & Payne 1965, p. 269). These easily transportable and traditional animals did not do well in the initial colonization, perhaps for lack of suitable local habitats around the sites reviewed.

Two highly prized animals, the horse for transportation and cattle for products such as meat, hides and tallow, present problems for introduction. They are both large and therefore difficult to transport and they reproduce slowly. Despite the difficulties of transport and relatively slow herd growth, cattle prospered remarkably well, particularly in Hispaniola, as evidenced by documentary reports and the large sample of large-sized cattle remains in locus 39 of Puerto Real (Reitz 1986). The early history of cattle in Florida is different. Although cattle were present in the 16th century, ranches were not established in what is now northern Florida until the second half of the 17th century (Arnade 1965). This may have been more because of Indian predation on the herds and their keepers than the absence of suitable habitats. Horses were particularly difficult to transport. Many succumbed in transit, particularly in the Horse Latitudes – named for their toll of horses (Crosby 1972, p. 80). Once safely landed in the New World, they are reported to have prospered along with cattle in the grazing lands, free of competitors and predators. Horse remains, however, are seldom encountered in the archaeological sites. This may be a function of patterns of disposal, in which remains of beasts of burden which were not usually consumed would not be incorporated in food or butchering refuse remains.

The impact of Spanish culture and animal introductions into the southeast and Caribbean was profound. The rapid adaptation of the introduced animals modified the New World environments. The introduction of Spanish traditions of animal procurement and uses, and the introduction of domestic animals changed the western hemisphere almost immediately and permanently.

Acknowledgements

In presenting these examples I lean heavily on the work of colleagues, friends, and students. I am particularly grateful for the scholarly production of Elizabeth J. Reitz who did the original research on the faunas from St. Augustine and locus 39 at Puerto Real (Reitz 1986). I am also indebted to Bonnie McEwan for her analysis of the fauna from locus 33/35 at Puerto Real (McEwan 1983). Thanks are due to Karla Bosworth, Laurie McKean, Erika Simons, and several other diligent students for their work on the identification of the faunas from Hontoon Island and En Bas Saline. I am grateful to Drs Barbara Purdy and Kathleen Deagan, the archaeologists who entrusted the faunal samples they so carefully recovered to us for our studies.

References

Arnade, C. W. 1965. *Cattle raising in Spanish Florida, 1513–1763*. Saint Augustine Historical Society Publication Number 21, 3–11.
Crosby, A. W. Jr. 1972. *The Columbian exchange: biological and cultural consequences of 1492*. Westport: Greenwood Publishing Company.
McEwan, B. G. 1983. *Spanish colonial adaptation on Hispaniola: the archaeology of Area 35, Puerto Real, Haiti*. Unpublished thesis, University of Florida, Gainesville, Florida.
Newsom, L. A. 1986. *Plant, human subsistence, and environment: a case study from*

Hontoon Island (8–VO–202), Florida. Unpublished thesis, University of Florida, Gainesville, Florida.

Reitz, E. J. 1985. Comparison of Spanish and aboriginal subsistence on the Atlantic coastal plain. *Southeastern Archaeology* **4**(1), 41–50.

Reitz, E. J. 1986. Cattle at Area 19, Puerto Real, Haiti. *Journal of Field Archaeology* **13**, 317–28.

Reitz, E. J. & C. M. Scarry 1985. *Reconstructing historic subsistence with an example from sixteenth-century Spanish Florida.* Society for Historic Archaeology Special Publication 3, 1–150.

Sauer, C. O. 1966. *The early Spanish Main.* Berkeley: University of California Press.

Williamson, G. & W. J. A. Payne 1965. *An introduction to animal husbandry in the tropics.* London: Longman.

8 Osteological evidence for the process of animal domestication

RICHARD H. MEADOW

Introduction

The domestication by humans of a small number of non-human animal species has taken place in various parts of the world at different times in the past (for summaries, see Zeuner 1963, Clutton–Brock 1981, Mason 1984). Records of these successes, as well as of unsuccessful attempts, have been recovered from the writings, iconographic representations, and material remains of ancient societies. In this chapter, focus is placed on one kind of evidence for animal domestication, namely that provided by animal bones and teeth recovered from archaeological sites. Most of what is discussed should be seen as referring principally to those social ungulate species that have become economically important to humans, namely cattle, sheep, goat, pig, horse, donkey, camel, water buffalo, and South American camelids.

Domestication

To most of us, animal domestication implies the development of special kinds of human–animal relationships that, intuitively, seem different from those between hunter and prey. Yet, as some researchers have emphasized, there is a continuum of conceivable relationships stretching from random hunting through intentional game-cropping, herd-following, animal-penning, and pet-keeping to the breeding of genetically isolated 'domestic' stock (e.g. Higgs & Jarman 1969, Jarman & Wilkinson 1972, Hecker 1982). Postulating slow, almost imperceptible changes based on rational, ecologically sound choices, these investigators contend that no discontinuity in human–animal relationships is to be expected between pre-pastoral and pastoral societies. Domestication of hunted species is thus seen as an elaboration of the predator–prey relationship, the ultimate result being that herding instead of hunting is used to guarantee a dependable and renewable source of animal protein with minimum risk and expenditure of energy.

While such views are compelling, Ingold (1980) has convincingly argued that they are deficient because they fail to find a place for the important social factors that help to shape human behaviour. As earlier investigators (notably V. Gordon Childe) have also maintained, times that see the beginnings of pastoralism are marked by a transformation in the infrastructure of society, with the source for this 'revolution' to be found in the area of social relations particularly as they relate to production. Therefore, it is to manifestations of changing socio-economic relations within human societies, as well as to changing relations between humans and various animal populations, that investigators must turn when seeking to study the domestication of economically important species (Compagnoni & Tosi 1978, Ducos 1978).

Following from these arguments, animal domestication can be defined as being a selective diachronic process of change in human–animal relationships involving, at the very least, a change of focus on the part of humans from the dead to the living animal and, more particularly, from the dead animal to the principal product of the living animal – its progeny. This process, while varying from culture to culture and from species to species, is manifest in two ways: first, in structural transformations in socio-economic dimensions of the human societies that interact with the living animals and, secondly, in changes in the behaviour and eventually in the morphology and physiology of the animals being domesticated (Clutton–Brock 1981, pp. 9–25, Meadow 1984, p. 310).

Although the remainder of this chapter is concerned with how animal (principally ungulate) domestication is reflected in faunal remains from archaeological sites, it is important to stress that concomitant changes in society are also manifest in other portions of the prehistoric record. For example, all three aspects of the 'Neolithic revolution' in the Middle East – the development of settled village life, plant cultivation, and animal husbandry – reflect a growing concern for property and for its maintenance through time. The development of this attitude was a prerequisite for the kind of social stratification and differentiation based on the accumulation of material wealth that formed the foundation for 'civilization' in the region. One must note, too, that the domestication of bovids and pigs in the Middle East took place in an already established context of cereal crop agriculture and permanent villages (see Gebel [1984] for a convenient summary of the data). One can thus argue that animal-keeping was almost forced upon those human groups as a means to permit them to maintain a supply of animal protein in the face of an overhunting of restricted territories, which was brought on by an absence of periodic movement and by the need to control crop-robbing herbivores (Uerpmann 1979). Thus, while herding can indeed be seen as a kind of predator–prey relationship, it reflects not the end of a continuum but a complete change in human attitudes towards animals.

When evaluating evidence of animal domestication, it is important to bear in mind that there is a conceptual difference between identifying the presence of domestic animals at a site and identifying the process of domestication. The latter implies documenting changes over time (however short) while the former involves identifying the end results of a process, and includes an implicit contrast between 'wild' and 'domestic'. Bökönyi, in his often-cited 1969 essay, recognized this difference and noted that it is first necessary to determine whether animals were being kept by peoples at a site and then to determine whether they were domesticated locally. His statements provide a useful summary of the kinds of evidence commonly used to support claims for animal domestication:

> . . .there is certain evidence for animal keeping in a prehistoric settlement if:
> (1) the proportion of age groups of a domesticable species is not the same as found normally in the wild population;
> (2) the proportions of the sexes of a domesticable species is not the same as found normally in the wild population;
> (3) domesticated species appear which have no wild ancestors in that particular region, at least since the Pleistocene;
> (4) morphological changes appear in domesticated animals;
> (5) there are artistic representations of domesticated animals;
> (6) there are objects associated with animal husbandry.

Note that, in all cases, Bökönyi makes at least an implicit contrast between a normative situation characteristic of wild animals and one characteristic of kept or domesticated forms. In other words, he is dealing with dichotomies and is not trying to identify a process. Unfortunately, this sense of dichotomy remains even in his answers to his own processual questions:

> . . .what is the evidence of local domestication in the material of a site?
> (1) Remains of both wild and domesticated forms on the site.
> (2) The existence of transitional forms between wild ancestor and the domesticated animal.
> (3) Changes in the proportions of age and sex groups in the wild form.
> (4) Representations of scenes of capture.

Of principal concern to us in this essay are points 1–4 of evidence for animal-keeping (Bökönyi 1969, pp. 220–1) and points 1–3 of evidence for local domestication (Bökönyi 1969, p. 223). It is important to point out, however, that the significance and meaning of artistic representations and objects associated with domestic animals, like the faunal evidence itself, are often open to multiple interpretations particularly when coming from periods early in the domestication process. Thus, such evidence is best used as only one line of support for an argument of animal domestication.

Demographic evidence for animal domestication

Age and sex ratios that change or are different from those 'found normally in the wild population' of a taxon are one kind of evidence that has been widely employed as evidence for domestication or for the keeping of domestic ungulates (e.g. Dyson 1953, Ducos 1969, 1978, Wright & Miller 1976; for a good summary discussion, see Hesse 1978, pp. 236–56). One of the most significant applications of the approach was its use to identify the presence of 'presumably domestic sheep' in 9th millennium levels of Zawi Chemi Shanidar (eastern Iraq) and to help support a claim for the local domestication of these animals, this on the basis of a high frequency of bones from sheep in general and from immature sheep in particular (Perkins 1964, revived by Reed 1983).

Considerable doubt has been cast on the validity of the demographic approach, however (Jarman & Wilkinson 1972, Simmons & Ilany 1975–77, Collier & White 1976). In the case of those ungulate species for which such data are even available, there are no such things as age and sex profiles 'normally found' in a wild population. Instead, the proportion of young and juveniles varies widely over time, as does the demography of individual social units which can even vary seasonally. These factors make average statistics of questionable value for understanding what sexes and ages might have been available to a group hunting in a particular area at a particular time. Furthermore, simulation studies carried out by Uerpmann (1979) show that, given a closed population of animals (gazelles, for example), increasing hunting pressure applied non-selectively over time will tend to create a younger overall population from which the hunters will draw their prey.

Before using arguments for domestication that rely heavily on age and sex ratios, therefore, it is necessary to have good modern data on the demography and behaviour of wild relatives of the taxon or taxa in question, and to have carefully examined the faunal materials for evidence of possible seasonal exploitation. Even with the best of information, however, it is almost impossible to use age and sex ratios to *rule out* selective hunting as an explanation of particular faunal configurations. Furthermore, given the same information, it is equally difficult to prove the existence of *intentional* selective hunting (game management), a phenomenon that some investigators feel must have preceded the domestication of ungulate species.

If there is no such thing as a 'typical' hunting pattern, the same is also true for herding patterns. A herder will tend to follow different culling practices for each of the various animal products desired (e.g. meat, milk, hair, traction, security; see Payne 1973, Redding 1984). Although these practices may come to be reflected in faunal assemblages, identification of one or more can be difficult for taphonomic reasons and because of the presence of mixed economic goals. In addition, social reasons *for* killing animals may override economic reasons *not* to kill them, and disease or injury can leave the herder with no choice at all.

Archaeologists deal with probabilities more than with clear-cut proof, and it is thus necessary to draw upon multiple lines of evidence when presenting an argument. Age and sex ratios are one kind of evidence that can provide potentially important information on human–animal relationships but one that generally should not be used as the sole support for an hypothesis of animal domestication, or for demonstrating the presence of domestic animals at a given site. As Dyson (1953, p. 662) has noted, however, 'Such objections may be minimized to a large extent by the repetition of the statistical pattern in a number of sites from a wide area, provided the sample in each case is of adequate size'.

In sum, documentation of demographic patterns over space, or trends over time, seems a more productive way to proceed than trying to identify particular configurations by comparing them with illusionary 'normal' patterns. The same is also true for the species composition of a faunal assemblage, that being another line of evidence that has been used to identify the presence of domestic animals at a site. A good example of this latter approach is provided for the southern Levant by Bar–Yosef (1981, Fig.11, based in large part on the dissertation research of Simon Davis). He notes that the 'frequencies of hunted ungulates in each [of 26 Epipalaeolithic and Pre-pottery Neolithic occupations] reflect the available game in its vicinity. . .' and that '[d]omestication is reflected in the shift to caprovines and the almost total abandonment of traditional game. This occurred with the transition from Pre-pottery Neolithic A to Pre-pottery Neolithic B in Israel . . . It seems therefore that during the Epipalaeolithic the hunting of available game continued uninterruptedly and no special man/animal relationship or "hunter's choice" can be discerned' (Bar–Yosef 1981, pp. 403–6).

Zoogeographic evidence for domestic animals

One of the potentially most reliable indicators of the presence of a domestic animal is finding its remains at a site in a region that is beyond the natural range of its wild relatives. Use of this criterion, however, requires knowledge of the distribution of past wild populations. More generally, it is necessary to know what the native wild species of an area were in order to demonstrate either the introduction of non-native taxa or the possibility of local domestication. One example is Corsica, where the question of whether the 'wild' sheep and pigs on the island are truly wild or merely feral has been debated for years (Groves, ch. 5, this volume). The latest opinions are that the free-ranging animals are feral (Poplin 1979, Franceschi 1980, Geddes 1985), having been introduced as domesticates on to the island as early as the 7th millennium BC (Vigne 1984).

Another example is the question of the true horse in Anatolia. Do the 4th and 3rd millennium remains reported from northwestern and eastern Anatolia (Boessneck & von den Driesch 1976, Rauh 1981, Bökönyi in press) represent relict wild populations that were hunted, or imported domestic stock? The answers to this question, and to others concerning the late Pleistocene and early Holocene distribution of various mammalian species, are far from clear (for the Middle East see Uerpmann 1987), with the result that, in some cases, it is impossible to use the apparent absence of wild relatives as clear evidence for the domestic status of a given taxon. Should there be morphological evidence for the presence of bones of wild stock in faunal collections, however, arguments for local domestication can be made, but such claims have to be evaluated against the possibility that local wild stock was being hunted at the same time that domestic animals imported from elsewhere were being kept.

Morphological evidence for animal domestication

The principal difficulty with the use of morphological evidence has been not so much in using it to define the presence of fully domestic stock, but with the conception that it is not possible to use such evidence to identify early stages in the process of animal domestication because changes would not have had time to take place. The key to this problem might be to distinguish between genetically based changes and those resulting from immediate environmental conditions, including primitive husbandry practices. A well-known example of the phenotypic manifestation of genotypic change is in the horncores of sheep and goats, which are very different in wild and domestic stock (Zeuner 1955, Reed 1960, Hole et al. 1969, Bökönyi 1975, Stampfli 1983). Even here, however, we have very little understanding of the range of variation occurring in wild populations, and of the number of generations of relaxed natural selection pressure or of intentional selection that would have been required to produce observable results. In addition, there is the problem of possible gene flow from domestic stock to wild populations. For example, to what degree is the occasional occurrence of hornlessness in modern wild female sheep due to such gene flow in the past?

With respect to the time factor in morphological change, Bökönyi (1976) has made the following observations:

> . . .well defined morphological changes do not occur before about 30 generations, according to modern experiments on domestication. The length of a generation is 2–3 yr in small species (dog, sheep, goat, pig) and 5–6 yr in large species (cattle, horse, and so on). Since the earliest domestication goes back as far as 8,000–14,000 yr, the delay of 60–90 or 150–180 yr in the occurrence of morphological changes is hardly significant.

While it may be true that, when taken together, the faunal remains from a multi-component site are unlikely to cover less than 200 years, individual strata can contain materials deposited in very short periods of time (Wright et al. 1980). To the degree that it is possible to deal with such deposits individually, the absence of evidence for morphological change could be an impediment to identifying the domestication process when it began occurring.

There are, however, at least two kinds of morphological change that can begin as early as the first or second generation. The first is the appearance of skeletal manifestations of pathological conditions brought on by keeping animals confined. Examples are Tepe Sarab (western Iran, early ceramic Neolithic) where 'a lot of chronic arthritis cases' and some periodontitis are reported for goats (Bökönyi 1977, p. 38) and, more significantly, Ain Ghazal (Jordan, PPNB) where a high proportion of morphologically 'wild' goat remains show pathologies (Köhler–Rollefson 1986 and pers. comm.). The second is size diminution, a phenomenon that is known to have accompanied the domestication of bovids and pigs in the Middle East and Europe (Bökönyi 1974, Boessneck & von den Driesch 1978, Uerpmann 1979, Stampfli 1983).

Although biological mechanisms for size change have rarely been discussed in the zooarchaeological literature, it appears that initial size decrease need not have been the result of genetic change. Instead, it may be related to a lower overall level of nutrition and lack of necessary diversity in the foods made available to early domestic stock whose mobility was restricted to increasingly overgrazed areas and whose supplementary diet, if any, was of low-quality forage. Elsie Widdowson has suggested that while genetics may be of primary importance in determining growth rates in the early stages of development of a foetus, nutrition had a major influence in later stages. She notes (1980, p. 7)

> . . .because the growth rate in the last part of gestation has been slow, the appetite after birth will be 'set' at a level appropriate to the size and rate of growth at the time before birth when the appetite centres in the hypothalamus were developing. The infant or animal that is small at birth takes less food than its larger counterpart and shows no sign of the 'catch-up growth' which is so characteristic of rehabilitation after undernutrition at older ages.

This condition resulting from malnutrition of the mother may have been compounded by the presence of parasitic infestations, or by drawing milk from the mother for human consumption. Indeed, it is conceivable that these latter two factors affecting the food intake and metabolism of the young animal directly could have led to size reduction without the presence of maternal malnutrition (proposals made by Noddle 1986, Köhler–Rollefson pers. comm.).

Accompanying nutrition-related changes would have been natural and even human selection for smaller females bearing smaller young that, just because they were smaller, would have a better chance of surviving lean periods in marginal environments (Jarman & Wilkinson 1972, Boessneck & von den Driesch 1978). At the same time, forces of natural selection that favoured large males would have been removed and this, combined with positive selection for smaller body size in the context of restricted populations, would have led to continuous size diminution until a lower plateau was reached. Such phenomena would be reflected in the faunal record by an increase in the variability of bone dimensions followed by an overall decrease in size, as shown by increasingly smaller extreme and median measurement values (Uerpmann 1979, Meadow 1984). Only with selective breeding for size, combined with good nutrition, would the trend be reversed (Boessneck & von den Driesch 1978).

The above discussion of size diminution is largely an *ex post facto* explanation of phenomena observed to have taken place in food species that were later known to be domestic and that were kept in relatively large numbers, probably to ensure a dependable and renewable source of animal products for settled agriculturalists in eastern Europe and the Middle East. Size change in mammals, however, need not be associated with domestication. Thus, Davis (1981) has discussed temperature-related size diminution in fox, wolf, boar, aurochs, wild goat, and gazelle at the end of the Pleistocene in Israel. Uerpmann (1978) has shown that size decreased up to the present in Middle Eastern wild sheep, while Pietschmann (1977) has documented the same pattern for red deer in Europe. Finally, Jordan (1975) discusses a case of decrease over time in size of red deer on the basis of remains recovered from prehistoric and early historic levels of Magula Pevkakia in Thessaly, a decrease likely to be related to the deer being confined to increasingly marginal areas as the result of the spread of human settlement. These examples make it clear that, just as with age and sex ratios, size diminution alone is not sufficient evidence for demonstrating the existence of the domestication process.

Conclusion

Arguments for animal domestication based on faunal remains from archaeological sites are likely to be more convincing if they employ multiple lines of evidence than if they are based on any one feature alone. Furthermore, examination in isolation of such traits as increasing proportions of specimens from selected species, morphological change, size diminution, or changing age and sex ratios, and faunal distribution patterns can provide little insight into coevolutionary aspects of animal and human societies. Where possible, all of these features are best examined together and trends documented and evaluated on the basis of large faunal collections from single sites covering significant spans of time, or from multiple sites within a limited region. In addition, interpretations are best made within the archaeological context of the site and region being examined, because only then can features of community and settlement patterning, site structure, and material culture be evaluated and related to the faunal remains. Such an integrated approach calls for archaeologically literate faunal analysts and zooarchaeologically literate archaeologists, each with a good understanding of the possibilities and limitations of the other's data. It also calls for an appreciation of biological and behavioural factors related to both the humans and the non–human animal taxa concerned.

References

Bar–Yosef, O. 1981. The Epi-Palaeolithic complexes in the Southern Levant. In *Préhistoire du Levant*. Colloques Internationaux du CNRS no. 598, 389–408. Paris: Editions du CNRS.

Boessneck, J. & A. von den Driesch 1976. Pferde im 4./3. Jahrtausend v. Chr. in Ostanatolien. *Saügetierkundliche Mitteilungen* **24**(4), 81–7.

Boessneck, J. & A. von den Driesch 1978. The significance of measuring animal bones from archaeological sites. In *Approaches to faunal analysis in the Middle East*, R. H. Meadow & M. A. Zeder (eds). Peabody Museum Bulletin no. 2, 25–39. Cambridge, Mass.: Harvard University.

Bökönyi, S. 1969. Archaeological problems and methods of recognizing animal domestication. In *The domestication and exploitation of plants and animals*, P. J. Ucko & G. W. Dimbleby (eds), 219–29. London: Duckworth.

Bökönyi, S. 1974. *History of domestic mammals in central and eastern Europe.* Budapest: Akadémiai Kiadó.

Bökönyi, S. 1975. Some problems of animal domestication in the Middle East. In *Domestikationsforschung und Geschichte der Haustiere*, J. Matolcsi (ed.), 69–75. Budapest: Akadémiai Kiadó.

Bökönyi, S. 1976. Development of early stock rearing in the Near East. *Nature* **264**, 19–23.

Bökönyi, S. 1977. *Animal remains from the Kermanshah Valley, Iran.* Oxford: BAR Supplementary Series 34.

Bökönyi, S. in press. Late Chalcolithic horses in Anatolia. In *Equids in the ancient world II*, R. H. Meadow & H.–P. Uerpmann (eds), Beihefte zum Tübinger Atlas der Vorderen Orients, Reihe A, Nr. 19/2. Wiesbaden: Dr Ludwig Reichert.

Clutton–Brock, J. 1981. *Domesticated animals from early times.* London: British Museum (Natural History) & Heinemann.

Collier, S. & J. P. White 1976. Get them young? Age and sex inferences on animal domestication in archaeology. *American Antiquity* **41**(1), 96–102.

Compagnoni, B. & M. Tosi 1978. The camel: its distribution and state of domestication in the Middle East during the third millennium B.C. in light of finds from Shahr-i Sokhta. In *Approaches to faunal analysis in the Middle East*, R. H. Meadow & M. A. Zeder (eds), Peabody Museum Bulletin no. 2, 91–103. Cambridge, Mass.: Harvard University.

Davis, S. J. M. 1981. The effects of temperature change and domestication on the body size of Late Pleistocene to Holocene mammals of Israel. *Paleobiology* **7**(2), 101–14.

Ducos, P. 1969. Methodology and results of the study of the earliest domesticated animals in the Near East (Palestine). In *The domestication and exploitation of plants and animals*, P. J. Ucko & G. W. Dimbleby (eds), 265–75. London: Duckworth.

Ducos, P. 1978. Domestication defined and methodological approaches to its recognition in faunal assemblages. In *Approaches to faunal analysis in the Middle East*, R. H. Meadow & M. A. Zeder (eds), Peabody Museum Bulletin no. 2, 53–6. Cambridge, Mass.: Harvard University.

Dyson, R. H. Jr. 1953. Archeology and the domestication of animals in the Old World. *American Anthropologist* **55**(5), 661–73.

Franceschi, P. F. 1980. *Essai de caractérisation génétique du porc corse.* Thesis 3ème cycle. Paris VI: l'Université Pierre et Marie Curie.

Gebel, H.–G. 1984. *Das Akeramische Neolithikum Vorderasiens.* Beihefte zum Tübinger Atlas des Vorderen Orients, Reihe B. Nr. 52. Wiesbaden: Dr Ludwig Reichert.

Geddes, D. S. 1985. Mesolithic domestic sheep in West Mediterranean Europe. *Journal of Archaeological Science* **12**(1), 25–48.

Groves, C. P. 1989. Feral mammals of the Mediterranean islands: documents of early domestication. In *The walking larder*, J. Clutton–Brock (ed.), ch. 5. London: Unwin Hyman.

Hecker, H. 1982. Domestication revisited: its implications for faunal analysis.

Journal of Field Archaeology **9**(2), 217–36.

Hesse, B. C. 1978. *Evidence for husbandry from the early Neolithic site of Ganj Dareh in western Iran.* Unpublished PhD dissertation, Faculty of Political Science, Columbia University. Ann Arbor: University Microfilms International no. 78–9905.

Higgs, E. S. & M. R. Jarman 1969. The origins of agriculture: a reconsideration. *Antiquity* **43**(169), 31–41.

Hole, F., K. V. Flannery & J. A. Neely 1969. *Prehistory and human ecology of the Deh Luran Plain.* Memoirs of the Museum of Anthropology, no. 1. Ann Arbor: University of Michigan.

Ingold, T. 1980. *Hunters, pastoralists, and ranchers.* Cambridge: Cambridge University Press.

Jarman, M. R. & P. F. Wilkinson 1972. Criteria of animal domestication. In *Papers in economic prehistory*, E. S. Higgs (ed.), 83–96. Cambridge: Cambridge University Press.

Jordan, B. 1975. *Tierknochenfunde aus der Magula Pevkakia in Thessalien.* Inaugural Dissertation. Munich: Institut für Palaeoanatomie, Domestikationsforschung und Geschichte der Tiermedizin der Universität München.

Köhler–Rollefson, I. 1986. *From sedentary animal herding and hunting to semi-nomadic pastoralism: the faunal evidence from Neolithic Ain Ghazal (Jordan).* Unpublished paper presented at 5th International Conference of the International Council for Archaeozoology, Bordeaux, France, 25–30 August.

Mason, I. L. (ed.) 1984. *Evolution of domesticated animals.* London: Longman.

Meadow, R. H. 1984. Animal domestication in the Middle East: a view from the eastern margin. In *Animals and archaeology.* Vol. 3: *Early herders and their flocks*, J. Clutton–Brock & C. Grigson (eds), 309–37. Oxford: BAR International Series 202.

Noddle, B. 1986. *Flesh on the bones. Productivity of animals in the past: speculations based on historical data and the performance of primitive livestock in Third World countries.* Unpublished paper presented at 5th International Conference of the International Council for Archaeozoology, Bordeaux, France, 25–30 August.

Payne, S. 1973. Kill-off patterns in sheep and goats: the mandibles from Asvan Kale. *Anatolian Studies* **23**, 281–303.

Perkins, D. Jr. 1964. Prehistoric fauna from Shanidar, Iraq. *Science* **144**, 1565–6.

Pietschmann, W. 1977. *Zur Grösse des Rothirsches* (Cervus elaphus *L.) in vor-und frühgeschichtlicher Zeit.* Inaugural Dissertation. Munich: Institut für Palaeoanatomie, Domestikationsforschung und Geschichte der Tiermedizin der Universität München.

Poplin, F. 1979. Origine du mouflon de Corse dans une nouvelle perspective paléontologique: par marronnage. *Annales de Génétique et de la Sélection Animale* **11**, 133–43.

Rauh, H. 1981. *Knochenfunde von Säugetieren aus dem Demircihüyük (Nordwestanatolien).* Inaugural Dissertation. Munich: Institut für Palaeoanatomie, Domestikationsforschung und Geschichte der Tiermedizin der Universität München.

Redding, R. W. 1984. Theoretical determinants of a herder's decisions: modelling variation in the sheep/goat ratio. In *Animals and Archaeology.* Vol. 3: *Early herders and their flocks*, J. Clutton–Brock & C. Grigson (eds), 223–41. Oxford: BAR International Series 202.

Reed, C. A. 1960. A review of the archeological evidence on animal domestication in the prehistoric Near East. In *Prehistoric investigations in Iraqi Kurdistan*, R. J. Braidwood & B. Howe (eds), Studies in Ancient Oriental Civilizations (SAOC) no. 31, 119–45. Chicago: University of Chicago Press.

Reed, C. A. 1983. Archeozoological studies in the Near East: a short history (1960–1980). In *Prehistoric archeology along the Zagros flanks*, L. Braidwood, R. J. Braidwood, B. Howe, C. A. Reed & P. J. Watson (eds), Oriental Institute Publications, Vol. 105, 511–36. Chicago: Oriental Institute of the University of Chicago.

Simmons, A. H. & G. Ilany 1975–77. What mean these bones? Behavioral implications of gazelles' remains from archaeological sites. *Paléorient* **3**, 269–74.

Stampfli, H. R. 1983. The fauna of Jarmo with notes on animal bones from Matarrah, the Amuq, and Karim Shahir. In *Prehistoric archeology along the Zagros flanks*, L. S. Braidwood, R. J. Braidwood, B. Howe, C. A. Reed & P. J. Watson (eds), Oriental Institute Publications, Vol. 105, 431–83. Chicago: Oriental Institute of the University of Chicago.

Uerpmann, H.–P. 1978. Metrical analysis of faunal remains from the Middle East. In *Approaches to faunal analysis in the Middle East*, R. H. Meadow & M. A. Zeder (eds), Peabody Museum Bulletin no. 2, 41–5. Cambridge, Mass.: Harvard University.

Uerpmann, H.–P.1979. *Probleme der Neolithisierung des Mittelmeerraumes*. Beihefte zum Tübinger Atlas des Vorderen Orients, Reihe B, Nr. 28. Wiesbaden: Dr Ludwig Reichert.

Uerpmann, H.–P. 1987. *The ancient distribution of ungulate mammals in the Middle East*. Beihefte zum Tübinger Atlas des Vorderen Orients, Reihe A, Nr. 27. Wiesbaden: Dr Ludwig Reichert.

Vigne, J.–D. 1984. Premières données sur le début de l'élevage du mouton, de la chèvre et du porc dans le sud de Corse (France). In *Animals and archaeology. Vol. 3: Early herders and their flocks*, J. Clutton–Brock & C. Grigson (eds), 47–65. Oxford: BAR International Series 202.

Widdowson, E. M. 1980. Growth in animals. In *Growth in animals*, T. L. J. Lawrence (ed.), 1–9. London: Butterworth.

Wright, G. A. & S. J. Miller 1976. Prehistoric hunting of New World wild sheep: implication for the study of sheep domestication. In *Culture change and continuity*, C. E. Cleland (ed.), 293–312. New York: Academic Press.

Wright, H. T., N. Miller & R. W. Redding 1980. Time and process in an Uruk rural center. In *L'archéologie de l'Iraq du début de l'époque Néolithique à 333 avant notre ère*. Colloques Internationaux du CNRS 580, 265–84. Paris: Editions du CNRS.

Zeuner, F. E. 1955. The goats of early Jericho. *Palestine Exploration Quarterly* April, 70–86.

Zeuner, F. E. 1963. *A history of domesticated animals*. London: Hutchinson.

9 Animal exploitation and the phasing of the transition from the Palaeolithic to the Neolithic

HANS–PETER UERPMANN

Introduction

The decision whether plant cultivation or animal husbandry was practised by a prehistoric population is usually based on surviving plant and/or animal remains. However, these two sources of information are neither equally available nor can they be evaluated scientifically to the same extent. Generally, animal bones are among the most common finds in archaeological contexts, whereas plant remains are often rare or absent. Moreover, animal bones are commonly left as a result of animal exploitation, whereas many plants of economic importance do not produce waste materials likely to become part of the archaeological record. In addition, bone remains are closely correlated to the type of use of the respective species. The skeleton of each taxon has a definitive number of elements and the survival of these remains is not too different from taxon to taxon. Thus, contrary to plants, fairly reliable estimations of the relative importance of different animals can be based on bone finds from archaeological sites. With plants, on the other hand, meaningful proportions can only be established between remains that are related in the same way to the useful parts as bones are related to flesh. This is the case, for example, with nutshells. However, any plant remains which are dependent on carbonization for preservation will always reflect their particular susceptibility to being burnt rather than their relative economic importance.

In the context of the beginnings of cultivation and domestication, there are morphological responses that enable us to recognize the processes. These morphological changes probably happen in animals faster than in plants, or, perhaps more correctly, it is less easy to imagine early forms of animal husbandry which would *not* produce morphological changes than it is to suggest early forms of cultivation which would leave the respective plants unchanged. Therefore, if an economic definition of the Neolithic versus the Palaeolithic period is applied, the insight into the animal part of the economy is potentially more distinct than the information available on the vegetal side.

It seems appropriate, therefore, to develop a scheme of definitions based mainly on animal remains through which the various stages of the transition can be classified from Palaeolithic hunters and gatherers to Neolithic farmers and herders, at least for the circum-Mediterranean areas and the adjacent parts of Europe and western Asia.

Terminology of the Palaeolithic–Neolithic transition

Rather than creating another new terminology, existing terms like 'Neolithic' and 'Mesolithic' will be assigned in this chapter to a more precise socio-economic meaning. This fairly conservative approach is based on accepted Childean criteria for the Neolithic. In the 1950s Gordon Childe took an economy based on plant cultivation and animal husbandry as the defining character for distinguishing the Neolithic from the Palaeolithic period. This definition provided a more meaningful subdivision of the Stone Age than did the former characterization of the Neolithic as the period of polished stone versus flaked stone tools. However, it would be unsatisfactory merely to seek evidence of 'plant cultivation' and/or 'animal husbandry' when attempting to characterize the actual processes of economic change in the Fertile Crescent of Southwest Asia between about 10 000 and 6000 BC. It is now clear that the domestication of meat-producing animals is only the last major step in the evolution of the Neolithic economy. There is evidence for plant cultivation prior to animal domestication, and it is likely that plant cultivation is only an intermediate step in the development of the early Neolithic. The initial stage of this process is best described as an economy based on 'intensive plant exploitation including long-term storage of vegetal food'. The village facies of the Natufian culture in the Levant is the prototype of this stage of socio-economic development.

Given the problems of the evidence about the exploitation of plants as outlined above, one cannot exclude the possibility that the Natufian population had already discovered methods of stimulating the growth of cereals and vegetables, and thus had begun plant cultivation. The succeeding culture, which in order to use a consistent terminology should be called 'Sultanian', evidently practised some agriculture. There are cereal remains from Sultanian contexts which differ sufficiently from their ancestors to indicate changes in reproductive selection due to manipulation by early agriculturalists. The still more widespread use of the term 'Pre-pottery Neolithic A' (PPNA) for this cultural unit should be abandoned for two reasons:

(a) this term is inconsistent with the use of type-site names for the designation of Stone-age 'cultures';
(b) the 'PPNA' is not a neolithic culture in the full sense.

Because it is lacking evidence of domesticated meat-producing animals, the Sultanian and similar cultural units should be grouped with the preceding Natufian into a stage called 'Protoneolithic'. Sites like Zawi Chemi, Karim Shahir, and Asiab in the northern part of the Fertile

Crescent are also included in this term, although evidence for 'intensive exploitation of plants including long-term storage of vegetal food' can sometimes only be inferred indirectly from structures which suggest some degree of sedentism. Incipient sedentism is here considered both as a result of, and a presupposition for, effective long-term storage of larger quantities of food. Sedentism is also instrumental in the evolution of some of the features of the fully developed Neolithic, including animal domestication. Therefore, sedentism based on intensive exploitation and long-term storage of vegetal food in the absence of domestic animals for meat-production should be the defining characteristic of the Protoneolithic stage of economic development.

Among the first civilizations to combine animal husbandry with the features already present in the Protoneolithic was the 'Pre-pottery Neolithic B' of the Levant. The PPNB is a real Neolithic culture, having both plant cultivation and animal husbandry as defining characters. However, one should also apply type-site names for this earliest horizon of the Neolithic period. The name 'Tahunian' has been used in this context (Prausnitz 1970), but one might also propose to use Beidha as an eponymic site because of the availability of an extensive description of its flint industry (Mortensen 1970). Good evidence for animal domestication at this chronological level has been produced for several sites in the southern Levant, including Beidha (Hecker 1974, 1982, Uerpmann 1979), Jericho (Clutton–Brock & Uerpmann 1974, Clutton–Brock 1979), and others. In the north of the Fertile Crescent, the upper part of the aceramic levels of Çayönü represent a real Neolithic (Lawrence 1980, 1982). Jarmo (Stampfli 1983) and Ali Kosh (Hole et al. 1969) are examples from the foothills along the northeastern border of Mesopotamia, whereas Ganj-e Dareh (B–D) is an early representative of a Neolithic economy in the Zagros mountains (Hesse 1978).

The first animals to be domesticated for meat production during the 7th millennium BC were sheep and goat, both in different parts of the Fertile Crescent (Uerpmann 1979, 1987). The full potential of Neolithic economy was only reached, however, more than half a millennium later when cattle and swine joined the domestic livestock. Their domestication could be used to differentiate between an early and a developed stage of the Neolithic economy. It is less easy, though, to trace domestication in cattle than in sheep or goat. Cattle bones are usually even more broken than those of the smaller ruminants, and the ecological shifts imposed on bovines during early domestication are potentially less severe than in sheep and goat. Nevertheless, there is at least a theoretical distinction possible between an 'early Neolithic' with an incomplete set of domestic animals, and a 'full Neolithic' with cattle, pig, sheep, and goat forming the complete list of protein-producing domesticates of this period. The later use of these animals for other products than meat adds further facets to the Neolithic economy (Sherratt 1981), but is not marked as a clear-cut 'revolution'. If there is evidence for the use of secondary products, however, it can be used to define a 'late Neolithic'.

The end of the Neolithic period can also be defined in terms of animal exploitation: the domestication of special animals for purposes of labour, like donkey, horse, and camel, is a reflection of the changes in overall economic patterns at the beginning of the metal ages (the formation of cities and states in the Middle East). Thus, the end of the Neolithic is marked by a further broadening of the exploited resources. Like all the economic shifts described above, it consists of additions of new subsistence technologies and of new schemes of exploitation of the environment to a previous pattern which was almost never completely replaced.

Terminology of other economic stages

Within the circum-Mediterranean area the data available at present indicate that an autochthonous development from the Epipalaeolithic through the Protoneolithic into the Neolithic economy only occurred in the Fertile Crescent of southwest Asia. The Epipalaeolithic is found almost everywhere in the Old World, and the Neolithic spread over most of it in later times, but the Protoneolithic is confined to the core area where natural stands of wild cereals and pulses favoured 'intensive plant exploitation including long-term storage of vegetal food' and the subsequent development of sedentism (Uerpmann 1979). However, a fairly high degree of sedentism also seems to have been achieved by some 'Mesolithic' civilizations without signs of an intensified use of plants. The economic basis of this sedentism is the exploitation of aquatic animal populations, mainly in the form of fish and shellfish. The sedentism reached by some fishing populations actually led to cultural developments diverging from a Palaeolithic lifestyle. There are similarities with Neolithic features which actually justify an intermediate term for this stage of development. The 'Mesolithic' as a term in economic prehistory should be restricted to groups with a non-productive economy which nevertheless allowed for some degree of sedentism based on the exploitation of aquatic resources.

The general hunters and gatherers of the post-Pleistocene, which are also often included in the term Mesolithic, do not need a special intermediate term between the Palaeolithic and the Neolithic. From their economy they are clearly part of the Palaeolithic and might best be grouped under the term 'Epipalaeolithic'. An economic definition of this term is difficult, although the use of long-ranging versus medium-ranging weapons (bow and arrows versus spear) for hunting might provide a basis for its discrimination from the Upper Palaeolithic, both in the tool kit and in the choice of prey. In particular, in those parts of the Middle East which had open vegetation, gazelles only became significantly available to prehistoric hunters after the invention of long-ranging weapons. The introduction of the bow and arrows may have been the reason for the apparent expansion of the Epipalaeolithic into semi-desert areas.

Obviously, Epipalaeolithic groups existed alongside Protoneolithic villages, and in particular the Natufian had an Epipalaeolithic and Protoneolithic facies. The limits of typologically defined 'cultures' do not necessarily coincide with economic units such as the 'Neolithic' and the 'Palaeolithic'. This is an important point, for a number of prehistoric groups, mainly in the deserts of Arabia and North Africa, are called 'Neolithic' because of the presence of arrow heads or pottery. So long as the economic characteristics of the Neolithic have not been demonstrated for these groups they should really be given a locally defined name which does not include them in one of the larger periods.

There are, however, prehistoric groups with known economic characters that do not fit the scheme developed above. For example, there existed so-called Mesolithic groups in the south of France which had domestic sheep (Geddes 1985). Apparently, the ascription of these groups to the Mesolithic is only based on the fact that no pottery was found in their sites. Since this is not a diagnostic feature, one could easily include the 'Castelnovien' within the Neolithic, if there were indications of the other defining characters of the Neolithic. These examples are most probably groups which have lost, or only adopted part of, the Neolithic economy during their distribution and diffusion outside of the central areas of Neolithic development in southwest Asia. The term 'Paraneolithic' seems appropriate to describe the peripheral situation of these groups which are in the process of economic change.

Plainly nomadic herders, who apparently did not practise plant cultivation, also cause terminological problems. Although there is no good evidence that there was a direct transition from hunters to herders without a Protoneolithic interlude, such a shift is at least a theoretical possibility. The term 'Semi-neolithic' might be appropriate with regard to the fact that these groups lack the plants of the full Neolithic. This term should, however, not be used for derived groups which have given up agriculture while adapting to particular environmental circumstances. In this case, terms like the notorious 'Shepherd Neolithic' more closely reflect actual developments.

Another common deviation from the more normal pattern are real Mesolithic economies with the added use of some domestic animals. In fact, the exploitation of aquatic resources may well have been much more significant for the economic status of the respective groups than was the contribution made by domestication (e.g. Biagi et al. 1984). In such cases, composite terms such as Meso-Neolithic, could be used and similar names for other mixtures of Palaeo-, Meso-, and Neolithic features could be developed.

The importance of having a precise terminology in common usage is to specify the socio-economic processes and changes which took place. The particular terminology suggested in this chapter cannot be directly applied to areas unconnected with Neolithic development in the Near and Middle East. However, it is hoped that the suggested precision in defining the relevant processes under consideration may aid comparative studies of the origins of productive economies in other parts of the world as well.

References

Biagi, P., W. Torke, M. Tosi & H.–P. Uerpmann 1984. Qurum: a case study of coastal archaeology in Northern Oman. *World Archaeology* **16**, 43–61.

Clutton–Brock, J. 1979. The mammalian remains from the Jericho Tell. *Proceedings of the Prehistoric Society* **45**, 135–57.

Clutton–Brock, J. & H.–P. Uerpmann 1974. The sheep of early Jericho. *Journal of Archaeological Science* **1**, 261–74.

Geddes, D. 1985. Mesolithic sheep in west Mediterranean Europe. *Journal of Archaeological Science* **15**, 25–48.

Hecker, H. M. 1974. *The faunal analysis of the primary food animals from pre-pottery Neolithic Beidha (Jordan)*. Unpublished PhD dissertation, Columbia University, New York.

Hecker, H. M. 1982. Domestication revisited: its implications for faunal analysis. *Journal of Field Archaeology* **9**, 217–36.

Hesse, B. C. 1978. *Evidence for husbandry from the early Neolithic site of Ganj Dareh in western Iran*. Unpublished PhD dissertation, Columbia University, New York.

Hole, F., K. V. Flannery & J. A. Neely, 1969. *Prehistory and human ecology of the Deh Luran Plain*. Memoirs of the Museum of Anthropology, University of Michigan no. 1. Ann Arbor: Museum of Anthropology.

Lawrence, B. 1980. Evidences of animal domestication at Çayönü. In *The joint Istanbul–Chicago universities' prehistoric research in southeastern Anatolia 1*, H. Çambel & R. J. Braidwood (eds), 285–308. Istanbul: Istanbul University, Faculty of Letters no. 2589.

Lawrence, B. 1982. Principal food animals at Çayönü. In *Prehistoric village archaeology in south-eastern Turkey*, L. S. Braidwood & R. J. Braidwood (eds), 175–99. Oxford: BAR International Series 138.

Mortensen, P. 1970. A preliminary study of the chipped stone industry from Beidha, an early neolithic village in southern Jordan. *Acta Archaeologica* **41**, 1–54.

Prausnitz, M. W. 1970. *From hunter to farmer and trader: studies in the lithic industries of Israel and adjacent countries*. Jerusalem: Sivan Press.

Sherratt, A. 1981. Plough and pastoralism: aspects of the secondary products revolution. In *Patterns of the past*, I. Hodder, G. Isaac & N. Hammond (eds), 261–305. Cambridge: Cambridge University Press.

Stampfli, H. R. 1983. The fauna of Jarmo with notes on the animal bones from Matarrah, the 'Amuq, and Karim Shahir. In *Prehistoric archaeology along the Zaqros flanks*, L. S. Braidwood, R. J. Braidwood, B. Howe, C. A. Reed & P. J. Watson (eds), Oriental Institute Publications Vol. 105, Chicago: Oriental Institute of the University of Chicago.

Uerpmann, H.–P. 1979. *Probleme der Neolithisierung des Mittelmeerraumes*. Beihefte zum Tübinger Atlas des Vorderen Orients, Reihe B, Nr.28. Wiesbaden: Dr Ludwig Reichert.

Uerpman, H.–P. 1987. *The ancient distribution of ungulate mammals in the Middle East*. Beihefte zum Tübinger Atlas des Vorderen Orients, Reihe A, Nr. 27. Wiesbaden: Dr Ludwig Reichert.

10 *A two-part, two-stage model of domestication*

FRANK HOLE

It has been customary to conceive of domestication as a 'revolutionary' event whose essential elements were in place throughout Southwest Asia as early as the 10th millennium cal. BC, and were fully established by the end of the 8th millennium BC. The events leading to domestication have recently been re-examined, leading to two different points of view (Hole 1984, Moore 1982, 1985). Both Hole and Moore argue for a long period of development of plant domestication, but Hole considers livestock to have been brought under control through entirely separate processes, perhaps without a lengthy period of development, and that it was the conjunction of these two economic adaptations that resulted in the Neolithic revolution. In this chapter I shall briefly outline the substance of this argument and then look specifically at evidence for domestication of sheep and goats.

Essentially, archaeological and zoological evidence suggests that sheep and goats were domesticated earlier in the Zagros Mountains of eastern Anatolia and western Iran than in the Levant. Although these species are present sporadically in the Levant and Anatolia, they do not constitute a consistently high proportion of remains in archaeological sites in these western regions until sedentism with cereal agriculture is manifest.

Howe (1983) has outlined the archaeological and zoological finds at the earliest settlements outside the coastal Levant, with special emphasis on the Zagros region. Here, a series of sites estimated to be between 8000 and 10 000 cal. BC exhibit the following: semi-sedentary camps lacking traces of permanent architecture, although some circular huts are present; widespread use of stones for pounding or grinding vegetable food; a mix of fauna but with an emphasis on sheep/goat. The locations and physical conditions of these sites suggest that they were used as seasonal camps rather than as permanent residences. These camps, in the foothills and mountainous regions, were probably used during the warmer months when cereals or other plants would have been ripe. In short, these sites – Shanidar B1, Zawi Chemi, Karim Shahir, M'Lefaat, Gird Chai, and Asiab – all suggest some form of seasonal transhumance, whether for herding and hunting, or collecting and harvesting of planted fields. At this stage – often called the Protoneolithic – there are no permanent settlements known.

Because of one radiocarbon date, Ganj Dareh E has often been considered as a Protoneolithic component in a site that otherwise fits comfortably into the early Neolithic. However, in a detailed review of the grounds for dating this and other sites, Hole (in press) prefers the later date in cal. years BC. This means that Tepes Qazemi and Ghenil are, likewise, 8th millennium. The dates for Shanidar and Zawi Chemi can also be questioned, particularly as they relate to possible evidence of sheep domestication, but the relative antiquity of these sites and the others mentioned seems correct. One hopes that the recent excavations by Stefan Kozlowski at Nemrik on the upper Tigris will yield dates for this era.

Sites pre-dating the Protoneolithic (generally known as Zarzian and Baradostian) were very much smaller and often located in caves and shelters (Hole & Flannery 1967). At none of the Pleistocene sites is there evidence of either the use of grinding stones or of domesticated livestock. Thus, the nature of the later Holocene sites and their distributions, coupled with faunal evidence, suggest that the first step in the process of sheep and goat domestication had been taken by 10 000 cal. BC, although there is debate among faunal specialists whether any or all of these sites actually evidences the use of domestic stock. The important issue is that the *pattern* of life essential to a herding economy had been taken with the seasonal reuse of these 'pastoral' camps. Although the evidence varies from site to site, it is clear that the processing of vegetable foods went hand in hand with these seasonal movements.

A critical gap in our knowledge of this era exists for most of Anatolia, particularly the eastern and southern portions which lie adjacent to the Zagros, and a similar gap exists on the Soviet side. But perhaps the most crucial gap is what happened on the rolling foothills at the base of the mountains in northeastern Iraq and on the adjacent north Mesopotamian steppe which today provides a transition to the desert. Here, where much attention has been placed upon mounded sites, one would expect to find the lowland counterparts to the upland Zawi Chemi type sites (Nemrik is probably one such site). If other camping debris from this era exists, it has gone unnoticed in previous surveys, and it may be that most of it is no longer visible at the surface.

The pattern in the Levant has been summarized by Bar–Yosef (1980) and Moore (1985). Here, evidence for the introduction of domestic stock, chiefly goats, appears only in the 8th millennium, during the Pre-pottery Neolithic B, in spite of a very long history of semi-sedentary and sedentary sites at which there is abundant evidence of the grinding and pounding of vegetable foods. Indeed, the appearance of domesticated sheep, goats, cattle, and pigs heralds a dramatic replacement of the indigenous gazelle, onager, and deer populations. Sedentism, as manifest at Jericho, Mureybat, Aswad (near Damascus), Abu Hureyra, and Mallaha, preceded the introduction of domestic livestock by as much as 2000 years. A similar pattern of abrupt replacement of wild stock is repeated at other sedentary sites, such as Çayönü in south–central Anatolia (Lawrence 1980).

The archaeological evidence thus suggests a differential timing in the development of the chief components of domestication: herding is earlier in the east and plant processing and sedentism are much older in the west. This assessment of the differences between the two areas is supported by cytological evidence of a Levantine origin for the domesticated races of wheat, peas, and lentils and their subsequent dispersal as an agricultural complex through the Near East and ultimately into Europe (Zohary pers. comm.). The independent development of these two lines toward fully effective domestication is Stage One of my rubric, in which the development of a herding economy is Part One and the slow separate evolution of an agricultural economy out of a long tradition of plant processing is Part Two. However, one must not ignore the deplorable state of our knowledge about these eras. If, as I argue, intensive plant use was a lowland adaptation, then we would not expect to find early evidence in the mountainous regions. One might then postulate seed-using economies developed as early in Mesopotamia as in the Levant, but present evidence is insufficient to test this hypothesis. In fact, although we know of a long tradition of grinding-stone technology in the Levant, we are still not able to state with assurance when cultivation began, so that even in this region the rapidity of domestication is in doubt. Nevertheless, in each case, patterns of behaviour – vertical transhumance in the case of sheep and goats, and storage/sedentism in the context of an intensive plant-gathering economy – must have been established for some time before a full transition to the use of domesticates came about.

Stage Two is represented by the 8th millennium in the Near East as the widespread appearance of permanent settlements which make use of both domestic animals and plants. It is important to note that sedentism in the west precedes Stage Two by as much as 2 millenniums, whereas sedentism (as opposed to seasonal occupancy) in the zone of animal domestication coincides with the introduction of cereal domestication.

Thus, an outline of the structure and timing of the processes of domestication is evident even though there are still many problems concerning the evidence and its interpretation (see Ducos & Helmer 1981). But perhaps the most important remaining problems are conceptual and access to relevant sites. Since Braidwood and Childe defined the problems in archaeological terms (Braidwood & Howe 1961, pp. 1–9), the search for suitable sites has concentrated on the 'nuclear' zones where many species converged. There are three problems with this:

(a) the environment of southwest Asia changed markedly during the time of concern;
(b) although there are broad zones where many species overlap in geographic distribution today, this was not necessarily the case at the time domestication began; and
(c) each species of animal, and to a lesser extent plant, has its own characteristics and behaviour which required different methods of exploitation by humans.

It is not surprising, therefore, to find that the species are differentially represented in sites, especially during the early stages of domestication. Thus, we find a preponderance of sheep over goats at Zawi Chemi and of goats over sheep at Asiab. These occurrences probably reflect different local environments. But we also find that cattle and pigs, which were probably distributed throughout the Near East, are rarely present as domesticates in early sites. Indeed, domestic cattle, *Bos taurus*, are thought to have originated in Greece or western Anatolia (but see Meadow 1984 for *Bos indicus*), and pigs are absent as demonstrable domesticates before the 8th millennium. These latter omissions are unlikely to have resulted from environmental considerations; rather, the explanation probably resides in the behaviours of the species and their interactions with humans. The picture that emerges, therefore, is one of a series of experiments by people who took advantage of the locally dominant fauna, which were both desirable as food and tractable to human control. Of these available potential domesticates, sheep and goats, both relatively benign, gregarious, herd animals, lent themselves best to human domination. We must look, therefore, not for one event – 'domestication' – but at a series of events that brought different animals and plants under control. The problem is thus more complicated than once conceived, but perhaps more approachable because we can now focus on separate aspects of the transition rather than having to deal with all of its parts simultaneously.

Because of environmental changes at the end of the Pleistocene and since (including man's dispersal of species), distributions of fauna may have differed substantially from the present so that today's maps do not accurately reflect the biogeographic landscape in which domestication took place. In particular, changes in the location and proportions of forest and forage will have had a determining effect upon the distributions of the browsers (goats) versus the grazers (sheep). In this regard, it is interesting, in view of the rapid spread of sheep after domestication, that they are seldom found in either Palaeolithic or early Neolithic sites, whereas goats are relatively common, along with various other hunted species such as red deer, gazelles, onagers, and oxen. Sheep occur in the Pleistocene caves of Shanidar, Palegawra, and Douara. In the early Holocene they are found at Zawi Chemi and Asiab in the Zagros, Abu Hureyra and Mureybet on the Euphrates, and in sites in the eastern and southern Levant, including G VIII in the Negev. These two distributions are strikingly different and suggest that sheep were not distributed in the past as they are today – that they were not abundant in the places where we have dug sites. As sheep are grazers who prefer rolling country, the upper margins of the Fertile Crescent would appear to have been the most likely places for sheep to have lived. However, as sheep do not penetrate forested regions, it is possible that their distribution lay south of what we consider today to be steppe.

According to pollen studies, at the end of the Pleistocene conditions were drier and cooler and began to grow warmer and wetter after 10 000 cal. BC, resulting in slow resurgence of forest and grasses over lands that previously had been arid, shrubby steppe (Van Zeist & Bottema 1982). That some forest and cereals remained in isolated refuges seems inescapable, and is supported by the presence of red deer in several Palaeolithic sites in the Zagros. Nevertheless, for the most part during the late Pleistocene, the mountainous regions would have been much less suitable to sheep than to goats, a conclusion that is supported by faunal data from sites in the same region – only Shanidar and Palegawra Caves hold any evidence of sheep in the Pleistocene (Turnbull & Reed 1974, Evins 1982).

Following the Pleistocene, it is generally thought that forests stretched well out onto the lowland steppe to approximately the 250 mm rainfall isohyet, a zone that today lacks trees owing largely to human interference. The grassy stretches that would have been preferred by sheep may, therefore, have been still farther south in the zone that we regard today as desert-steppe. Evidence to support this is meagre, largely because sheep are only minor components of early archaeological faunal assemblages. For example, in the Epipalaeolithic levels at Abu Hureyra 85 per cent of the fauna is gazelle, and sheep/goat compose only 8 per cent. Interestingly, here *both* sheep and goat are present, although one would not expect to find the latter so far from mountainous terrain (Moore *et al.* 1986). At Mureybet, just 35 km north of Abu Hureyra, also on the Euphrates, a slightly later occupation contains a few sheep but not goats, and the majority of fauna consists of ass and gazelle. Ducos (1978) offers a behavioural reason for the relative lack of sheep in these sites: gazelles are easier to hunt, and sheep were taken only incidentally during gazelle hunts.

It is clearly established that sheep occupied the upper Mesopotamian lowlands at the time of the earliest Holocene settlements and perhaps the western Levant as far south as the Negev (Ducos 1978, Payne 1983, Davis 1982); it remains problematic whether they also occupied the mountain zone to the north. Their entry into this habitat may have been more a response to the decline of woody and shrubby vegetation following human uses of the land, than to the 'naturalness' of that particular habitat for sheep.

That the mountain plains held some forest during the late Pleistocene (in spite of some pollen indications) is suggested by the ubiquity of red deer and cattle in Palaeolithic sites (e.g. Shanidar, Gar Arjeneh, and Bisitun). That much open ground also existed seems indicated by the presence of onager and gazelles, but the important point is that routes of migration between grazing areas may have been closed during the early Holocene by forest and scrub, both of which would impede the travel of sheep more so than of goats. I argue, therefore, for a spatial separation of the major concentrations of sheep from goats at the time of early domestication.

According to cytological studies, the wild ancestor of the domestic sheep is *Ovis orientalis* (Armenian variety) whose present distribution is west of a line from the Caspian Sea to the Arabian Gulf (Valdez *et al.* 1978). Although I have remarked above on the problems of using modern distributional data to infer past situations, it is interesting to note that the cytological data support an origin in northeastern Mesopotamia, as would the Shanidar and Zawi Chemi fauna. However, on modern environmental grounds, Ducos & Helmer (1981, p. 525) propose that sheep were domesticated in Central Anatolia, and Hilzheimer (1936), Zeuner (1963), and Ryder (1982, 1986) have argued that sheep and goats may have originated in Central Asia. Meadow (1984, p. 324) finds sheep in the earliest aceramic levels of Mergharh in Baluchistan, but both their absolute ages and genus are in doubt. Clearly, the issue has not been settled but present evidence suggests that sheep were much more restricted geographically than were goats, and that the distributions of the two species did not overlap much, if at all.

By contrast, there is evidence of goats throughout the mountainous regions of southwest Asia and eastward into central Asia dating well back into the Pleistocene, a reflection of their ability to withstand much more rugged topography and poorer forage, and perhaps an indication that they were easy to hunt. As browsers they are able to thrive on a variety of foods, but their density in the Levantine region must have been quite low and the *Capra ibex* of the southern Levant was apparently never domesticated.

Concluding remarks

My model is based on evidence of variable quantity and quality, which presently suggests that plant domestication was essentially a lowland adaptation that had a long period of technological development, beginning with grinding-stones from some 18 000 years ago. The species that were eventually domesticated are native to the grassy steppes and lightly forested regions. Potentially, adaptations of this type could be found on both sides of the Mesopotamian plain where rainfall or surface runoff stimulated the growth of the relevant species. However, present archaeological and cytological evidence suggests that the development was much older in the Levant and along the Nile than in Mesopotamia, which it may have reached through slow diffusion.

In contrast, we find no evidence of the use of domesticated livestock before about 9000–10 000 cal. BC and it occurs in a narrow zone along the front of the Zagros mountains in seasonal camps where caprine herders also engaged in some form of food collecting or cultivation. The picture with animals is more complicated than with plants in that the four major species: goats, sheep, cattle, and pigs, were probably domesticated separately and at different times by different processes. The reasons for this separation may have to do both with environmental distributions of the animals and with their different behaviours. The earliest species, sheep and goats, have habits that place them seasonally in the same locales as the ripening plants that people wished to harvest, so that propinquity was an important factor. However, the fact that these species readily lent themselves to herding and to multiple human uses, as contrasted with gazelles or onagers that also moved with the ripening plants, may have had more to do with the domestication than their mere presence in the environment.

In the future, as well as looking into plant use during the late Pleistocene as Moore has suggested, it will be important to acquire much more precise data on the changing vegetation of the north Mesopotamian plain, and to seek possible herding camps as well as seed-processing camps in this lowland steppe.

References

Bar–Yosef, O. 1980. Prehistory of the Levant. *Annual Review of Anthropology* **9**, 101–33.

Braidwood, R. J. & B. Howe 1961. *Prehistoric investigations in Iraqi Kurdistan*. The University of Chicago Oriental Institute, Studies in Ancient Oriental Civilization, no. 31, Chicago: The University of Chicago Press.

Davis, S. J. M. 1982. Climatic change and the advent of domestication: the succession of ruminant artiodactyls in the late Pleistocene–Holocene in the Israel region. *Paleorient* **8**(2), 5–15.

Ducos, P. 1978. *Tell Mureybet étude archéozoologique et problèmes d'écologie humaine*, Vol. 1. Lyon: CNRS.

Ducos, P. & D. Helmer 1981. Le point actuel sur l'apparition de la domestication dans le Levant. In *Préhistoire du Levant*, J. Cauvin & P. Sanlaville (eds), Actes du Colloque Internationale no. 598, 523–8. Paris: Edition du Centre de la Récherche Scientifique.

Evins, M. A. 1982. The fauna from Shanidar Cave: Mousterian wild goat exploitation in Northeastern Iraq. *Paleorient* **8**, 37–58.

Hilzheimer, J. 1936. Sheep. *Antiquity* **10**, 195–206.

Hole, F. 1984. A reassessment of the Neolithic Revolution. *Paleorient* **10**, 49–60.

Hole, F. in press. Chronologies in the Iranian Neolithic. Colloque International du CNRS, *Chronologies relatives et chronologie absolu dans le Proche Orient de 16000 à 4000 B P.* Lyon.

Hole, F. & K. V. Flannery 1967. The prehistory of Southwestern Iran: a preliminary report. *Proceedings of the Prehistoric Society* **33**, 147–206.

Howe, B. 1983. Karim Shahir. In *Prehistoric archeology along the Zagros flanks*, L. S. Braidwood, R. J. Braidwood, B. Howe, C. A. Reed & P. J. Watson (eds), Oriental Institute Publications, Vol. 105, 23–154. Chicago: Oriental Institute of the University of Chicago.

Lawrence, B. 1980. Evidences of animal domestication at Çayönü. In *Prehistoric research in southeastern Anatolia*, H. Cambel & R. J. Braidwood (eds), 285–308. Istanbul: Istanbul University, Faculty of Letters No. 2589.

Legge, A. J. 1975. The fauna of Tell Abu Hureyra: preliminary analysis. In The excavation of Tell Abu Hureyra: a preliminary report, A. M. T. Moore. *Proceedings of the Prehistoric Society* **41**, 74–7.

Meadow, R. H. 1984. Animal domestication in the Middle East: a view from the eastern margin. In *Animals and archaeology*. Vol. 3: *Early herders and their flocks*, J. Clutton–Brock & C. Grigson (eds), 309–37. Oxford: BAR International Series 202.

Moore, A. M. T. 1982. Agricultural origins in the Near East – a model for the 1980s. *World Archaeology* **14**, 224–36.

Moore, A. M. T. 1985. The development of Neolithic societies in the Near East. *Advances in World Archaeology* **4**, 1–69.

Moore, A., J. Gowlett, R. Hedges, G. Hillman, A. Legge & P. Rowley–Conwy 1986. Radiocarbon accelerator (AMS) dates for the epipalaeolithic settlement at Abu Hureyra, Syria. *Radiocarbon* **28**, 1068–76.

Payne, S. 1983. The animal bones from the 1974 excavations at Douara Cave. In *Paleolithic site of Douara Cave and paleogeography of Palmyra Basin, Syria*, K.

Hanihara & T. Akazawa (eds), Museum of the University of Tokyo, Bulletin 21, 1–108. Tokyo: University of Tokyo.

Ryder, M. L. 1982. Sheep–Hilzheimer 45 years on. *Antiquity* **56**, 15–23.

Ryder, M. L. 1983. *Sheep and man*. London: Duckworth.

Turnbull, P. F. & C. A. Reed 1974. The fauna from the terminal Pleistocene of Palegawra Cave. *Fieldiana Anthropology* **63**(3), 81–146.

Valdez, R., C. F. Nadler & T. D. Bunch 1978. Evolution of wild sheep in Iran. *Evolution* **32**, 56–72.

Van Zeist, W. & S. Bottema 1982. Vegetational history of the eastern Mediterranean and the Near East during the last 20,000 years. In *Palaeoclimates, palaeoenvironments and human communities in the eastern Mediterranean region in later prehistory*, J. L. Bintliff & W. Van Zeist (eds), 277–319. Oxford: BAR International Series 133.

Zeuner, F. E. 1963. *A history of domesticated animals*. London: Hutchinson.

11 *The domestic horse of the pre-Ch'in period in China*

CHOW BEN–SHUN

Introduction

Somewhat more than 9000 years ago, certain human populations in China changed their subsistence patterns from hunting to a more settled way of life and began to cultivate plants and to domesticate animals. The earliest Neolithic culture so far discovered in northern China is the 'Cishan-Peilikang culture', of which the radiocarbon dates are around 8000 years BP. Remains of *Sus domesticus* and *Canis familiaris* have been found at both Cishan and Pei-li-kang (Chow Ben–Shun 1984). The excavations show that animal husbandry and crop-raising occurred together. The pig and the dog were the first animals to be domesticated in Neolithic China. Pigs especially were of economic importance throughout the period, since they could provide meat rapidly, ensuring a regular food supply for large groups of people, and organic fertilizer for farming. As a supplier of meat, the horse was of little importance in the earliest Neolithic cultures.

In the past three decades more than 7000 Neolithic sites have been found on the mainland of China, and over 100 of them have been excavated. Thousands of animal bone fragments have been found in the debris of these sites, but there are almost no reliable reports of horses associated with the early Neolithic culture in China. The earliest locality that has yielded horse remains is the Yangshao culture (4800–3000 BC) Panpo Site at Sian, from which two molars and a first phalanx are regarded by Li You-heng & Han Defen (1959) as *Equus przewalskii*. Most investigators have suggested that the domesticated horse was established during the Luangshan period (3000–2300 BC), but even at this period osteological remains of horses are rare and have been identified only as *Equus* sp., making it impossible to say if the animals were really domesticated. Horse remains collected from sites of Machayao culture in Kansu Province (which is a later regional culture, *c.* 3000 BC, that shares a common origin with the Yangshao) and from a Chichia culture (2000 BC) cemetery at Chin-Wei-Chia in Yung-Ching County, Kansu Province (Kansu Archaeological Team 1975) indicate that the first domestication of horse took place around 2000 BC in northwestern China. Before the early Shang Dynasty, around 1300 BC, there is no archaeological evidence for domesticated horse in the Central Plain.

The horse of the Chinese Bronze Age

The Chinese Bronze Age, lasting for at least 1500 years from almost 2000 BC until the 3rd century BC, was a formative stage in Chinese civilization. The use of the horse in the wars between states and in transport gained it a special position in the exercise of political power. The horse became an instrument of power in Eurasia from Europe in the west to China in the east at almost the same time.

In the past half century, many Bronze-age chariot burials with horses have been excavated in China, indicating the rapid progress of horse-raising in this period. During the past few years the author has made an intensive study of horse skeletons from Bronze-age sites. Nearly 100 horse skeletons, dating from the Shang Dynasty Anyang phase (13th–11th century BC), Western Chou Dynasty (11th–8th century BC), down to the Spring and Autumn Period (770–746 BC), have been studied to trace the development of Chinese horses, with special attention to the height at the withers. Complete long bones of these horse skeletons were measured and the sizes of their withers were calculated. The height of the withers of domesticated horses of the Shang Dynasty Anyang phase, found in the sacrificial pits at the northern part of Wuguan village, Yin Hsu, is 133–143 cm. As recent Przewalski's horses had a withers-size of 134 cm on average, it would appear that the late Shang horses were related to the Mongolian wild horse. The horses from a Yen cemetery of the Western Chou Period at Liulihe, near Beijing (The Joint Archaeological Team & the Municipal Archaeological Team of Beijing 1984) reached a height of 135–146 cm. Horses of the Spring and Autumn Period from chariot pits belonging to a crown prince of the state of Kuo at Shang-ts'un-ling near the city of San-Men Gorge, Honan Province (The Institute of Archaeology, Academia Sinica 1959) grew to a withers height of 139–149 cm. Thus, the existing osteological evidence suggests that in the Chinese Bronze Age, the height of the horses gradually increased from ponies of 133 cm to medium-sized horses of 149 cm. They retained throughout this time the characteristics of the Przewalski's horse, including relatively heavy heads, thick necks, upright manes, short bodies, and relatively short legs, and can be regarded as a homogeneous population. The artists of the Chinese Bronze Age illustrated these features in their works, as, for instance, in the jade figure of horses from the tomb of Lady Hao, Anyang period c. 1300–1030 BC (Institute of Archaeology 1980), and the Western Chou bronze receptacle (tsun) in the shape of a foal (11th–8th century BC) unearthed at Meixian, Shensi Province (Kuo Mo-jo 1957). Down to the Ch'in Period the Chinese horse was clearly related to the Mongolian pony. The life-size terracotta war-horses of the First Emperor of China, Ch'in Shih-huang-ti, discovered at Lintong in Shaanxi Province, near the Emperor's Mausoleum, indicate that both chariot and saddle horses belonged to the same stocky breed, characterized by compact bodies, short legs, and broad necks. After the Western Han Dynasty the Chinese horses received some inflow of genes from the Ferghana horses which were derived from the tarpan, and new and larger breeds of graceful horses with small heads made their appearance. The Eastern Han bronze horses, excavated from Kansu (The Kansu Provincial Museum 1974), illustrate perfectly the graceful carriage of the tarpan breed in which the horse is shown as flying with its hoof on a falcon.

References

Chow Ben–Shun 1984. Animal domestication in Neolithic China. In *Animals and archaeology*. Vol. 3: *Early herders and their flocks*, J. Clutton-Brock & C. Grigson (eds), 363–9. Oxford: BAR International Series 202.

Institute of Archaeology, Academia Sinica 1959. *The cemetery of the State of Kuo at Shang Tsun Ling*. Archaeological excavations at the Yellow River reservoirs, Report no. 3, 1–85. Peking: Science Press.

Institute of Archaeology, Chinese Academy of Social Sciences 1980. *Tomb of Lady Hao at Yinxu in Anyang*. Beijing: Cultural Relics Publishing House.

Joint Archaeological Team of Institute of Archaeology, Chinese Academy of Social Sciences & The Municipal Archaeological Team of Beijing 1984. Excavation of a Yan cemetery of the Western Zhou Period in Liulihe in 1981–1983. *Kaogu (Archaeology)* **200**, 405–26.

Kansu Archaeological Team, Institute of Archaeology, Academia Sinica 1975. Excavation of Chichia culture cemetery at Chin-Wei-Chia in Yung-ching County, Kansu Province. *Kaogu Xuebao* **43**, 57–69.

Kansu Provincial Museum 1974. The Han Tomb at Lei-Tai in Wuwei County, Kansu Province. *Kaogu Xuebao* **41**, 87–110.

Kuo Mo-jo 1957. Notes on inscription of bronze vessels made by the Family Li. *Kaogu Xuebao* **16**, 1–6.

Li You–heng & Han Defen 1959. Animal bones from Pien-Po Neolithic site near Sian. *Paleovertebrata et Paleoanthropologia* **1**, 173–88.

12 *Utilization of domestic animals in pre- and protohistoric India*

P. K. THOMAS

In India the history of domestication of animals goes as far back as 7000 years BP, from the Mesolithic cultural phases at Bagor, district Bhilwara in Rajasthan (Misra 1973) and Adamgarh, district Hoshangabad in central India (Joshi & Khare 1966). A series of radiocarbon dates is available from Bagor, and the earliest date for the Mesolithic cultural phase is 6245 ± 200 BP, TF 786 (Agrawal *et al.* 1971). At Adamgarh the two enigmatic dates are 2765 ± 105 BP, TF 116, from 1.90 m below the surface on charred bones, and 7240 ± 125 BP, TF 120 on shells from 0.15–0.20 m depth (Agrawal & Kusumgar 1968). The discrepancy in the ^{14}C dates of Adamgarh is dealt with in more detail elsewhere (Thomas 1975). Although these two sites belong to the same cultural period, the composition of the fauna represented in them varies considerably. At Bagor, sheep/goat has been identified as the principal domestic animal (Thomas 1975, 1977); while at Adamgarh cattle (*Bos indicus*), buffalo (*Bubalus bubalis*), Sheep (*Ovis aries*), goat (*Capra hircus*), pig (*Sus domesticus*), and ass (*Equus asinus*) are reported as domestic animals (Joshi 1968).

The humped Indian cattle (*Bos indicus*) appears to be the most predominant domestic animal in the early cultures of India, with only a few exceptions. The Mesolithic site at Bagor and the Late Jorwe phase at Inamgaon, a Chalcolithic site in Maharashtra, have yielded more sheep/goat bones (about 65 per cent of the total faunal assemblage) than those of cattle. From all prehistoric levels, and even up to some of the early historic cultural periods, cattle were killed for meat and were a major source of subsistence for the early populations. The age group studies of the cattle killed in a majority of archaeological sites suggest that young animals around the age of three years were preferred for food. In order of preference in the food economy, the next important domestic animals were sheep, goat, pig, and buffalo. However, in some sites like Navdatoli (Chalcolithic) in district Nimar in central India, instead of sheep and goat, pigs were the second most important animals in the food economy (Clason 1977). Dog (*Canis familiaris*) is associated with most of the prehistoric cultures and may have served as watch animal for the settlement and domestic herd, and also in the hunting pursuits of early man. However, a solitary evidence of dog being killed for meat is reported in the Late Jorwe phase (*c.*1000–700 BC) at Inamgaon (Thomas 1984a).

The role of domestic animals further increased in the economy of the early populations. Alur identified ossified hock joints of cattle from the Neolithic sites at Hallur, district Dharwar, and T. Narasipur, district Mysore, both in the Karnataka State, and suggests that these animals must have been used for heavy traction or draught purposes (Alur 1971a, 1971b). The presence of a comparatively large number of aged bulls in sites like Ramapuram (Chalcolithic) district Kurnool, Andhra Pradesh (Thomas 1981) and Veerapuram (Neolithic) district Kurnool, Andhra Pradesh (Thomas 1984b) suggests that these animals were probably used in agricultural operations, such as traction or even for running irrigation devices. Also, the use of this animal in threshing the harvested crops is a possibility, as it is practised even today in Indian villages. The evidence of agriculture comes from the grains, cereals, and pulses excavated from sites, and also the impressions of some of these on different materials. The prehistoric agricultural tools made of wood may not have survived in the Indian archaeological sites as they have not been reported from the vast majority of sites. The strong antlers with definite shapes found in the excavations must have also been used in agricultural operations. The large storage jars and silos found in most of the Neolithic and Chalcolithic sites may also add to the agricultural evidences. Along with warfare equipment, the iron hoes, saddle querns, and pounders unearthed from the Vidarbha megalithic sites such as Naikund (Deo 1982), Mahurjhari (Deo 1973), Borgaon, and Bhagimohari (Deo in press), all in the Nagpur district of Maharashtra, point out the agricultural activities of the early inhabitants.

Terracotta wheeled toy carts are reported from some of the protohistoric sites in India. The earliest record is at Mohenjodaro and Harappa, dating back to 2300 BC (Allchin & Allchin 1968). The cart from Mohenjodaro is shown pulled by bullocks. Inamgaon, a Chalcolithic site in Maharashtra, has yielded an engraving of a cart drawn by bullocks on a storage jar from the Jorwe phase dated to 1400–1000 BC (Dhavalikar 1974, Sankalia 1974). These representations suggest the use of cattle as draught animals from very early times in India.

The cattle by-products must have been of great importance to the people. There is no direct evidence for the consumption of milk until early historic times. One of the earliest evidences for milking is depicted in a stone relief at Mahabalipuram dated to about 7th–8th century AD (Zeuner 1963). That cowdung was extensively used for plastering the floors and walls of the early houses, and also as a fuel, is known from the prehistoric settlements. Neolithic 'ash-mounds' consisting of cow dung have been reported at Kupgal, district Bellary in Karnataka and Utnur, district Mahabubnagar in Andhra Pradesh (Allchin & Allchin 1968). Cattle dung is the main source of domestic cooking fuel even today in Indian villages (Harris 1966). This must have also been used as a manure in agriculture. Another important cattle by-product is hide, for which archaeological evidences are sparse.

The horse (*Equus caballus*) was introduced late into the early cultures of India in an already domesticated form. It was during the Iron Age (c.1000–400 BC) that horse-breeding became prevalent (Thomas in press a). Horse ornaments and equipment retrieved from the megalithic sites in the Vidarbha region of Maharashtra suggest the use of this animal for riding. A few scanty representations of horse are noticed in the late phases of the Neolithic and Chalcolithic cultures (Nath 1968, Alur 1971a, Thomas 1984a) which may tally chronologically with the Iron Age cultures.

The close man–animal relationship can be further traced from the various activities conducted by man involving his domestic stock. Terracotta figurines of humped cattle are reported from a majority of protohistoric sites, and these animals are also depicted on pottery and rock-shelters. At present it may not be appropriate to suggest that these animals were worshipped. One thing that goes against the worship is the profuse killing of these animals in the pre- and protohistoric periods. However, the depiction itself suggests the significance of these animals to the early human populations.

Food offered to the dead is an age-old custom in Indian history. At Inamgaon, bones of cattle, sheep, and goat are found along with urn and extended burials. Such offerings are also reported from a number of other sites. Animal sacrifices were also common in the early cultures. The important sacrificed animals were cattle, sheep, goat, dog, and horse. At Burzahom, a Neolithic site in Kashmir, dog/wolf heads were buried with a human dead body (Allchin & Allchin 1968). The Chalcolithic site, Ramapuram, in Andhra Pradesh has yielded partial and complete burials of goat in the human graves. In partial burials, skulls and lower parts of the limb bones of goat are represented. In some cases complete burial of a goat has also been reported (Thomas 1981).

The megalithic stone circles in the Vidarbha region of Maharashtra, belonging to the Iron Age period, have brought to light partial burials of horses in the form of skulls and lower extremities of limb bones. A number of megalithic stone circles are being excavated by S. B. Deo of Deccan College, Pune, and animal remains from these graves and the habitation sites are being studied by the present author. From these studies it has been proved that horse sacrifice was not a regular custom, but it might have been a status symbol among the Iron-age populations of Vidarbha. So far, the only exception to the sacrifice of another animal in place of horse is in one of the megalithic circles at Naikund, where cattle bones are identified (Thomas in press b). From the recent excavations in the megalithic habitations at Naikund and Bhagimohari (Thomas in press c) the bones of horse with cut marks are reported. A similar find has been reported from the Late Jorwe phase at Inamgaon. Probably after the sacrifice and subsequent offerings to the dead persons, the rest of the body of the horse might have been consumed by the inhabitants. Another interesting evidence of animal sacrifice is found at Khanpur, a Harappan site in Gujarat, where a ventrally pierced atlas vertebra (probably executed with a spear-like instrument) is reported (Chitalwala & Thomas 1978). This probably suggests the sacrifice of the animal, involving the collection of hot, spurting blood as in the case of bull sacrifices performed in Crete (Fabri 1937).

Sharing of cattle meat between people of two different houses belonging to the same settlement is found at Inamgaon (Thomas 1984a). Two equal halves of a three-year-old bull are represented in two different houses. The degree of charring and the metrical homogeneity of bones of the left and right side of the body have been studied in detail. Probably on some occasion people of these two different houses shared this animal.

The economic outlook of the people and the varied uses of cattle might have been the reasons for the abundance of cattle in the early cultures of India. Man now felt a closer affinity and emotional feeling towards the animal which provided almost everything for him. Simultaneously, a taboo was introduced on the slaughter of cattle and this animal procured a place in the Hindu religion. An exact date for the ban on cow slaughter cannot be postulated at present; however, this development in India must have taken place during Buddhism and Jainism in the beginning of the Christian era.

References

Agrawal, D. P. & S. Kusumgar 1968. Tata Institute Radiocarbon Date list V. *Radiocarbon* **10**, 131–43.

Agrawal, D. P., S. K. Gupta & S. Kusumgar 1971. Tata Institute Radiocarbon Date list IX. *Radiocarbon* **13**, 442–9.

Allchin, B. & R. Allchin 1968. *The birth of Indian civilization*. London: Penguin.

Alur, K. R. 1971a. Skeletal remains. In *Protohistoric cultures of Tungabhadra valley (a report on Hallur excavation)*, M. S. N. Rao (ed.), 107–24. Dharwar: M. S. Nagaraja Rao.

Alur, K. R. 1971b. Report on the animal remains. In *Report on the excavation at T. Narasipur*, M. Seshadri (ed.), 19–104. Mysore: Department of Archaeology, Government of Karnataka.

Chitalwala, Y. M. & P. K. Thomas 1978. Faunal remains from Khanpur and their bearing on culture, economy and environment. *Bulletin Deccan College Research Institute* **38**, 11–14.

Clason, A. T. 1977. Wild and domestic animals in prehistoric and early historic India. *The Eastern Anthropologist* **30**(3), 241–89.

Deo, S. B. 1973. *Mahurjhari excavations*. Nagpur: Nagpur University.

Deo, S. B. 1982. *Excavations at Naikund (1977–78: 79–80)*. Bombay: Government Press.

Deo, S. B. in press. *Excavations at Borgaon and Bhagimohari*. Bombay: Government Press.

Dhavalikar, M. K. 1974. Subsistence pattern of an early farming community of western India. *Puratattva* **77**, 39–56.

Fabri, C. L. 1937. The Cretan bull-grappling sports and bull sacrifice in the Indus Valley civilization. In *Archaeological survey of India Annual Report, 1934–35*, J. F. Blakistan (ed.), 93–101. Delhi: Manager of publications.

Harris, M. 1966. The cultural ecology of India's sacred cattle. *Current Anthropology* **7**(1), 51–66.

Joshi, R. V. 1968. Late Mesolithic culture in Central India. In *La préhistoire: problèmes et tendances*, F. Bordes & D. de Sonneville Bordes (eds.), 245–54. Paris.

Joshi, R. V. & M. D. Khare 1966. Microlithic bearing deposits of Adamgarh rock-shelters. In *Studies in prehistory, Robert Bruce Foote memorial volume*, D. Sen & A.

K. Gosh (eds.), 90–5. Calcutta: Firme K. L. Makhopadhyay.

Misra, V. N. 1973. Bagor – a late Mesolithic settlement in north-west India. *World Archaeology* **5**(1), 92–110.

Nath, B. 1968. Advances in the study of prehistoric and ancient animal remains in India, a review. *Records of the zoological survey of India* **61**(1 & 2), 1–63.

Sankalia, H. D. 1974. *Prehistory and protohistory of India and Pakistan*. Poona: Deccan College Postgraduate and Research Institute.

Thomas, P. K. 1975. Role of animals in the food economy of the Mesolithic culture of western and central India. In *Archaeozoological studies*, A. T. Clason (ed.), 322–8. Amsterdam: North Holland Publishing Company.

Thomas, P. K. 1977. *Archaeozoological aspects of the prehistoric cultures of western India*. Unpublished PhD dissertation, Poona University.

Thomas, P. K. 1981. A preliminary report on the animal remains from Ramapuram. *Indian Archaeology: A Review 1981*.

Thomas, P. K. 1984a. The faunal background of the chalcolithic culture of western India. In *Animals and archaeology*. Vol. 3, J. Clutton–Brock & C. Grigson (eds), 355–61. Oxford: BAR 202.

Thomas, P. K. 1984b. Faunal assemblage of Veerapuram. In *Veerapuram: a type site for cultural study in the Krishna Valley*, T. V. G. Sastri, M. Kasturi Bai & J. Vara Prasad Rao (eds), Appendix C, i–xi. Hyderabad: Bhavani Printers.

Thomas, P. K. in press a. Subsistence and burial practice based on animal remains at Khairwada. In *Excavations at Khairwada*, S. B. Deo (ed.). Nagpur: Government Press.

Thomas, P. K. in press b. Animal remains at Naikund. In *Excavations in Vidarbha*, S. B. Deo (ed.). Nagpur: Government Press.

Thomas, P. K. in press c. Faunal background of the megalithic habitation at Bhagimohari. In *Excavations at Bhagimohari*, S. B. Deo (ed.). Nagpur: Government Press.

Zeuner, F. E. 1963. *A history of domesticated animals*. London: Hutchinson.

PASTORALISM

Introduction to pastoralism

JULIET CLUTTON-BROCK

Patterns of subsistence based on hunting, herding, farming, pastoralism, and nomadism all depend on the exploitation of herd animals and were all established during the prehistoric period yet they still survive at the present day. The cultural and environmental determinants of these social systems and the transitions between them have intrigued anthropologists (e.g. Evans–Pritchard 1940) since the beginning of this century, but it has only been within the past decade or so that comparative studies have been carried out on pastoral economies. Notable amongst these have been the publication (in English translation) of the work on pastoral nomadism by Khazanov (1984), and the reviews of Ingold (1980, 1984).

As discussed in Chapter 8 by Meadow, it might be logical to assume that there could be a continuity from the hunting of wild animals to the following of herds, and hence to pastoralism. Khazanov (1984) and other authors claim, however, that this has seldom, if ever, occurred. As discussed by Ingold (1980), the pressures induced by human social systems have resulted in a sequence of cultural changes that have followed the same pattern throughout Eurasia. This began with broad-spectrum hunting towards the end of the Pleistocene, which was replaced by dependence on a few species of large mammals as resources diminished and the human population increased. Change of climate, over-hunting, or the immigration of agriculturalists then resulted in settlement and the cultivation of plants. The generally accepted thesis is that it is only after agriculture and animal husbandry became well established that pastoral nomadism developed as a social system in Europe and Asia. However, as discussed later, plant cultivation as an essential intermediate stage between hunting and a pastoral economy is contended for southern Africa and South America, as well as, with perhaps less evidence, for the reindeer-herders of northern Europe.

Khazanov (1984) maintains that the sources of pastoral nomadism in the Old World are now clear and that they can be directly linked to a food-producing economy. He holds, as is now the general belief, that only people who were leading a relatively sedentary way of life and who had surpluses of vegetable food could domesticate hoofed animals. These food-producing economies spread from their core areas into new habitats, some of which could support settled husbandry but some, especially in arid regions, could support only mobile pastoralism. The work of Khazanov has been centred on the East European steppes, and, indeed, the basis for some of his ideas has been provided by the work of Shilov who discusses nomadism in Chapter 13. Zarins, in Chapter 14, presents archaeological evidence for the beginnings of pastoralism in Arabia, which was one outward extension from the Near Eastern centre of domestication, and he shows how, as the climate became progressively more arid, sheep and goats were replaced by camel-herding. Dhavalikar, with a similar theme but from a later period, in Chapter 15, discusses how successive droughts and famines forced the settled communities of the Deccan plateau of India to resort to pastoral nomadism in order to survive.

The traditions of herd management of reindeer in Lapland, together with some fascinating anecdotes, are described in Chapter 16 by Aikio. While accepting that the general view of reindeer-herding is that it is of relatively recent origin, perhaps dating to the end of the Middle Ages, Aikio personally believes that its beginnings are far more ancient and probably go back to the end of the Ice Age.

In Chapter 17, Tani describes his exhaustive study of how the shepherds of the transhumant flocks of sheep and goats in the Mediterranean countries train certain animals to lead the flocks. These flock-leaders may or may not be castrated and Tani discusses the origins and functions of castration in this context as well as drawing parallels with the castration of human males in the ancient world.

Khazanov (1984) discussed only briefly the origins of pastoralism in Africa and, indeed, they are only just beginning to be investigated. Cattle-herding in North and East Africa is discussed in Chapters 18 and 19. Whereas sheep and goats were undoubtedly brought into Africa from western Asia, there does seem to be some slight evidence to suggest that cattle may have been locally domesticated in North Africa (Clutton–Brock in Ch. 18). Robertshaw, in Chapter 19, gives the evidence for the first pastoralism in East Africa as being found from the late 3rd millennium BC. From the diversity of animal remains from archaeological sites and ethnographic accounts, he discusses the relationships between hunters and herders, attitudes of cultural superiority by pastoralists, and the question of stock-thieving by hunters; a topic that has also been discussed by Davidson in Chapter 6 and mentioned by Clutton–Brock in the Introduction to Section I of this book.

The insistence by Khazanov (1984) and other anthropologists that pastoralism can only develop from a society that practises agriculture may not apply in Africa, perhaps because there was no shortage of wild plant foods for both animals and people. Robertshaw claims that the pastoralists of East Africa were reluctant cultivators, and the evidence from the western Cape of South Africa indicates that the aboriginal Khoi (Hottentots) were never settled and never cultivated plants, with the exception of a narcotic (Klein 1986). The Khoi were originally hunter-gatherers who adopted, sometime after 2000 years ago, the sheep and cattle that had moved south through the continent with migrating people over the preceding millennium. Again, as discussed by Klein (1986) the Khoi may have obtained their cattle by raiding from Bantu immigrants. They used oxen as draught animals and sheep to provide milk and meat (Smith 1986).

In Chapter 20 Galaty describes the systems used by the Maasai for the naming and classification of their cattle. The mastery of pastoral cognition is gained through learning and experience which are today inevitably decreasing as young people turn towards school education.

The remaining four chapters in this section of the book are about camelid pastoralism in South America, a subject that has been little investigated until very recently. Chapters 21 and 22 by McGreevy and Brotherston, respectively, discuss the role of camelids before the Spanish Conquest and, in particular, the importance of the llama in Inca society. Browman, in Chapter 23, presents a review of the earliest evidence for camelid exploitation and the shift from hunting to pastoralism, which does appear to have preceded the cultivation of plants. These three chapters demonstrate how modern data on animal distributions and behaviour, literary sources, and the investigation of animal remains from archaeological sites can be combined to elucidate the history of ancient social systems and palaeoeconomies.

Finally, in these studies on camelid pastoralists, Rabey in Chapter 24 describes fieldwork that he has carried out amongst llama-herders in the hills and plains of the south central Andes. With the hill herdsmen Rabey describes what could be the most primitive form of pastoralism, which is merely an extension of hunting with very little control or ownership of the animals. If this is indeed a relic of an anciently established system of exploitation, it appears to support Aikio (Ch. 16) and to belie the theories of Khazanov (1984) that pastoralism only develops from true domestication. On the other hand, like much modern reindeer nomadism in northern Eurasia, this form of herd-following may involve the secondary use of feral animals, which were anciently domesticated, within a modern hunting economy, as discussed by Ingold (1980, p. 123).

At this point it may be helpful to explain some of the terms used in studies of pastoralism, using the definitions of Ingold (1980) and Khazanov (1984):

Hunters are food-extractors who are only interested in the dead animal, as discussed by Meadow in Chapter 8. Like other carnivores, the hunter interacts with prey only when it is about to be killed. All other social systems of food-producing protect the living animal until it is ready for slaughter.

Herd-following may apply to a human population that ranges over the same area in its annual cycle as the animal population, or it may apply to particular humans that are associated with particular herds of animals, in which case it is equivalent to *ranching*.

The *rancher* loosely owns herds of animals for exploitation of meat and other resources that are often marketed. In origin the animals may be wild, feral, or domestic but they live as wild animals except that their territory is usually restricted.

Nomads may be wandering hunters and gatherers or mobile pastoralists. However, the reasons for mobility in these two groups are so different that Khazanov considers that hunter-gatherers should be termed 'wandering' and the term 'nomadism' should be reserved for mobile pastoralists.

Pastoralists are divided by Khazanov into a number of different categories within two main groups. *Pastoral nomadism proper*, which is characterized by the absence of agriculture and is exemplified by, say, the pastoralists of the Sahara, or the 'pure pastoralists' of McGreevy (Ch. 21), and *semi-nomadic pastoralism* in which there is periodic changing of the pastures during the year but the cultivation of crops is also practised. This is the most common form of pastoralism.

Transhumants are agriculturalists living in the Mediterranean and southern Europe who move their livestock between mountain and lowland pastures.

These somewhat simplistic definitions are given here to clarify some of the terms used in this section of *The walking larder*: for further discussion see Ingold (1980, 1984) and Khazanov (1984). The intention of the chapters on pastoralism is to present a picture of the complicated and fascinating relationships between herders and their flocks from many different parts of the world in the past and at the present day.

References

Evans–Pritchard, E. E. 1940. *The Nuer*. Oxford: Clarendon Press.

Ingold, T. 1980. *Hunters, pastoralists and ranchers*. Cambridge: Cambridge University Press.

Ingold, T. 1984. Time, social relationships and the exploitation of animals: anthropological reflections on prehistory. In *Animals and archaeology*. Vol. 3: *Early herders and their flocks*, J. Clutton–Brock & C. Grigson (eds), 3–12. Oxford: BAR International Series 202.

Khazanov, A. M. 1984. *Nomads and the outside world* (translated by J. Crookenden). Cambridge: Cambridge University Press.

Klein, R. G. 1986. The prehistory of stone-age herders in the Cape province of South Africa. *South African Archaeological Society Goodwin Series* **5**, 5–12.

Smith, A. B. 1986. Competition, conflict and clientship: Khoi and San relationships in the western Cape. *South African Archaeological Society Goodwin Series* **5**, 36–41.

13 The origins of migration and animal husbandry in the steppes of eastern Europe[1]

VALENTIN PAVLOVICH SHILOV

(translated by Katharine Judelson)[*]

There are two points of view on the origins of nomadic animal husbandry in the literature. Some researchers (Hahn 1886, pp. 132–3, 1891, p. 484, 1905, pp. 96, 99–100, Golmsten 1933, p. 89, Gryaznov 1955, p. 24, 1957, p. 2, Chernikov 1957, pp. 31–2, Rudenko 1961, p. 2) consider that nomads emerged as a separate group from settled communities engaged in both arable farming and animal husbandry at the end of the 2nd and the beginning of the 1st millennium BC. Others (Rousseau 1775, Smith 1776, Jselin 1786, Schmidt 1915–16, pp. 593, 610, 1924, p. 193, 1951, pp. 213–17) are of the opinion that nomadic animal husbandry emerged earlier, at the time of hunting tribes. For the development of the nomadic economy, there are also two points of view: first that it came into being suddenly (Gryaznov 1957), and secondly that it took shape over several centuries (Smirnov, K. F. 1964, pp. 45–7, Smirnov, A. P. 1966, p. 13).

What objective data do we have to resolve these questions? First of all let us consider evidence from written sources. In the 7th and 6th centuries BC the territory to the north of the Black Sea was colonized by the Greeks, who built a dense network of towns, the inhabitants of which established close links with the local people. From these colonists there have come down to us quite valuable, albeit fragmentary, observations concerning life in this territory, which provided the original source for the works of Greek and Roman historians. The Greek colonists and travellers from this area were struck first and foremost by the, for them, unusual nomadic way of life in the steppe, where families and their herds moved from one place to another 'depending upon where they could find an abundance of grass and water'. The poet Hesiod (Strabo 1st century AD, p. 7) who lived in the 8th century BC, pointed out in his *Theogony* that the Scythians milked horses. Homer, whose works are usually held to date from between the 10th and 8th centuries BC, referred to the 'amazing horse-milkers' in the steppes bordering on the Black Sea, who milked horses and drank their milk (Homer 1936, p. 355). It can also be assumed that the Cimmerites, who settled the territory adjoining the Black Sea before the arrival of the Scythians, were already nomads, since they succeeded in

[*] Modern Languages Department, Totton Sixth Form College, Hampshire.

carrying out long expeditions, for instance as far as Asia Minor. Finally, the fact that the Scythians in the 7th and 6th centuries BC engaged in a highly developed form of nomadic animal husbandry shows that it must have existed long before that time. The study of archaeological monuments, in particular that of Scythian burial mounds, has revealed marked property differentiation. Side by side with graves for the poorest strata of the population can be found rich graves, the furniture of which includes costly gold ornaments (the Litoi burial mound, the burial site on the Kalitva River, the Kelermessky burial mounds, etc.), and also graves where together with the deceased were buried hundreds of horses (the Ulsky burial mound and others). Furthermore, the burial sites of the Scythian nobility were enormous monuments, the erection of which demanded gigantic manpower resources. This already reflects a complex picture of the social life of the nomadic people, and does not in any way point to merely preliminary steps towards the development of nomadic animal husbandry, but rather to an advanced stage in that development, when tremendous wealth, in the form of livestock and expensive luxury objects in gold, was concentrated in the hands of the leaders.

Another early emergence of nomadic animal husbandry is to be observed in desert regions, where it can still be found nowadays: for instance, the ancient population to the west of the Nile was leading a nomadic life in the neighbouring deserts as early as the 4th century BC. Egyptian hieroglyphic inscriptions inform us as to the large numbers of livestock the Egyptians used to obtain from the Libyan deserts, where the people were engaged in nomadic animal husbandry (Struve 1941, p. 146).

A typical example of a nomadic people was that of the Hyksos who conquered Egypt in 1710 BC (Struve 1941, pp. 169–70). In the Bible it is written (*Chronicles* I v. 39–41) that the tribe of Judah went 'to the entrance of Gedor, even unto the east side of the valley, to seek pasture for their flocks. And they found fat pasture and good, and the land was wide, and quiet, and peaceable; for they of Ham had dwelt there of old. And these written by name came . . . and smote their tents, and the habitations that were found there, and destroyed them utterly unto this day and dwelt in their rooms: because there was pasture there for their flocks.'

The date at which herdsmen first moved into the open steppe is a question that my colleagues and I have studied in the examination of close on 800 burial sites in the open steppe dating from various periods. In addition, palaeogeographers and soil scientists working with us have demonstrated, on the basis of their study of deposits in the lakes of the steppes and the soils lying beneath the burial mounds that have been excavated, that the climate and the steppe vegetation found in the Bronze Age in the steppes of eastern Europe have changed little since then up to the present day, although there may have been fluctuations in aridity (Chuguryaeva 1960, p. 282, Neishtadt 1967, pp. 198–9). In 478 Bronze-age burial sites the animal remains were distributed as shown in Table 13.1. In so far as there were found in these burial sites the remains of men, women, children, and infants we can assume that at the time of the Pit-grave culture there were whole families living in the steppes. It was

Table 13.1 The remains of domestic animals collected from burial sites of Bronze-age cultures in the lower reaches of the Volga valley.

Name of archaeological culture	domestic animals Total No.	sheep and goats No. %	cattle No. %	horses No. %	camels No. %	dogs No. %	uniden-tified No. %
Pit-grave[a]	40	26 (65.0)	6 (15.0)	3 (7.5)	—	2 (5.0)	3 (7.5)
Poltavkin[b]	48	37 (77.1)	6 (12.5)	2 (4.2)	—	1 (2.1)	2 (4.2)
North-Caucasian[c]	152	106 (69.7)	21 (13.8)	11 (7.2)	1 (0.7)	1 (0.7)	12 (7.9)
Chamber[d]	238	128 (53.8)	60 (25.2)	27 (11.3)	—	1 (0.8)	21 (8.8)

Number of graves with bones of:

 a The Pit-grave culture spread throughout the steppes of eastern Europe from the River Danube in the west to the Urals in the east, extending on its southern edges into the wooded steppe belt. It is dated, in the main, as belonging to the 3rd millennium BC or to the first centuries of the 2nd millennium BC.
 b The Poltavkin culture was to be found in the steppes between the Volga and Ural rivers. Some graves relating to this culture have also been encountered on the eastern bank of the Volga and in the steppes between the Volga and the Don. It dates from the end of the 3rd millennium BC to the 17th century BC.
 c The North-Caucasian culture was found throughout the steppes of eastern Europe between the Don and the Ural rivers, stretching as far as the foothills of the Caucasus. It dates from the 21st to the 14th centuries BC.
 d The Chamber culture was to be found in the steppes of eastern Europe over the same territory as the Pit-grave culture. Chamber-culture tribes settled in the chernozem areas as well as in the wooded steppe zone, where long-term settlements are also found. It dates back to the period between the 16th and the 10th centuries BC. The Sauromatian–Sarmatian tribes were living as nomads in the Volga–Ural steppes between the 7th and 4th centuries BC, and in the 4th century BC they began to oust the Scythians from the steppes of the northern shores of the Black Sea. Later, nomads – the Huns, Avars, Torks, Hazaras, Pechenegs, Polovtsi, and Tartars – seized control of the steppes of eastern Europe after the 5th century BC, the so-called 'wild field'.

formerly held that the people of the Pit-grave culture were hunters and gatherers. Only more recently has the hypothesis emerged that they had a producer economy (Latynin 1957, p. 31, Merpert 1961, Lagodovskaya *et al.* 1962, p. 178). In 40 of the 263 burial sites (15.2 per cent) investigated in the steppes of the area between the Don, Volga and Ural rivers, bones of domestic animals, mainly sheep, were found.

The custom of placing meat in graves testifies undoubtedly to the fact that animal husbandry was widespread in the valley of the lower reaches of the Volga. In the Bronze Age the animals bred were for the most part sheep, cows, horses, and camels. Dogs were also kept and these were at that time buried with the humans.

Animal husbandry, hunting, and fishing were the principal activities in the economy of the Bronze-age population in the lower part of the Volga valley. Moreover, the role of animal husbandry increased considerably between the age of the Pit-grave culture (47 per cent) and the Chamber culture (94.1 per cent).

Sheep and goats predominated in the herds of the Bronze-age population in the lower part of the Volga valley, an arid land with saline and sandy soils. In such areas the sheep and goats would graze in the plains during the winter and in the mountains in the summer.

In pastures where a mixture of grasses predominated there would be more horses and cows in the herds. When the animal husbandry included different kinds of animals then the best winter pastures were those where both steppe and meadow vegetation were to be found, i.e. where there would be fodder both for sheep and camels, on the one hand, and for horses and cattle on the other (Geins 1897, pp. 60–1).

Therefore, on the basis of analysis of the bones found in burial sites, it can be said that in the economy of the Pit-grave culture and subsequent cultures animal husbandry was the predominant form of economic activity. Hunting and fishing did not play a prominent part. Up until the present time it has not proved possible to find permanent settlements of the population of the Pit-grave culture in the lower part of the Volga valley. So far over 400 settlements dating from the Neolithic Age, Eneolithic, and Early Bronze Age have been recorded. They are situated in terrain unsuitable for cultivation, in windswept sandhills (Rykov 1928, pp. 20, 23–4, 26, Minaeva 1929, p. 1, Sinitsyn 1931, pp. 81–2, 1933, p. 89, 1948, p. 151–60, Filipchenko & Kurochkin 1960, p. 272ff.).

It is also impossible to link bone remains with specific cultural horizons, in so far as they are encountered in a dispersed cultural stratum. This was borne out by archaeological excavations in the Ryn-Sands near Astrakhan undertaken by the Leningrad Division of the Institute of Archaeology under the USSR Academy of Sciences led by A. N. Melentiev. At that site archaeologists have found dozens of settlements between sandhills in terrain unsuitable for cultivation. In these settlements pottery and microlithic tools from the Neolithic, Aneolithic, and Bronze Ages were found. These settlements resemble in character temporary camps for herdsmen, rather than sites of permanent habitation. Approximately 300 wooden carts have also been discovered by the members of this expedition.

Thus, the archaeological finds from the steppes of the lower Volga valley testify to the nomadic way of life of the people from the Pit-grave, Poltavkin, and North-Caucasian cultures. This was determined by the natural geographical habitat, which did not favour extensive land-cultivation. There are large tracts of saline land containing many obstacles in the form of sandhills, between which there were temporary camps, but no permanent settlements.

Despite systematic research, nothing has been found so far which would confirm substantial development of land cultivation before the age of the Chamber culture. Long-term settlements in the lower Volga valley are in those areas where chernozem (black earth) predominated, in the northern part of the Great Irgiz Basin and further north.

Many researchers consider that the predominance of sheep and goats in herds testifies to a nomadic way of life (Herre 1949, p. 317). This is also borne out by the small number of graves in the burial grounds of the Pit-grave culture age (see Table 13.1), and also probably by the emergence of the custom of making burial mounds over the graves. In the monuments of the Pit-grave and North-Caucasian cultures covered carts have been found which constitute houses on wheels (Sinitsyn 1948, pp. 151–60, Kaposhina 1962, pp. 41, 48, Sinitsyn & Erdniev 1963, p. 14, 1966, pp. 32–5).

Ethnological data indicate that animal husbandry in the lower Volga valley would not have been possible without migration. In order to supply food for one Kalmyk group, or *khoton*, (consisting of four tents each with 13 people) it would have been necessary to keep 100 sheep, four cows, two horses, and four camels. The leader of one such *khoton* from the Bogutov Clan of the Yandyk nomad camp, or *ulus*, was considered the 'last of the wealthy men', i.e. the owner of that minimum below which comes poverty. Even given such a small quantity of animals, the *khoton* would have been obliged to be on the move. A study of a 19th-century *khoton* revealed that for about six weeks in the winter and spring the *khoton* stayed in the Mashtyk terrain. In mid-May it moved from there to the Khosh terrain some 18 km away, where it spent five days. Here the water had a salty-bitter taste and the *khoton* was obliged to move on to the Kiubedin–Ksentse terrain 10 km further on, where it spent 20 days. From there it advanced 30 km to the Dabst terrain where it spent 15 days. After that it moved 5 km to the Per–Marzlyk terrain where it spent 10 days and where it was encountered on 10 July by I. A. Zhitetsky (Zhitetsky 1892, pp. 95–7).

Thus the *khoton*, which began its migration in mid-May, changed pastures five times and covered a distance of 73 km. Moreover, the animals fed on the fodder to hand in the local pastures. No stocks of hay were laid in.

A large-scale Kalmyk economy in 1885 kept 300 horses, 1500 sheep, 50 camels, and 60 head of cattle. The families of the chief, the herdsmen, and the hands who saw to the wells lived in ten covered tents. This *khoton* would spend 60 days in its winter pasture. From 15 February over a period of 144 days it would move on four times and cover a distance of 80 km. Moreover, after May all the animals, apart from the cattle, would move on a further five times and cover a distance of 85 km. The camels and sheep would have an extra move involving another 25 km. The cattle moved four times and covered a distance of 80 km, the horses moved nine times over 109 km, the camels 10 times over 194 km, and the sheep ten times over 169 km. In May the *khoton* would divide into two groups, and during periods of drought into four groups.

This ethnographic information demonstrates convincingly that in conditions similar to those of the lower Volga valley it is impossible for people to engage in animal husbandry as their main economic activity without migration. The reasons for this are the shortages of grass and water. In order to feed one cow or horse in the conditions found in the Lenin District of the Volgograd Region, where the semi-desert steppeland is covered with tipchak and wormwood, 4.9 ha of degraded pasture, or 8–9 ha of very degraded pasture, is necessary (Terenozhkin 1937).

So archaeological and ethnographic evidence testifies that the peoples of the ancient Pit-grave culture, and subsequent cultures found in the lower Volga valley, led a nomadic way of life, breeding sheep and goats.

Nomadism did not prevail over the whole steppeland, however. Our research in the lower part of the Dnieper valley has clearly shown that here there was settlement on the open steppes in the age of the Pit-grave and Kemiobinsk cultures (the Dmitrievsky burial ground and the Orlov burial ground with burial mounds). A similar picture was observed in the northern Caucasus from monuments of the Maikop culture, where bones of cattle and pigs predominate. Bones of these animals in sites of the Maikop culture in Meshoko and in the Ukrainian site of the Pit-grave culture at Mikhailovka point to a settled way of life (Lagadovskaya et al.1962, pp. 168, 207).

Thus, it can be shown from study of the animal remains that in the areas with saline soil in the lower Volga valley and in the open steppes to the north of the Black Sea (the Dmitrievsky burial ground, the Orlov grave and others) sheep were kept, together with some goats. In the mountains and the foothills of the northern Caucasus, rich in alpine meadows and oak groves, cattle and pigs were kept: in the Don feathergrass steppes (on the eastern bank of the Don) horses were kept, and in the chernozem (black earth) areas of the lower part of the Dnieper valley land cultivation and cattle-breeding were practised. So during the time of the Pit-grave culture we can pick out regions with all these specialized economies, the development of which was determined by the natural and geographical environment. Finally, in the steppes of southern Russia during the time of the Pit-grave culture (3rd millennium BC) there appeared the first elements of nomadism, which, according to the testimony of classical writers, followed a long and complex path.

Note

1 Translation of an abbreviated version of Shilov (1985).

References

Chernikov, S. S. 1957. The role of the Andronov culture in the history of central Asia and Kazakhstan [in Russian]. *Kratkiye soobshcheniya instituta etnografii, Moscow* **26**, 31–2.

Chuguryaeva, V. A. 1960. Vegetation of the Trans-Volga region in the Bronze Age [in Russian]. *Materialy i issledovaniya po arkheologii SSSR, Moscow* **78**, 282.

Filipchenko, V. A. & Y. V. Kurochkin 1960. Flint tools from the island of Kulaly [in Russian]. *Sovietskaya Arkhaeologiya, Moscow* **3**, 272 ff.

Geins, A. I. 1897. *Collection of literary works*, Vol. I, 60–1 [in Russian]. Saint Petersberg.

Golmsten, V. V. 1933. On the question of ancient animal husbandry in the USSR [in Russian]. *Proiskozhdeniye domashnikh zhivotnykh, Moscow* **1**, 89.

Gryaznov, M. P. 1955. Aspects of the formation and development of early nomadic societies [in Russian]. *Kratkiye soobshcheniya instituta etnografii, Moscow* **24**, 24.

Gryaznov, M. P. 1957. Stages of development in the economy of the animal-breeding tribes of Kazakhstan and southern Siberia in the Bronze Age [in Russian]. *Kratkiye soobshcheniya instituta etnografii, Moscow* **26**, 21.

Hahn, E. 1886. *Die Haustiere und ihre Beziehungen zur Wirtschaft der Menschen*, 132–3. Leipzig.

Hahn, E. 1891. Waren die Menschen der Urzeit zwischen Jägerstufe und stufe Ackerbaus Nomadem? *Das Ausland* **64**, 484.

Hahn, E. 1905. *Das Alter der Wirtschaftskultur der Menschheit*, 96, 99–100. Heidelberg.

Herre, V. 1949. Zur Abstammung und Entewicklung der Haustiere. *Verhandlungen der deutschen Zoologen in Kiel 1948*, 317. Leipzig.

Homer 1936. *Illiad*, Vol. XIII, 5, 355. Moscow.

Isaak, E. 1971. On the domestication of cattle. In *Prehistoric agriculture*, 446. New York.

Jselin, I. 1786. *Uber die Geschichte der Menschheit*, Vol. 2. Basel.

Kaposhina, S. I. 1962. Findings from the work of the Kobyakov Expedition [in Russian]. *Kratkiye soobshcheniya instituta arkheologii An SSSR, Moscow–Leningrad* **103**, 48.

Lagodovskaya, O. V., A. G. Shaposhnikova & M. A. Makarevich 1962. *The Mikhailovskoye Settlement* [in Russian], 178. Kiev.

Latynin, V. A. 1957. The level of development of productive forces in the Early Bronze Age [in Russian]. *Kratkiye soobshcheniya instituta arkheologii, Moscow–Leningrad* **70**, 3.

Liberov, P. D. 1960. On the history of animal husbandry and hunting in the territory to the north of the Black Sea [in Russian]. *Materialy i issledovaniya po arkheologii SSSR, Moscow* **50**, 134, Table 6.

Merpert, N. Y. 1961. The Eneolithic Age in the steppe belt of the European part of the Soviet Union [in Russian]. Lecture delivered at a symposium in Prague, issued as a separate offprint.

Minaeva, T. M. 1929. *The flint industry in the lower reaches of the Volga* [in Russian]. Trudy Nizhne–Volzhskogo oblastnogo nauchnogo obshchestva kraevedeniya, Saratov p. 31.

Neishtadt, M. I. 1967. *The history of forests and palaeogeography in the USSR in the Holocene Epoch* [in Russian]. 198–9. Moscow.

Rousseau, J. J. 1775. *Le discours sur l'origine et les fondements de l'inégalité parmi les hommes*. Paris.

Rudenko, S. I. 1961. On the question of forms of animal-breeding economies and nomads [in Russian]. *Materialy po etnografii, Leningrad* **1**, 2. Geographical Society, USSR.

Rykov, P. S. 1928. Archaeological surveys and excavations in the lower reaches of the Volga, carried out in 1928. *Izvestiya Saratovskogo Nizhne–Volzhskogo instituta kraevedeniya im. M. Gorkogo, Saratov* **IV**, 20, 24, 26, 33.

Schmidt, W. 1915–16. *Anthropos* **593**, 610.

Schmidt, W. 1924. *Anthropos* **193**, 193.

Schmidt, W. 1951. Zu den Anfängen der Herdentierzucht. *Zeitschrift für Ethnografie* **76**, 213–17.

Shilov, V. P. 1964. Problems relating to the opening up of the steppes in the Bronze Age [in Russian]. *Arkheologicheski sbornik Gosudarstvennogo Ermitazha, Leningrad* **VI**, 102.

Shilov, V. P. 1985. Ancient finds in the Kalnyk area of the USSR. *Elista*, pp. 23–33.

Sinitsyn, I. V. 1931. Flint tools from the dune sites in the Kalmyk Region [in Russian]. *Izvestiya Saratovskogo Nizhne–Volzhskogo instituta kraevedeniya im. M. Gorkogo, Saratov* **IV**, 81–92.

Sinitsyn, I. V. 1933. Ancient monuments of the coastal area of the Kalmy Region [in Russian]. *Izvestiya Saratovskogo Nizhne–Volzhskogo instituta kraevedeniya im. M. Gorkogo, Saratov* **IV**, 89.

Sinitsyn, I. V. 1948. Monuments of the Pre-Scythian Epoch in the lower Volga steppes [in Russian]. *Sovietskaya Arkheologiya, Moscow–Leningrad* **10**, 151–60.

Sinitsyn, I. V. & U. E. Erdniev 1963. Archaeological excavations in the Kalmyk ASSR in 1961 [in Russian]. *Trudy Kalmytskogo respublikanskogo kraevedcheskogo muzeya, Elista* **XVII**, 14.

Sinitsyn, I. V. & V. E. Erdniev 1966. New archaeological monuments in the territory of the Kalmyk ASSR [in Russian]. *Trudy Kalmytskogo nauchnoiss-ledovatelskogo instituta yazyka, literatury i istorii i Kalmytskogo respublikanskogo kraevedcheskogo muzeya, Elista* **II**, 32–5.

Smith, A. 1776. *An inquiry into the nature and causes of the wealth of nations*, Vols. I & II. London.

Smirnov, A. P. 1966. *The Scythians* [in Russian], 13. Moscow.

Smirnov, K. F. 1964. The production and nature of the economy of the Early Sarmatians [in Russian]. *Sovietskaya Arkheologiya*, **3**, 45–7.

Strabo, 1st century AD. S **VII**, 3, 7.

Struve, V. V. 1941. *History of the East in ancient times* [in Russian],146. Moscow.

Terenozhkin, I. I. 1937. *Wild pastures and hay-yielding grasses in the Stalingrad Region* [in Russian]. Stalingrad.

Zhitetsky, I. A. 1892. Astrakhan Kalmyks [in Russian]. Sbornik trudov chlenov Petrovskogo obshchestva issledovatelei Astrakhanshogo kraya, 95–7.

14 Pastoralism in southwest Asia: the second millennium BC

JURIS ZARINS

Introduction

The question of pastoral nomadism in the Middle East has been increasingly re-examined in the light of historical, archaeological, and ethnographic work (Irons & N. Dyson–Hudson 1972, Rowton 1974, Dyson–Hudson & Dyson–Hudson 1980, Eph'al 1981, Galaty & Salzman 1981, Lancaster 1981, Briant 1982, Khazanov 1984, Zarins 1988, Chang & Koster 1986). Archaeological work within a defined arid zone in Southwest Asia over the past decade has particularly emphasized the complex pastoral pattern which has emerged since the 7th millennium BC (Fig. 14.1). This chapter seeks to shed light on a particularly interesting period of this development, namely the pivotal 2nd millennium BC. It is known that a basic pastoral way of life developed along certain characteristic lines during the period 6000–1800 BC (Zarins 1988) and that this society contrasted with a quite different type developed during the period following 800 BC (Bulliet 1975, p. 76ff., Eph'al 1981, Briant 1982). The earlier stage was characterized by the reliance on domesticated ovicaprids and bovids and the construction of stone structures which, by their particular nature, define the associated culture to roughly 4 millenniums. The later culture is characterized by the reliance on domesticated camels and much more ephemeral archaeological remains. The principal region in question (Fig. 14.2) covers the northern part of the Arabian peninsula and the adjoining Levant. The nature of the transition period will be dealt with here in some detail. It covers the thousand years from 1800–800 BC, which is the period between the end of the better-documented Early Bronze Age and the beginning of the Iron Age throughout much of the Middle East (Edens 1986).

The archaeological evidence for this millennium has long been thought to be absent or poorly known, particularly in such specialized discussions as the EB–MB transition in Palestine, the MB–LB Period of Transjordan, the origins of camel nomadism, the history of Mesopotamia, the nature of the Second Intermediate Period in Egypt, and other related concerns. Three lines of evidence will be dealt with here. First, the historical documentation includes the Middle Assyrian and Kassite cuneiform records and the Egyptian hieroglyphic texts such as the Execration Texts of the late Middle Kingdom, the New Kingdom Papyri Harris I, Anastasi VI, and the Tell el Amarna correspondence. Secondly, we have increasing

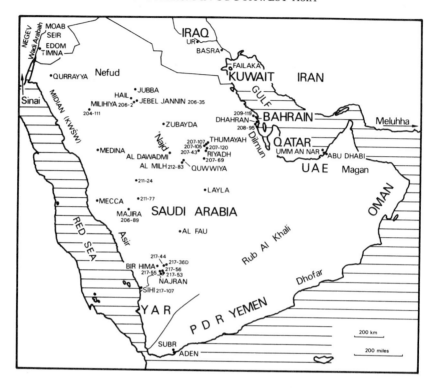

Figure 14.1 Principal site localities in the Arabian peninsula and adjoining regions.

archaeological evidence from Eastern Arabia/Bahrain, the greater Nejd, southern Arabia, Midian (including southern Jordan), and the Negev/ Sinai. Thirdly, the status of camel domestication during this millennium will be reviewed.

The early transition (1800–1400 BC)

The nature of human occupation within areas marginal to the Fertile Crescent has only been examined within the past two decades. While our knowledge of the 'Agricultural Revolution' has increased dramatically, we are only now beginning to understand the probable course of human adaptation in the arid areas south of the primary agricultural zone itself. In the northern part of the Arabian peninsula the pattern of human occupation can be divided into two phases: (a) the Early Pastoralist Phase, 6000–3500 BC; and (b) the Mature Phase, 3500–1900 BC. These pastoralists followed a multi-resource food procurement strategy but principally depended on the herding of ovicaprids and, to a lesser extent, cattle. The second phase saw the introduction of equids, such as horses and donkeys,

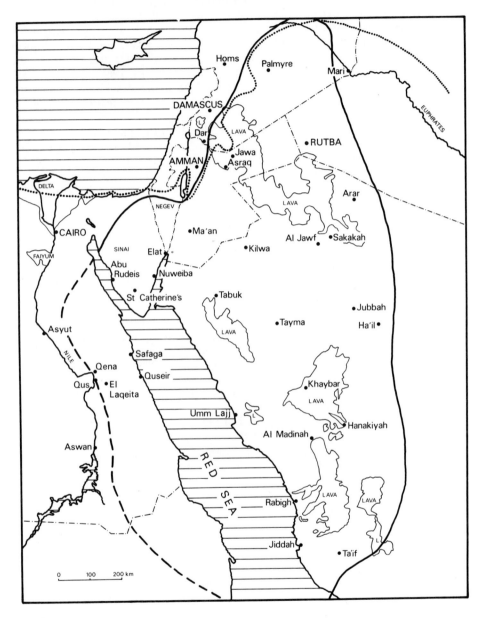

Figure 14.2 The early pastoral technocomplex (after Zarins 1988).

and increased mobility. By the EB IV period (*sensu lato*) *c.* 2200–1900 BC, in the arid Negev/Sinai, there is a definite change in the number of sites and population density. Sites become more numerous and smaller and I suggest that this basic shift may be correlated in large part with deteriorating climatic conditions. Without exception, the archaeological sites of the technocomplex 'disappear' around 1900 BC never to be reoccupied. (For details of this record, see Zarins 1988. Avner [1984, p. 119] has also commented on the increase in standing pillars in settled communities only after the decline of this typical desert feature in the early 2nd millennium BC.)

This shift may be paralleled in the historic records of late Old Kingdom/First Intermediate Period Egypt and the Ur III and I/L periods of southern Mesopotamia (2100–1900 BC). Simpson (1971, pp. 238–40) states that the Asian Bedouin occupied the Delta nomes at the end of the Old Kingdom and the well-known *Instructions of Merykare* describes these pastoral nomads rather vividly. The involvement of the Amorites in redirecting the political fortunes of southern Mesopotamia is equally well known (Edzard 1957). To what extent these pastoral nomads contributed to bringing about the collapse of the EB civilization in Palestine is still being debated (Prag 1984). Two exceptions to this overall picture need to be mentioned here. First, the Amorite pastoral nomads of the Diyala region are well attested in the Eshnunna documentation of the Ur III period down to *c.* 1750 BC, but these records have not been fully published (Whiting 1983). In addition, understanding of the relationship between the pastoral nomads and the established political system in the Diyala remains vague. Secondly, there is voluminous data for the Mari nomads of the period 1825–1759 BC (Kupper 1957, Luke 1965, Prag 1985) but they represent the latest example of this way of life in the northern tip of the arid region. Neither of these groups is, in fact, part of the '2nd millennium BC' transition problem.

Historically, the period 1800–1400 BC is rather a curious mixture, and references to pastoral peoples of the larger Arabian peninsula are virtually non-existent. For example, the Old Babylonian period is dominated by an overlapping series of three dynasties (Isin, Larsa, and Babylon) stretching from *c.* 1850 to 1600 BC. However, these three dynasties (as well as the contemporary one from Assyria) are derived from the earlier well-known Amorite pastoral nomads of the Ur III period – Kinglist A records them as part of the *palû MAR.TU.*, the Amorite reign (Hallo & Simpson 1971, Fig. 20). Thus, the evidence for pastoral populations in the southern arid regions adjacent to Babylonia for this period remains essentially unknown. Rowton (1969, pp. 68–9) suggests that they had vanished from the documentary sources because the sedentary population had become bi-ethnic during the Old Babylonian period and distinctions of foreign populations were not necessary. It may also be plausible that the local Semitic rulers of this background would not distinguish between the settled populations of southern Mesopotamia and the nomadic sections or tribes still found in the adjacent arid zones. Such a distinction appears only in the poorly understood succeeding Kassite period (see below). Yet even at the time of Ammiṣaduga (c. 1600 BC), such honorific titles as *Abu Amurrim* may have pointed to the continued presence of a tribal organization of the Amorites (Rowton 1969, p. 69, Matthews 1979, p. 129). Similarly at Sippar, there is the 'fossilized' presence of pastoral nomadic terms, such as *rabianum* and *babtum*, used in urban context (Harris 1975, p. 38ff.). Even within the context of Old Babylonian records, there is a rather abrupt cessation of records in the southern plain by 1740 BC and 1720 BC in Middle Babylonia (Stone 1977, p. 271). The Assyrian documentation from the north is similar in pattern. There is abundant documentation from the *karum* rulers of the Old Assyrian period, beginning c. 1940 BC, and the Assyrian rule of Shamši-Adad and Ishme–Dagan (1740 BC) covering 200 years. But nothing of importance concerning pastoral nomads appears until Adad–Nirari I (1307–1275 BC).

Concerning the Kassite records, historians generally assign a total length of rule from 1595 to 1155 BC, covering some 440 years. But even here the early portion of the period is very poorly attested textually (Brinkman 1976). Hallo & Simpson (1971, pp. 105–9) remark that 'for the first two centuries of their rule we have virtually no contemporaneous documentation' and label the early part of the period as a 'Dark Age'. Economic and historical data of relevance here only appear at the time of Burnaburiaš II, c. 1350–1333 BC (Brinkman 1976, *passim*).

The archaeological data from southern Mesopotamia itself reflect this condition. In the Uruk area Adams & Nissen (1972, p. 39) note that 'the stable configurations of settlement . . . of the ED–OB period . . . drew to a close in the Warka region with the OB period.' Widespread abandonment followed as an outcome of a slow protracted process. In the Kassite period, sites were smaller and there were a number of disruptive breaks. By the end of the period, a nadir in settled life had occurred. In the Ur area, commercial life ceased around 1740 BC and did not revive until approximately 1400 BC, 250 years after the beginning of the Kassite period (Wright 1981, p. 332). Sites show extensive abandonment both in the late Larsa–OB period and OB–Kassite times (Ibid. p. 331). Adams (1981, pp. 167–9) suggests that a sharp economic and demographic entrenchment took place in the OB period in the entire southern alluvial plain, and in the Kassite period geographical fragmentation took place.

From Egypt, a somewhat similar analogy can be suggested. The Middle Kingdom evidence as derived from the Sinuhe account, the Execration Texts, *The Instructions of Amenemhet I*, *The Instructions of Merykare*, and the Brooklyn Papyrus suggest that Asians (Aamu) were both present in Egypt and found in a wide range of contexts in Southwest Asia itself. By the time of Amenemhet IV (1798–1790 BC) they had begun to arrive in Egypt in large numbers (Van Seters 1966, p. 78, Prag 1985, p. 85). The problem is to attempt a separation between those Asians who belonged to the pastoral tribes of the Negev/Sinai and the sedentary populations of the Levant. Certainly portions of the Sinuhe account and *The Instructions of Merykare* refer to peoples of the Sinai, because of geographical proximity to the eastern delta and defined lifestyle. This interpretation is confirmed by the following passage from *The Prophecy of Nefer–Rohu* (in reference to the fortifications of the northern part of the eastern delta): '. . . the Asiatics will not be permitted to come down into Egypt that they might beg for water in the customary manner, in order to let their beasts drink' (Wilson 1958, p. 257). (For a frontier fort of the new kingdom dated to Seti I in the northern Sinai at Bir el Abd, see Oren 1973a.) The Execration Texts (Sethe 1926, Posener 1940) are a particular source of controversy. The dating is still debated (Middle Kingdom c. 1880–1790 BC?), and defining the geographical location of the names is very difficult. Van Seters (1966, p. 80) is quite adamant that the mentioned princes and inhabitants are those of cities and 'that there is no reason to regard them as nomads.' However, Brussels texts E50–E51 mentioning 'chief of the tribes of Kwšw' may parallel the Amarna correspondence references to northwest Arabia (Posener 1940, pp. 88–9). In addition, the Egyptian use of 'Aamu' as a term describing nomads or settled groups is open to debate.

Similarly, the Pharaonic inscriptions found at Serabit el Khadim in the eastern Sinai and dated principally to Amenemes III and IV (1842–1790 BC) (Černý et al. 1955) mention and depict Asians (Černý 1935) in a role consistent with the Egyptian historical evidence discussed above. We may also lay to rest the idea that the Asians mentioned in the Middle Kingdom texts at Serabit el Khadim did not come from the Negev/Sinai (contra Van Seters 1966, pp. 88–90). These arguments are now rendered obsolete by the archaeological evidence from the Negev/Sinai. This region was occupied by a series of peoples from at least 6000 BC, which culminated in a definite pastoral lifestyle by at least 4000 BC. The archaeological record for this way of life is both abundant and distinctive, and shows clear ties to both the southern Levant, Arabia, and Egypt. The disappearance of this archaeological phenomenon during the MB I period c. 1850 BC, has caused continuous debate and may be closely related to Asian influx into the delta by late Middle Kingdom times (Albright 1955, Hayes & Watkins 1955).

The Second Intermediate Period (Hyksos domination, c. 1750–1550 BC) again spans a critical period of the 2nd millennium BC. Here, as in Mesopotamia, the Semitic populations of the Negev/Sinai and the southern Levant dominated the Egyptian state at least in the eastern delta (see Weinstein 1981, p. 10) and references to Asian populations of the Negev/Sinai, southern Jordan, and Arabia would of course be non-existent and irrelevant. How were the Asians perceived in the delta by the Egyptians themselves? According to *The Admonitions of Ipuwer* and other sources, it appears that the Asians, (1) settled in the delta and took over its affairs, (2) assimilated Egyptian culture and displaced Egyptians in places of authority, (3) caused the northeast frontier of the delta to be open to all Asians, and (4) apparently even became pharoahs in the XIII Dynasty (Van Seters 1966, pp. 115–16, Redford 1970). As Van Seters (1966, p. 126) suggested, the Hyksos analogy is to the Amorite and Kassite control of southern Mesopotamia.

The later transition (1400–900 BC)

This is the key period in terms of historical data, both from Mesopotamia and Egypt. It is within this time that we see the appearance of pastoral nomadic groups in the arid zones which once again threaten the *status quo* of established and revitalised urban states, such as New Kingdom Egypt, the Levantine city states, and Kassite Babylonia/late Middle Assyria. The relevant historical data are summarized below by ethnic name, beginning with Mesopotamia.

Ahlamu/Aramu

The earliest occurrence of the term Ahlamu/Ahlame (discounting the earlier references to Aram) is to be found in the second half of the 2nd millennium BC. The term seems to refer to tribal confederations found on the fringes of the Syrian desert. The first question to be asked is if this term refers generically to 'bedouin'. If it does, what is the relationship to the earlier Amorite pastoral nomads decribed in the 3rd and early 2nd millenniums BC?

In Assyria, the earliest royal reference comes from Adad–Nirari I (1307–1275 BC) in an apparent mention of his father's exploits (Arik-din-Ili, 1318–1307 BC). Here the 'hordes of Ahlamu' are mentioned in the context of the northern Euphrates region (Grayson 1972, p. 58). Shalmaneser I (1274–1245 BC) mentions 'an army of the Ahlamu' who are assisting the Hittites in the northern Euphrates region (Grayson 1972, p. 82). Tukulti–Ninurta I (1244–1208 BC) states that the Ahlamu were to be associated with one of the traditional regions of the Amorite nomads, the hilly arid zone west of Mari in the Jebel Bishri area. He boasts that he controlled 'the lands of Mari, Hana, Rapiqu and the mountains of Ahlamu' (Grayson 1972, p. 119). The last royal reference to the Ahlamu is that of Ashur-resha-ishi I (1133–1116 BC) who again mentions the defeat of 'an extensive army of the Ahlamu' in the northern Euphrates area (Grayson 1972, p. 147).

In the Tell el Amarna texts, EA 200 provides the sole example where the term Ahlamu (*lú ah-la-ma-i*, or *lú ah-la-ma-ú*) is used. This text, dated to Amenophis IV (*c.* 1389–1358 BC), depicts the Ahlamu as agitating for food in a context that may mention the king of Babylon (Brinkman 1968, p. 389 & note 2185). If the term denotes 'bedouin' the Ahlamu must have been widespread from the eastern desert of Arabia, well north to the Syrian steppe. It seems reasonably clear that the Ahlamu are to be identified with the distinct nomadic group later called Arameans who enter history with the inscriptions of Tiglath–Pileser I and maintain a distinct ethnic and social identity well through the 1st millennium BC (for summaries of their characteristics, see Kupper 1957, p. 117ff., Brinkman 1968, p. 267ff. Albright 1975b, p. 532ff.). Organized along tribal lines, the units appeared to be rather small, i.e. segmented social patrilineages, with numerous elders or sheikhs (*nasiku*). Of principal concern here is the lack of mention of camels as domesticated animals.

Shosu/Kashu

The earliest clear reference to the Shosu (*Š3św*) comes from Thutmose II, 1500–1490 BC. The references reach a peak during the reign of Ramses II (1304–1237 BC) and the last mention is under Ramses III (1198–1166 BC). A compilation of the data principally by Giveon (1971, see also the review of Ward 1973) suggests that they were to be found in the Nile delta, southern Palestine (Negev), and in the Transjordan. Helck (1968, p. 477) states that they were located south of the Dead Sea. More specifically, the Shosu are often associated with Seir, a mountainous region located in southern Transjordan. The term Seir occurs as early as the Amarna period (1389–1358 BC) (EA = Tel el Amarna text 288 line 26: *a-di KUR.MEŠ Se-e-ri-ki*) and there are at least four documents to link the Shosu with this region (Giveon 1971, p. 235). At the time of Ramses II, they were said to inhabit six districts, among which is *Św Ša-ʿ-r-r* 'Shosu in the land of Seir' (Helck 1968, p. 477). The Tanis stele of Ramses II states that he 'laid waste to the land of the Asiatic nomads, who has plundered Mount Seir with his valiant men.' (Bartlett 1969, p. 1). The papyrus Anastasi VI associated with Merneptah, also links the Shosu with Seir (Giveon 1971, pp. 131–4) as does the papyrus Harris I of Ramses III (Bartlett 1969, p. 2, Giveon 1971, p. 136). According to Helck, another name for Seir was *Ja-ha-wa* (one of the districts in Shosu land at the time of Ramses II) and we see this term also in an inscription of Amenophis III at the Temple of Amon at Soleb (*t3 Św yhw* – 'Yahwe in the land of Shosu') (Giveon 1971, pp. 26–8).

A number of New Kingdom Egyptian objects and inscriptions found in Transjordan suggests the local inhabitants were in contact with Egypt during this period. The best known of these is the Balua Stela which may depict local Shosu with Egyptians (for the latest summary and references, see Dornemann 1983, p. 22ff., 153–4). Other objects include an Amenophis III scarab found near Petra (Ward 1973), two Ramesside scarabs from central Jordan (Dornemann 1983, p. 27), and an inscription of Queen Twosret at Deir Alla (ibid.). (For the Wadi Timna material, see below.)

According to Giveon (1971, p. 240), the Shosu were seen principally as pastoralists. According to the papyrus Harris I quoted above, Ramses III destroyed the Shosu along with their tents, goods, and herds (Bartlett 1969, p. 2). The Merneptah papyrus Anastasi VI quotes an Egyptian frontier official who states: 'We have finished letting the Shosu tribes of Edom [ÍDM] pass the fortress "Merneptah-hotep-her-maat" which is in Tjekou, to the pools of Pi-tum [Pe-atoum] of M. which is in Tjekou, in order to sustain them and sustain their flocks through the good pleasure of pharoah.' (Redford 1963, p. 406, Bartlett 1969, p. 2, Giveon 1971, pp. 131–4, Ward 1973, p. 52). The site of Tjekou is generally identified with the eastern delta town of Tell Maskhoutah (Redford 1963, p. 406).

The titles of the Shosu also indicate that they were pastoral nomads.

A direct outgrowth of defining the Shosu is the question of the inhabitants of the Sinai/Negev in the New Kingdom. During the Old and Middle Kingdoms there was a wealth of written information about these people from the Wadi Magharah/Serabit el Khadim localities. However, as researchers have noted, the New Kingdom inscriptions tend to become almost devoid of information, heavily ritualized and stylized (Černý et al. 1955, p. 19). Since archaeological survey in the Sinai has suggested that no new ethnic groups arrived there during the Late Bronze Age (Beit–Arieh 1984, p. 52), we are left with the impression that the earlier Middle Kingdom inhabitants continued to be present in the region, albeit in a changed economic and social pattern. It thus appears likely that the Shosu are the best candidates for the inhabitants of the Negev/Sinai in the Late Bronze Age.

If we associate the Shosu with Edom and Moab (for the term *Mw-i-b* under Ramses II, see Kitchen 1964, p. 50), which group of people can be assigned to the Midianites, located further to the south, but unattested in New Kingdom contemporary accounts? (The term Midian/Madyan is unknown outside of the Biblical context until classical times, see Knauf 1983, p. 148.) This controversy centres around the term *Kašu*. The earliest possible attestation of the term occurs in the late Middle Kingdom Execration Texts. The historical clues to *Kašu* are derived from the well-known accounts of Genesis and Judges. According to recent re-evaluation, the Biblical material suggests the presence of a large confederation or tribe, sections of which were nomadic (Dumbrell 1975, Payne 1983, Mendenhall 1984). Whether the population functioned in a way similar to later bedouin tribes is open to debate (Knauf 1983, pp. 150–1). Combining the Egyptian evidence of *Kašu* and the Biblical Midianites, we feel that a powerful Semitic group in the area of Midian complemented the more northerly group identified with the Shosu and Edom.

The archaeological evidence

The archaeological record for the Ahlamu will be discussed first, secondly that of the Shosu/Kashu, and thirdly the evidence from the southern Nejd.

Concerning the archaeology of western Iraq and eastern Syria, virtually nothing is known from the 2nd millennium BC. Work to date has been confined to the Euphrates river valley and upper tributaries, and very little in the way of comparable material of the 3rd millennium BC and earlier is known from the desert (Zarins 1988). From northeastern Saudi Arabia, survey work in the tributary wadis to the Euphrates revealed the presence of 3rd millennium BC stone circle complexes, but nothing definitely attributable to the middle of the 2nd millennium BC (Parr et al. 1978, p. 37ff., Potts et al. 1978, p. 9, Gilmore et al. 1982, p. 16ff.). City III at the Qala on Bahrain defines the Kassite period (Larsen 1983, pp. 80–1, 249–51 & Fig. 54), but this distinctive ceramic horizon is also attested at other occupation sites (Salles no. 2068) and tumuli (al-Hajjar, Diraz, Sar)

(McNicoll & Roaf n.d., pp. 19–20, Rice 1972, pp. 69–72, Salles 1981, p. 2, Mughal 1983, p. 400). It should be noted, however, that in each case the number of Kassite occurrences is minimal. Finally, the record for the Kassite period clearly suggests a political presence related to shipping and commerce involving the Mesopotamian state to the north (including Failaka), and not pastoral nomads (Edens 1986).

For the meagre evidence suggestive of a pastoral nomadic presence such as the Ahlamu, there are the tumuli found in the Dhahran vicinity on the adjoining Arabian mainland. Based on the evidence of the Nippur letters, Cornwall (1952, p. 111) suggested that some of the more modest tumuli at Dhahran (and Bahrain) were built by the Ahlamu. From the excavations conducted in the Dhahran area, it appears that the tombs of the 3rd millennium BC were re-used during the middle to latter part of the 2nd millennium BC, in a number of which (B–7, B–6, A–5, and B–2) there were aceramic burials. The associated grave goods consisted of copper and occasionally iron arrowheads, copper bracelets, finger rings, beads, large seashells, small stone palettes, ear plugs, and in A–5, an apparent game made of a small white bowl and five stone balls (four black and one white) (Zarins et al. 1984, pp. 38–9). A single supporting ^{14}C date comes from the 3rd millennium BC tomb A–4 which was disturbed sometime in the mid-2nd millennium BC. The date taken from the fill yielded 1270 BC ± 345 (GX- 9590).

The archaeological evidence from the southern Nejd has filled in a long-neglected gap in our attempt to delineate the human occupation record of the Arabian peninsula. It is this record which may throw light on the archaeological and historical record already presented above. The human occupation of the Nejd in the Holocene began in the 6th millennium BC. It seems likely that in the northern part of the Nejd a pastoral way of life was present which was allied to a larger technocomplex described elsewhere (Zarins 1988), while in the central and southern Nejd hunting and gathering were probably dominant. By the end of the 3rd millennium BC radical climatic and ecological changes took place in the Nejd. With the advent of the 2nd millennium BC, there is a number of significant structural changes in the human settlements of the Nejd as well.

The later sites have little in common with the earlier ones. For example, site location preference is restricted to two basic locations. First, many sites are situated at the base of small, granitic outcrops in a 50 m arc. Secondly, sites can be found within small embayments or coves on low-lying outcropping strata. There are no site clusters and population density was clearly low. The most numerous structures are hearths, usually constructed of small slabs and often packed with small cobbles. Living structures are also common and are radically different from the earlier stone circle complexes so familiar from northern Arabia, being 'horseshoe'-shaped, about 2 m wide and 3 m on each side, and distinctly outlined by small boulders or cobbles (Zarins et al. 1980, p. 22 & Pl. 31B, Whalen et al. 1981, p. 51). In the Bir Hima region of southwest Arabia, there are 25 such structures aligned in a single row approximately 95 m long (Zarins et al. 1981, p. 30). Long, rectangular organic superstructures define a second

(a)

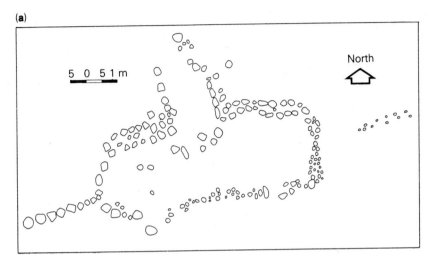

(b)

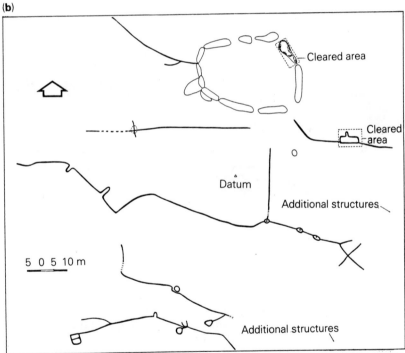

Figure 14.3 (a) Cleared habitation structure at 210–3, near Muwayh, Saudi Arabia (after Zarins *et al.* 1980, Pl. 10A); (b) plan of site 210–3, near Muwayh, Saudi Arabia (after Zarins *et al.* 1980, Pl. 9B).

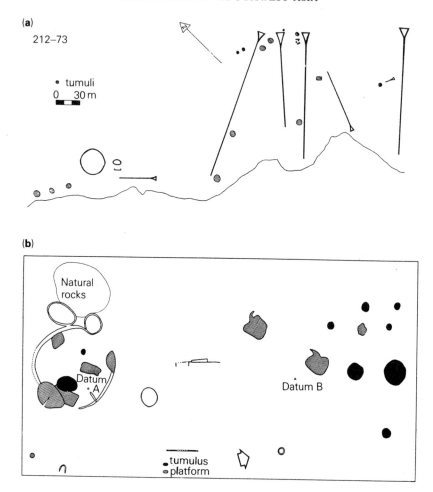

Figure 14.4 (a) Tapered structure site complex at 212–73 near Layla, Jebel Tuwayq, Saudi Arabia (after Zarins *et al*. 1979, Pl. 13A); (b) site plan at Waqir, 210–49, near Taif, Saudi Arabia (after Zarins *et al*. 1980, Pl. 9A).

type whose presence is attested only by an outline in small cobbles (Fig. 14.3). These structures often contain internal partitions and a well-defined entrance (Zarins *et al*. 1980, Pl. 10A, 31A, see Pl. 8C for re-used example). Troughs are a third type which are very prevalent (Zarins *et al*. 1980, Pl. 1B). They are, however, more common on wadi floors and junctions. These are 3–5 m long, double lines of slabs about 30 cm apart, filled with small pebbles. In many cases, they are 'hooked' on both ends with the extensions about 50 cm long. In some cases, the structures have an attached extension in the centre protruding about 0.5 m and also filled

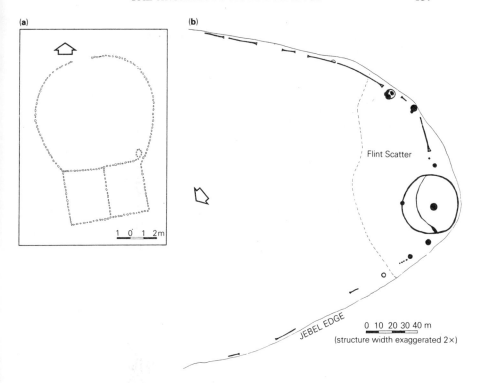

Figure 14.5 (a) Outline of enigmatic stone structure at site 211–24 near Bisha, Saudi Arabia (after Zarins *et al.* 1980, Pl. 8B); (b) Site plan of Safra Huqayl, 206–60B west of Riyadh, Saudi Arabia (after Zarins *et al.* 1980, Pl. 5).

with pebbles (Zarins *et al.* 1981, P1. 42; for similar structures identified as 'open sanctuaries' in the Negev/Sinai, see Avner 1984, p.119ff.). These structural types appear to be mutually exclusive at the sites surveyed to date.

Another group of structural remains may indicate a funerary or ritual complex. The most noteworthy is the 'tapered structure' which looks like an elongated wedge (Figs 14.4a & 14.5b). Already noted by researchers in the al Fau area of central Arabia in the 1960s (Field 1971, p. 44 & Pl. 21–5), hundreds of these structures have been examined in the Nejd region. A very narrow 'tail' is built of vertical slabs about 50 cm apart and filled with rubble. Depending upon the length of the tail, the structure gradually widens and expands in size until the maximum width is achieved at the 'head'. Here the walls taper outward and the centre of the structure remains hollow. Excavation of these structures has confirmed that they were not utilized for habitation or burial (Zarins *et al.* 1980, p. 32). Sometimes the tapered structures are associated with 'platforms' which were first noted on survey in 1979 (Zarins *et al.* 1980, p. 19) (Fig.14.4b).

The platforms are circles up to 10 m in diameter, completely filled with stones. Two such structures were excavated in 1979 and this confirmed the fact that they were deliberately built in such a fashion (ibid. p. 32, Gilmore *et al.* 1982, p. 16).

Pillars and aligned slabs represent the final structural category. Pillars as cultic objects are attested from the 6th to the 3rd millenniums in Arabia (Kirkbride 1969, Rothenberg 1973, Zarins 1979, Avner 1984, pp. 115–19), but aligned slabs may be manifestations of the 2nd and 1st millennium BC. Several of the latter in the Nejd have been examined.

Organic remains from the sites are rare since, with a few exceptions, they are essentially surface sites that have undergone major deflation. Bone awls have been recovered but animal remains are mostly fragmentary. From a number of sites ostrich shell is known, and from 211–55 and 211–56 identifiable *C. dromedarius* fragments have been found (Zarins *et al.* 1980, p. 32 n.6). From 2nd millennium BC Red Sea middens there are equid, bovid, ovicaprid, and *C. dromedarius* remains (Zarins & Badr 1986 in press). The best examples of *C. dromedarius* come from pit excavations at site 217–44 see (see Fig. 14.6).

The ethnoarchaeology of bedouin sites confirms the gap in time between the modern bedouin and the structures described above. It is unfortunate that detailed ethnoarchaeological work on the Arabian bedouin is lacking, although a few preliminary studies have been undertaken in the Negev (Rosen n.d.), the Sinai (Kozloff 1981), southern Jordan (Banning & Kohler–Rollefson 1983), Qatar (Montigny 1978, pp. 197–9 & Figs. 8–9) as well as in the central Nejd (Zarins *et al.* 1980, p. 23 & Pl. 11; for Iran, see Hole 1978).

In the Nejd, at the present day there is a minimal use of rock material to steady major tent poles, outline hearths, and batten down tent sides. In Qatar, cobbles are used to completely outline the tent structures almost in identical fashion to that reported from site 210–3A (see above). At Beidha in Jordan, Banning & Kohler–Rollefson (1983, p. 377) have seen stone

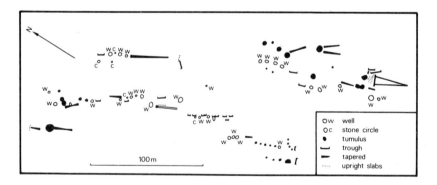

Figure 14.6   Site plan of 217–44 near Bir Hima, Saudi Arabia (after Zarins *et al.* 1981, Pl. 11).

hearths in use and also bed and storage platforms and small stone pens. In the Negev, Rosen (n.d.) identifies two types of settlements, the ephemeral campsites characterized by the presence of large stones which serve as benches, lined or unlined hearths, and black Gaza ceramics; and campsites consisting of rebuilt stone circles generally of the Roman–Byzantine period or earlier. All agree that bone, dung, and perishable debris predominate at an encampment. In contrast to their older archaeological counterparts, stone tools are entirely absent, ceramics were noted only at the Qatar and Negev sites, and it is unlikely that recent bedouin build, recognize, or understand the function of stone platforms, tapered structures, pillars, erect slabs, or troughs.

The dating of the ancient sites discussed above presents a number of difficulties, but probably most of them fall between 1900 BC and AD 650. Clues to possible dates remain minimal, due principally to lack of excavation.

The recognition of relevant archaeological data in both northern Arabia and the southern Levant to complement the findings in the Nejd remains elusive. There are virtually no 2nd millennium BC sites or structures allied to those mentioned above. From central and southern Sinai, with the exception of Serabit el Khadim itself, no sites are attributed to the Late Bronze or Early Iron I ages (1550–1000 BC) (Beit–Arieh 1984, p. 52 & Map, p. 49). In northern Sinai, with the exception of the formal Egyptian fort, Bir el Abd attributed to Seti I, no sites of the Late Bronze or Early Iron Age have been reported (Oren 1973a, 1973b, p. 200). From the Negev, a perusal of the Archaeological Emergency Project reveals no Late Bronze Age sites (Cohen 1979, pp. 250–4) although some of the open shrines near Eilat are attributed to the 13th–11th centuries BC (Avner 1984, p. 124). From southern Jordan, in the Wadi el Hasa and the Maan–Aqaba areas, again no sites are attributed to the Late Bronze Age I period (Jobling 1981, p. 109, 1982, MacDonald *et al.* 1982, p. 126). However, it is known that pastoral people were present in the area, from the historical data and from the isolated New Kingdom Egyptian finds discovered in the Negev/Sinai and as far east as the Transjordan. This point is driven home particularly well by the surveys and excavations carried out in the Timna area of the Wadi Arabah. Sites such as the Hathor shrine and rock engravings attest to the presence of Egyptians from at least the time of Seti I well into the XXth Dynasty (Rothenberg 1972, pp. 129–207). In the Sinai, what people invented and used the proto-Sinaitic script? Rothenberg (1980, p. 164) and Beit–Arieh (1983, p. 48) suggest they were semi-nomadic native inhabitants, descendants from the earlier centuries.

Most likely, the answer to why archaeological sites are 'missing' in these areas lies in the nature of the archaeological record because the structural remains are not diagnostic in themselves, and continued re-use makes attribution to a specific period difficult (Avner 1984, p. 124). Lithics are nondescript and ceramics do not readily match known cultural types found in the settled zones. Thus, in a recent survey of northwestern Arabia, no sites were found that could be linked to post-Neolithic materials, but 'the date of the enigmatic stone enclosures and cairns . . .

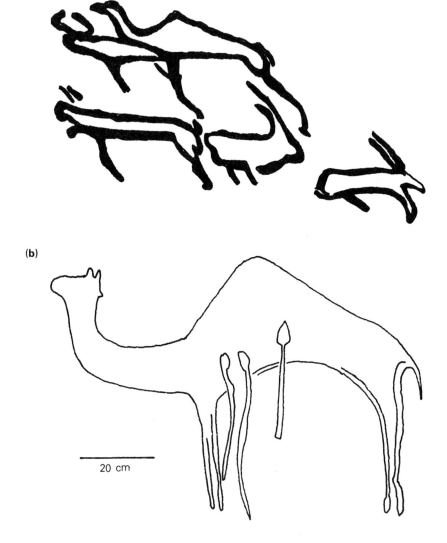

(a)

(b)

20 cm

Figure 14.7 (a) Phase I 'outline' rock art scene at Jebel Kawkab, north of Najran, Saudi Arabia (after Anati 1968*a*, p. 138 and Fig. 91); (b) Speared camel of Phase I or II from Bir Hima, site 217–35C (after Zarins *et al.* 1981, p. 34).

(a)

(b)

Figure 14.8 (a) Phase II camel hunting scene in the Bir Hima region (after Anati 1968*b*, p. 54 & Fig. 4); (b) Phase II camel hunting scene in the Bir Hima region (after Anati 1986b, p. 64 & Fig. 15).

may span the second and first millennia BC' (Fig. 14.5a) (Ingraham *et al.* 1981, p. 71 & Tables 1–2). Similar statements could be made for the adjoining arid parts of Jordan, Israel, Syria, Iraq, and the Sinai. It appears that in the north, particularly the northwest, culturally recognizable societies tied to camel-based caravans and the spice trade did not arise until after 1500 BC (Bulliet 1975). The crucial period here, then, is the time span between the end of the earlier pastoral technocomplex and the arrival of the domesticated camel, *c.* 1500 BC (Hakker–Orion 1984, p. 209).

Support for this hypothesis comes from a number of sites. The earliest bedouin site in the Bir Hima region has a date of 1215 BC; at Zubayda in the Burayda oasis, excavations yielded a series of ^{14}C dates which begin at 1315 BC (GX 7097, 3265 ± 150 BP for excavation unit III 7; Parr & Ghazdar 1980). From northwestern Arabia, in the Midian region, archaeological investigations at Qurrayya suggest that the Midianite culture began no earlier that the 13th century BC (Parr *et al.* 1970). During the 1980 survey, investigations confirmed that the distinctive Midianite ware was made locally at Qurrayya (Ingraham *et al.* 1981, pp. 71–5). Studies of the ware's distribution suggest a region which includes northwest Arabia, southern Jordan, the Wadi Arabah, and more rarely in the adjoining northern Sinai and southern Levant (ibid., Rothenberg & Glass 1983). Perhaps, in respect to cultural development, the Midianites paralleled the Nabateans of a millennium later. Both societies began as pastoral groups but attained domination by controlling the spice trade with adjoining formal states. For the Midianites, this can be seen at the Hathor shrine in the Wadi Arabah.

Camel domestication

The final focus of this study deals with the animal which transformed the pastoral societies of the Arabian peninsula, the camel. Two lines of evidence can be used, artistic and osteological. Contrary to earlier suggestions, the camel was known to the early Holocene inhabitants of the peninsula. Osteologically, the remains are widespread but not common (Ripinsky 1975, Zarins 1978, Grigson 1983, Hakker–Orion 1984). Artistic evidence, principally from the southwestern part of the peninsula, indicates that the camel was hunted as a game animal until sometime in the 2nd millennium BC. Based on rock art sequences developed by Anati (1962, 1968, 1970, 1972, 1974, 1979) and Tchernov (1974) and modified here, in the southwestern part of the peninsula a number of depictions can be fitted into the schema shown in Table 14.1. From southwestern Arabia, there are at least five camel depictions (from Phase 1 ('outline style') Fig. 14.7a) Anati 1968a, p. 110 & Fig. 74, (138 & Fig. 91, 1974, p. 234 & Fig. 243). In one case, the camel underlies a scene with a speared bovid (Zarins *et al.* 1981, Pl. 36B, site 217–36). In another, direct confirmation of hunting is to be seen in a depiction of a speared camel (ibid., p. 35 & Fig. 34E) (Fig. 14.7b). From Phase II, there is a number of additional scenes depicting hunted camels (Figs 14.8 & 9) (Anati 1968b, p. 11 & Fig. 2, p. 60 & Fig. 29, p. 54 & Fig. 4, p. 64 & Fig. 15, 1974, p. 132 & Fig. 248, p. 232 & Fig. 271). In one scene, the hunted camel is attacked by a man using a transverse arrowhead, a type common in the 5th–3rd millenniums

Table 14.1 The rock art sequence from southwestern Arabia.

Phase	Period	Date
V	Islamic	AD 650 to present
IV	Literate (Thamudic/South Arabic)	500 BC to AD 650
III	Pre-literate	1900–500 BC
II	early pastoralists	3500–1900 BC
I	early hunters	6000–3500 BC

Figure 14.9 (a) Camel hunting scene from Phase II at Bir Hima, site 217–23B (after Zarins *et al*. 1981, Pl. 34F); (b) Camel hunting scene from the Jebel Kawkab area (after Anati 1968b, p. 11 & Fig. 2).

Figure 14.10 (a) Phase I or II lifesize outline camel from Jubba Lake, Saudi Arabia (view sideways); (b) Hooked trough from site 207–46 in the Riyadh environs.

(a)

(b)

BC (Fig. 14.9a) (Zarins *et al.* 1981, p. 35 & Pl. 34F). Rock art from these two periods is known from other areas, including northern Arabia, Jordan, and the Negev/Sinai. However, rock art depicting the camel is only rarely attested outside the southwestern zone. One outstanding example is a Phase I outline camel from Jubba in the Nefud desert (Fig. 14.10a). From the United Arab Emirates, a camel is depicted in relief on several cairns at Umm an Nar. These can be dated to the 3rd millenium BC (Thorvildsen 1962, Figs 7–8).

Following Phase II, the clearest evidence for camel depictions comes from Phase IV. The principal theme in this period is riding and hunting or razzia raids. The rock art evidence for this period is widespread and common. From a brief analysis of the rock art, it seems clear that camel domestication began sometime in the late 3rd millennium BC, probably in southern Arabia, supporting the ideas of several scholars (Mikesell 1955, p. 242ff., Dostal 1959, 1979, Bulliet 1975). The strongest evidence is to be gleaned from the number of depictions in which human association with the camel is attested. These are to be found in the southwestern portion of the Arabian peninsula.

To turn now to the osteological evidence: identified remains of camel at archaeological sites are rare in the early Holocene, but they were part of the hunted game. For example, their remains have been recovered at Ain al Assad in southern Jordan in PPNB context (Kohler 1984, p. 201), at the Pottery Neolithic site of Shar ha Golan, *c.* 5000 BC (Stekelis 1951, p. 16), at Early Bronze Age I Arad, *c.* 2900 BC (Lernau 1978, p. 87), and at the Early Bronze Age IV site of Bir Resisim in the Negev, *c.* 2000 BC (Hakker–Orion 1984, p. 209). It is not possible, however, to delineate domestic from wild camels on the basis of morphological change in the skeleton alone (Hoch 1979, p. 607, Hakker–Orion 1984, p. 209). Thus, the question of camel domestication is extremely complicated and frustrating. Our best evidence to date for this process comes from eastern Iran. From the site of Shahr-i-Sokhta, the excavator recovered not only osteological remains but also hair and dung, found in a context datable to 2700 BC. This suggests that camel domestication began in Turkmenia and spread south (Compagnoni & Tosi 1978, pp. 95–9). The domestic camel was apparently known to the inhabitants of the Indus Valley civilization by 2300 BC, although the species utilized remains open to question (Meadow 1984, p. 134 and references).

Close trade contacts existed between the Indus Valley civilization and the eastern Arabian peninsula, and there is tentative evidence that the camel was merely one of many items traded during the late 3rd millennium BC. At the site of Umm an Nar, analysis of the osteological remains suggests a step towards domestication, deduced from the unusual number of camel bones, the age distribution, and the cultural context (Hoch 1979, p. 613). This stimulus may well have come from the Indus valley (Zarins 1978).

Another animal involved in this southern trade was the zebu (*Bos indicus*). Apparently domesticated by the mid-4th millennium BC on the Indian subcontinent (Allchin 1969, pp. 318, 322), from the Arabian eastern

Table 14.2 Suggested developmental model for camel bedouin.

	Phase	Camel utilization	Date		Cultural evolution
The later transition	V	North Arabian saddle (Shadad); Thamudic/ South Arabic	500 BC		rectangular goat hair tent; minimal use of stone
	IV	cushion saddle	1000 BC		rectangular stone outline; tapered structures
The earlier transition	III	South Arabic saddle (Hawlani/Hadaja) pack camels; overland incense trade; change in camel status	1500 BC	south to north	troughs; horse-shoe shape
	II	non-riding; herds for milk; little group movement	2200–1500 BC		Umm-an-Nar, Subr, Sihi; Phase II rock art in southwest Arabia
	I	the wild camel	6000–2200 BC		Phase I rock art in southwest Arabia

littoral the animal is attested both by artistic and osteological remains dating to the late 3rd millennium BC (Hoch 1979, pp. 567, 614–17, Cleuziou 1982, p. 19 n. 2, Zarins & Badr 1986). As in the case of the domesticated camel, the zebu is not attested in the Levantine littoral until the 15th century BC (Clason 1978).

Camel remains from southern Arabia supporting the thesis that the centre of domestication lay in the south are not common, but this may be due to a lack of survey and excavation. From Sihi, a shell midden on the southern Red Sea coast, camel remains have been recovered in a 2nd millennium BC context. These include large fragments of the left and right maxillae of a juvenile individual, as well as fragments of ribs, femur, and mandibles of adults (Zarins & Badr 1986). The likelihood that camel remains will turn up at other coastal sites of the 2nd millennium BC is great (e.g. see the site of Subr in South Yemen, Doe 1961, and the Italian Survey of the Yemen Red Sea coast, Tosi pers. comm.). Several bedouin sites of the 2nd millennium BC also yielded camel remains. From a series of pits excavated at site 217–44 in the Bir Hima region, there were a number of articulated camel fragments (Zarins *et al.* 1981, Pl. 43A–B). Other remains were found at 217–56 (Zarins *et al.* 1980, p. 23 n. 6).

From North Arabia and the southern Levant, the occurrence of camel remains follows the development of cultures involved in the South Arabian overland spice trade. With direct Midianite association, there is a sherd depicting a camel (Ingraham *et al.* 1981; Pl. 79/14). In the Wadi

Arabah, at Site 2, a copper smelting camp dated to the Ramessid period, *c.* 1350–1150 BC, 'several camel bones' were found with other faunal remains (Rothenberg 1972, p. 105). In a later report, the excavators mention that 'a large quantity of camel bones' was uncovered at the 13th–12th century BC sites of Timna (Rothenberg & Glass 1983, p. 122 n. 50). From Tell Jemmeh on the Gaza strip, Wapnish (1982, p. 2, 1984, p. 171) identifies only seven camel bones from levels attributable to the 14th–10th centuries BC. Similarly, at Heshbon in the Transjordan, camel remains are very infrequent from the earliest levels, 1230–1150 BC (Unit E 04–05, Weiler 1981, Table 4).

The osteological remains are supported by the historical evidence. The earliest attestation of the camel in the cuneiform literature occurs in the Assyrian context in the reign of Assur-Bel-Kala, 1074–1057 BC (Grayson 1976, p. 55). Camels were rare animals that were herded together and put on display in Nineveh. From the Biblical context there are the well-known accounts of the camel-riding Midianites (Judges 6–8), perhaps attributable to the 13th–12th centuries BC. Finally, from Egypt there seems to be no early term for the camel. The papyrus Anastasi VI, describing the herds of the Shosu and other Asians, uses the term *I3wt*, which usually denotes herds of small animals like sheep and goat, not larger cattle or camels (C. van Siclen pers. comm.).

All this information may be combined in a chronology for the use of the camel in Arabia (Table 14.2), integrating the ideas of Dostal (1959, 1979), Bulliet (1975), Compagnoni & Tosi (1978), and Zarins *et al.* (1980, 1981).

In conclusion, it seems that the camel pastoral nomads of the Arabian peninsula developed along a distinct path over the course of at least 2–3 millenniums. Lexically, the earliest evidence points to groups in the northern part of the peninsula such as the Ahlamu/Aramu, the Shosu and *Kašu*. These peoples must have undergone some type of radical change in cultural adaptation before the historical context beginning *c.* 1400 BC. Unfortunately, the earlier historical record covering the period 1800–1400 BC is largely irrelevant to the topic of pastoralism. Archaeologically, there was a slow development in the southern and central parts of the Arabian peninsula that can be interpreted from a number of distinctive building structures and it does appear that earlier ideas concerning the spread of camel domestication from south to north are largely correct.

References

Adams, R. Mc. 1981. *Heartland of cities*. Chicago: University of Chicago Press.

Adams, R. Mc. & H. J. Nissen 1972. *The Uruk countryside*. Chicago: University of Chicago Press.

Albright, W. F. 1955. North-west Semitic names in a list of Egyptian slaves from the eighteenth century BC. *Journal of the American Oriental Society* **74**, 222–33.

Albright, W. F. 1975a. Ch. XX, The Amarna letters from Palestine. In *The Cambridge ancient history*, 3rd edn. Vol. 11/2, I. E. S. Edwards, C. J. Gadd, N. G. L. Hammond & E. Sollberger (eds), 98–116. Cambridge: Cambridge University Press.

Albright, W. F. 1975b. Ch. XXXIII, Syria, the Philistines, and Phoenicia. In *The Cambridge ancient history*, 3rd edn. Vol. 11/2, I. E. S. Edwards, C.J. Gadd, N. G. L. Hammond & E. Sollberger (eds), 507–36. Cambridge: Cambridge University Press.

Allchin, F. R. 1969. Early domestic animals in India and Pakistan. In *The domestication and exploitation of plants and animals*, P. Ucko & G. Dimbleby (eds) 317–22. London: Duckworth.

Anati, E. 1962. *Palestine before the Hebrews*. New York: Alfred E. Knopf.

Anati, E. 1968a. *Rock art in central Arabia*. Vol. I: *The 'oval-headed people of Arabia'*. Institut Orientaliste. Louvain: Université de Louvain.

Anati, E. 1968b. *Rock art in central Arabia*. Vol. II: *'Fat tailed sheep in Arabia'* and *'The realist-dynamic style of rock art in the Jebel Qara'*. Institut Orientaliste. Louvain: Université de Louvain.

Anati, E. 1970. The rock-engravings of Dahtami Wells in central Arabia. *Bollettino del Centro Camuno di Studi Preistorici* **5**, 99–158.

Anati, E. 1972. *Rock art in central Arabia*. Vol. III: *Corpus of the rock engravings, sectors A–H*. Institut Orientaliste. Louvain: Université de Louvain.

Anati, E. 1974. *Rock art in central Arabia*. Vol. IV: *Corpus of the rock engravings, sectors J–Q*. Institut Orientaliste. Louvain: Université de Louvain.

Anati, E. 1979. *L'Arte rupestre del Negev e del Sinai*. Milano: Jaca Book.

Avner, U. 1984. Ancient cult sites in the Negev and Sinai deserts. *Tel Aviv* **11**, 115–31.

Banning, E. B. & I. Kohler–Rollefson 1983. Ethnoarchaeological survey in the Beidha area, southern Jordan. *Annual of the Antiquities Department of Jordan* **27**, 375–83.

Bartlett, J. R. 1969. The land of Seir and the Brotherhood of Edom. *Journal of Theological Studies* **20**, 1–20.

Beit–Arieh, I. 1983. Central-Southern Sinai in the Early Bronze Age II and its relationship with Palestine. *Levant* **15**, 39–48.

Beit–Arieh, I. 1984. Fifteen years in Sinai. *Biblical Archaeology Review* **10**(4), 26–54.

Briant, P. 1982. *Etat et pasteurs au Moyen–Orient ancien*. Cambridge: Cambridge University Press.

Brinkman, J. A. 1968. *A political history of post-Kassite Babylonia, 1158–722 B.C.* Rome: Pontificium Institutum Biblicum.

Brinkman, J. A. 1976. *Materials and studies for Kassite history*, Vol. 1. Chicago: The Oriental Institute.

Bulliet, R. W. 1975. *The Camel and the wheel*. Cambridge, Mass.: Harvard University Press.

Černý, J. 1935. Semites in Egyptian mining expeditions to Sinai. *Archiv Orientalní* **7**, 384–9.

Černý, J., T. E. Peet & A. Gardiner 1955. *The Inscriptions of Sinai* (2 vols). London: Egypt Exploration Society.

Chang, C. & H. A. Koster 1986. Beyond bones: towards an archaeology of pastoralism. In *Advances in archaeological method and theory*. Vol. 8, M. Schiffer (ed.), 97–148. San Francisco: Academic Press.

Clason, A. T. 1978. Late Bronze Age–Iron Age zebu cattle in Jordan? *Journal of Archaeological Science* **5**, 91–3.

Cleuziou, S. 1982. Hili and the beginning of oasis life in eastern Saudi Arabia. *Proceedings of the Seminar for Arabian Studies* **12**, 15–22.

Cohen, R. 1979. The Negev Archaeological Emergency Project. *Israel Exploration Journal* **29**, 250–4.

Compagnoni, B. & M. Tosi 1978. The camel: its distribution and state of domestication in the Middle East during the third millennium B.C. in light of finds from Shahr-i Sokhta. In *Approaches to faunal analysis in the Middle East*, R.M. Meadow & M. Zeder (eds), 91–103. Cambridge, Mass.: Harvard University Press.

Cornwall, P . B. 1952. Two letters from Dilmun. *Journal of Cuneiform Studies* **6**, 137–45.

Doe, D. B. 1961. Notes on pottery found in the vicinity of Aden. *Department of Antiquities Annual Report 1960–61*, 3–20.

Doe, D. B. 1977. Gazeteer of sites in Oman. *Journal of Oman Studies* **3**(1), 35–57.

Dornemann, R. H. 1983. *The archaeology of the Transjordan in the Bronze and Iron Ages*. Milwaukee: Milwaukee Public Museum.

Dostal, W. 1959. The evolution of bedouin life. In *L'Antica Societa Beduina*, F. Gabrieli (ed.), 1–34. Rome: Studi Semitici.

Dostal, W. 1979. The development of bedouin life in Arabia seen from archaeological material. In *Studies in the history of Arabia*, Vol. I, A. al-Ansary (ed.), 125–44. Riyadh: University of Riyadh Press.

Dumbrell, W. J. 1975. Midian – a land or a league? *Vetus Testamentum* **25**, 323–37.

Dyson–Hudson, R. & N. Dyson–Hudson 1980. Nomadic pastoralism. *Annual Review of Anthropology* **9**, 15–61.

Edens, C. 1986. Bahrain and the Arabian Gulf during the second millennium BC: urban crisis and colonialism. In *Bahrain through the ages*, Shaikha A. al Khalifa & M. Rice (eds), 195–216. London: Routledge & Kegan Paul.

Edzard, D. O. 1957. *Die 'Zweite Zwischenheit' Babyloniens*. Wiesbaden: Otto Harrasowitz.

Eph'al, I. 1981. *The ancient Arabs*. Leiden: E.J. Brill.

Field, H. 1971. *Contributions to the anthropology of Saudi Arabia*. Miami: Field Research Projects.

Galaty, J. G. & P. C. Salzman (eds) 1981. *Change and development in nomadic and pastoral societies*. Leiden: E.J. Brill.

Gilmore, M., M. al-Ibrahim & A. S. Murad 1982. Preliminary report on the Northwestern and Northern Region Survey 1981 (1401). *Atlal* **6**, 9–23.

Giveon, R. 1971. *Les Bedouins Shosu des Documents Egyptiens*. Leiden: E.J. Brill.

Grayson, A. K. 1972. *Assyrian royal inscriptions*. Wiesbaden: Otto Harrasowitz.

Grayson, A. K. 1976. *Assyrian royal inscriptions*, Part 2, Wiesbaden: Otto Harrasowitz.

Grigson, C. 1983. A very large camel from the Upper Pleistocene of the Negev Desert. *Journal of Archaeological Science* **10**, 311–16.

Hakker–Orion, D. 1984. The role of the camel in Israel's early history. In *Animals and archaeology*. Vol. 3: *Early herders and their flocks*, J. Clutton–Brock & C. Grigson (eds), 207–12. Oxford: BAR International Series 202.

Hallo, W. W. & W. K. Simpson 1971. *The ancient Near East, a history*. New York: Harcourt, Brace & Jovanovich.

Harris, R. 1975. *Ancient Sippar*. Istanbul: Nederlands Historisch–Archaeologisch Instituut.

Hayes, W. C. & J. B. Watkins 1955. *A papyrus of the late Middle Kingdom in the Brooklyn Museum*. New York: The Brooklyn Museum.

Helck, W. 1968. Bedrohung Palastinas durch Einwandernde Gruppen am Ende der

18. und am Anfang der 19. Dynastie. *Vetus Testamentum* **18**, 472–80.

Hoch, E. 1979. Reflections on prehistoric life at Umm-an-Nar (Trucial Oman) Based on faunal remains from the third millennium B.C. In *South Asian Archaeology 1977*, M. Taddei (ed.), 589–638. Naples: Istituto Universitario Orientale.

Hole, F. 1978. Pastoral nomadism in western Iran. In *Explorations in ethnoarchaeology*, R. A. Gould (ed.), 127–68. Albuquerque: University of New Mexico Press.

Ingraham, M., T. Johnson, B. Rihani & I. Shatla 1981. Preliminary report on a reconnaissance survey of the North-western Province (with a brief survey of the Northern Province). *Atlal* **5**, 59–84.

Irons, W. & N. Dyson–Hudson (eds) 1972. *Perspectives on nomadism.* Leiden: E.J. Brill.

Jobling, W. J. 1981. Preliminary report on the archaeological survey between Ma'an and 'Aqaba January to February, 1980. *Annual of the Antiquities Department of Jordan* **25**, 105–11.

Jobling, W. J. 1982. Aquaba-Ma'an Survey, January–February, 1981. *Annual of the Department of Antiquities of Jordan* **26**, 199–209.

Khazanov, A. M. 1984. *Nomads and the outside world.* Cambridge: Cambridge University Press.

Kirkbride, D. 1969. Ancient Arabian ancestor idols. *Archaeology* **22**, 116–21, 188–95.

Kitchen, K. A. 1964. Some new light on the Asiatic Wars of Ramses II. *Journal of Egyptian Archaeology* **50**, 47–70.

Knauf, E. A. 1983. Midianites and Ishmaelites. In *Midian, Moab and Edom*, J. F. A. Sawyer & D. J. A. Clines (eds), 147–62. Sheffield: JSOT Press.

Kohler, I. 1984. The dromedary in modern pastoral societies and implications for its process of domestication. In *Animals and archaeology*. Vol. 3: *Early herders and their flocks*, J. Clutton–Brock & C. Grigson (eds), 201–6. Oxford: BAR International Series 202.

Kozloff, B. 1981. Pastoral nomadism in the Sinai: an ethnoarchaeological study. *Bulletin de l'équipe et écologie et anthropologie des sociétés pastorales.*

Kupper, J.–R. 1957. *Les nomades en Mésopotamie au temps des rois de Mari.* Paris: Société d'Edition 'Les Belles Lettres'.

Lancaster, W. 1981. *The Rwala bedouin today.* Cambridge: Cambridge University Press.

Larsen, C. E. 1983 *Life and land use on the Bahrain islands.* Chicago: Chicago University Press.

Lernau, H. 1978. Faunal remains, Strata III–I. In *Early Arad*, R. Amiran (ed.), 83–113. Jerusalem: Israel Exploration Society.

Luke, J. T. 1965. *Pastoralism and politics in the Mari Period: a re-examination of the character and political significance of the major west Semitic tribal groups on the Middle Euphrates, ca. 1828–1758 BC.* Unpublished PhD dissertation, University of Michigan, Ann Arbor.

MacDonald, B., G. Rollefson & D. Roller 1982. The Wadi el-Hasa Survey 1981: a preliminary report. *Annual of the Antiquities Department of Jordan* **26**, 30–41.

McNicoll, A. & M. Roaf n.d. *Archaeological investigations in Bahrain, 1973–1975.* MS. on file, State of Bahrain, Ministry of Information.

Matthews, V. 1979. The role of the Rabi Amurrim in the Mari Kingdom. *Journal of Near Eastern Studies* **38**, 129–33.

Meadow, R. 1984. A camel skeleton from Mohenjo–Daro. In *Frontiers of the Indus civilisation*, B. B. Lal & S. P. Gupta (eds), 133–9. New Delhi: Indian Archaeological Society.

Mendenhall, G. E. 1984. Qurayya and the Midianites. In *Pre-Islamic Arabia*.Vol. II, A. al-Ansary (ed.), 137–45. Riyadh: King Saud University.

Mikesell, M. W. 1955. Notes on the dispersal of the dromedary. *Southwestern Journal of Anthropology* **11**, 231–45.

Montigny, A. 1978. Étude anthropologique au Qatar. In *Mission archéologique française à Qatar*, J. Tixier (ed.), 181–200. Paris: CNRS.

Mughal, M. R. 1983. *The Dilmun burial complex at Sar*. Manama: Ministry of Information.

Oren, E. D. 1973a. Bir el-Abd (northern Sinai). *Israel Exploration Journal* **23**, 112–13.

Oren, E. D. 1973b. The overland route between Egypt and Canaan in the Early Bronze Age. *Israel Exploration Journal* **23**, 198–205.

Parr, P. J. & M. Ghazdar 1980. A report on the soundings at Zubaida (al-'Amara) in the al-Qasim region: 1979. *Atlal* **4**, 107–17.

Parr, P. J., J. Harding & J. E. Dayton 1970. Preliminary Survey in N.W. Arabia, 1968. *Bulletin of the Institute of Archaeology, University of London* **8 & 9**, 193–242.

Parr, P. J., J. Zarins, M. Ibrahim, J. Waechter, A. Garrard, C. Clarke, M. Bidmead & H. al-Badr 1978. Preliminary report on the second phase of the Northern Province Survey 1397/1977. *Atlal* **2**, 29–50.

Payne, E. J. 1983. The Midianite Arc in Joshua and Judges. In *Midian, Moab and Edom*, J. F. A. Sawyer & D. J. A. Clines (eds), 163–72. Sheffield: JSOT Press.

Posener, G. 1940. *Princes et pays d'Asie et du Nubie*. Bruxelles: Fondation égyptologique Reine Élisabeth.

Potts, D., A. S. Mughannum, J. Frye & D. Sanders 1978. Preliminary report on the second phase of the Eastern Province Survey 1397/1977. *Atlal* **2**, 7–28.

Prag, K. 1984. Review article, continuity and migration in the South Levant in the late third millennium: a review of T. L. Thompson's and some other views. *Palestine Exploration Quarterly* **116**, 58–68.

Prag, K. 1985. Ancient and modern pastoral migration in the Levant. *Levant* **17**, 81–8.

Redford, D. B. 1963. Exodus I 11. *Vetus Testamentum* **13**, 401–18.

Redford, D. B. The Hyksos invasion in history and tradition. *Orientalia* **39**, 1–51.

Rice, M. 1972. The grave complex at Al-Hajjar, Bahrain. *Proceedings of the Seminar for Arabian Studies* **2**, 66–75.

Ripinsky, M. 1975. The camel in ancient Arabia. *Antiquity* **49**, 295–8.

Rosen, S. A. n. d. *Observations on bedouin archaeological sites near Ma'aleh Ramon*. Unpublished MS.

Rothenberg, B. 1972. *Were these King Solomon's mines?* New York: Stein & Day.

Rothenberg, B. 1973. Sinai Explorations III. *Museum Ha'aretz Yearbook* **15/16**, 16–34.

Rothenberg, B. 1980. *Sinai*. Bern: Kummerly & Frey.

Rothenberg, B. & J. Glass 1983. The Midianite pottery. In *Midian, Moab and Edom*. J. F. A. Sawyer & D. J. A. Clines (eds), 65–124. Sheffield: JSOT Press.

Rowton, M. B. 1969. The Abu Amurrim. *Iraq* **31**, 68–73.

Rowton, M. B. 1974. Enclosed nomadism. *Journal of the Economic and Social History of the Orient* **17**, 1–30.

Salles, J.-F. 1981. Bahrain: introduction et état des questions. In *Fouilles à Umm Jisr (Bahrain)*, S. Cleuziou, P. Lombard & J. T. Salles (eds), 1–12. Paris: Edition ADPF.

Sethe, K. 1926. *Die Ächtung Feindlicher Fürsten, Völker und Dinge auf Altägyptischen Tongefasscherben des Mittleren Reiches*. Berlin: Walter de Gruyter.

Simpson, W. K. 1971. Egypt. In *The ancient Near East, a history*, W. W. Hallo &

W. K. Simpson (eds), 185–302. New York: Harcourt, Brace & Jovanovich.

Stekelis, M. 1951. A new Neolithic industry: the Yarmukian of Palestine. *Israel Exploration Journal* **1**, 1–19.

Stone, E. C. 1977. Economic crisis and social upheaval in Old Babylonian Nippur. In *Mountains and lowlands*, L. D. Levine & T. C. Young, Jr. (eds), 267–90. Malibu: Undena Press.

Tchernov, E. 1974. A study of the fauna from Sectors A–Q. In *Rock art in Central Arabia*. Vol. IV: *Corpus of the rock engravings, Sectors J–Q*, E. Anati (ed.), 209–52. Louvain: Université de Louvain.

Thorvildsen, K. 1962. Gravroser pa Umm-an-Nar. *Kuml*, 191–219.

Van Seters, J. 1966. *The Hyksos, a new investigation*. New Haven: Yale University Press.

Wapnish, P. 1982. Camel caravans and camel pastoralists at Tell Jemmeh. *Journal of the Ancient Near Eastern Seminar* **13**, 101–21.

Wapnish, P. 1984. The dromedary and bactrian camel in Levantine historical settings: the evidence from Tell Jemmeh. In *Animals and archaeology*. Vol. 3: *Early herders and their flocks*, J. Clutton–Brock & C. Grigson (eds), 171–87. Oxford: BAR International Series 202.

Ward, W. W. 1973. The Shasu 'Bedouin', notes on a recent publication. *Journal of the Economic and Social History of the Orient* **15**, 35–62.

Weiler, D. 1981. *Säugetierknochenfunde vom Tell Hesban in Jordanien*. Inaugural Dissertation. München: Detlev Weiler.

Weinstein, J. M. 1981. The Egyptian Empire in Palestine: a reassessment. *Bulletin of the American Schools of Oriental Research* **241**, 1–28.

Whalen, N., A. Killick, N. James, G. Morsi & M. Kamal 1981. Saudi Arabian archaeological reconnaissance 1980. Preliminary report on the Western Province Survey. *Atlal* **5**, 43–58.

Whiting, R. 1983. *Early Old Babylonian letters from the Diyala*. Unpublished PhD thesis, University of Chicago, Chicago.

Wilson, J. A. 1958. Egyptian prophecy. In *The ancient Near East*. Vol. I, *an anthology of texts and pictures*, J. B. Pritchard (ed.), 252–7. Princeton: Princeton University Press.

Wright, H. 1981. Appendix: the southern margins of Sumer. In *Heartland of cities*, R. Mc. Adams (ed.), 295–345. Chicago: University of Chicago Press.

Zarins, J. 1978. The camel in ancient Arabia: a further note. *Antiquity* **52**, 44–6.

Zarins, J. 1979. Rajajil: a unique Arabian site from the fourth millennium BC. *Atlal* **3**, 73–8.

Zarins, J. 1986. MAR–TU and the land of Dilmun. In *Bahrain through the ages, the archaeology*. Shaikha H. al Khalifa & M. Rice (eds), 233–50. London: Routledge & Kegan Paul.

Zarins, J. 1988. Archaeological and chronological problems within the Greater southwest Asian Arid Zone: 8500–1850 BC. In *Absolute chronologies in Old World archaeology*, R.W. Ehrich (ed.). Chicago: University of Chicago Press.

Zarins, J. & H. Badr 1986. Archaeological investigations in the Southern Tihama Plain II (including Sihi, 217–107 and Sharja, 217–172) 1405/1985. *Atlal* **10**, in press.

Zarins, J., M. Ibrahim, D. Potts & C. Edens 1979. Saudi Arabian archaeological reconnaissance 1978: the preliminary report on the third phase of the Comprehensive Archaeological Survey Program – The Central Province. *Atlal* **3**, 9–38.

Zarins, J., A. S. Mughannum & M. Kamal 1984. Excavations at Dhahran South – the Tumuli Field (208–91), 1403 AH/1983. A preliminary report. *Atlal* **8**, 25–54.

Zarins, J., A. Murad & K. al-Yish 1981. Comprehensive archaeological survey program – a. The second preliminary report on the southwestern Province. *Atlal* **5**, 9–42.

Zarins, J., N. Whalen, M. Ibrahim, A. Morad & M. Khan 1980. Comprehensive Archaeological Survey Program – preliminary report on the central and southwestern provinces survey. *Atlal* **4**, 9–36.

15 *Farming to pastoralism: effects of climatic change in the Deccan*

M. K. DHAVALIKAR

It was generally thought by early historians that in the development of mankind, the hunting–gathering stage was followed by pastoral nomadism after which man began to produce his own food by domesticating plants and animals. This was the accepted view until the middle of the 19th century when Hahn suggested that pastoral nomadism was an offshoot of sedentary agriculture, and it is now generally agreed that pastoral nomadism came after crop cultivation (Khazanov 1984, p. 85). One reason for this view is that nowhere in Eurasia have there been pastoral nomads without the knowledge of agriculture. An excellent illustration of this view is provided by the archaeological evidence in the Deccan, which roughly comprises the State of Maharashtra and parts of Karnatak and Andhra Pradesh. Exploration followed by selective excavation in the Krishna valley has provided convincing evidence of flourishing agricultural communities in the latter half of the 2nd millennium BC, whose successors had to resort gradually to pastoral nomadism because of the drastic change in climate. This was confirmed in the course of our intensive excavations at Inamgaon District, Poona, Maharashtra (lat. 18°35′N, long. 74°32′E). Thirteen seasons' work has enabled us to study the culture process and culture change throughout the nine centuries of occupation at the site, from 1600 BC to 900 BC. This chapter attempts to explain the culture process during the Late Jorwe phase of occupation in the Deccan, which marks a shift from farming to pastoralism.

The first quarter of the 2nd millennium BC marks the appearance of early farming communities in the Deccan, the region between the Tapi and the Krishna rivers in the Indian peninsula. This distribution, however, was sporadic, but from the middle of the 2nd millennium BC the region was dotted with several self-sufficient villages, some of which, such as Prakash, Daimabad, and Inamgaon, developed rapidly and became large regional centres in the Tapi, the Godavari, and the Bhima valleys, respectively (Fig. 15.1). They appear to have been organized into chiefdoms, as the evidence from excavations at Daimabad and Inamgaon demonstrates (Dhavalikar 1981–83). This prosperity was in a large measure due to the congenial environment during 1500–1000 BC, which was a relatively wet phase (Krishnamurthy *et al.* 1981). The prosperity, however, did not last long and came to an abrupt end by the close of the 2nd millennium BC.

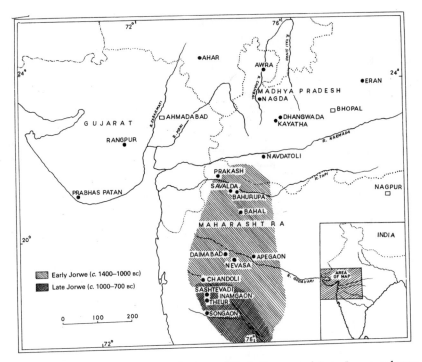

Figure 15.1 Map showing distribution of Early Jorwe and Late Jorwe cultures.

The opening of the 1st millennium BC witnessed large-scale desertion of settlements by the early farming communities in the northern Deccan. Everywhere in the Tapi and the Pravara–Godavari valleys human activity came to a halt (Dhavalikar 1984). We have, as yet, no satisfactory explanation of this abandonment of habitations, but a possible clue is furnished by the soil analysis of the sterile layer encountered in the excavations at Nevasa (District Ahmednagar, Maharashtra) which, according to Mujumdar & Rajaguru (1965, p. 152), belongs to the brown soil group and suggests a less humid climate than when the virgin black soil was formed. They observe:

Our pedological investigations have revealed that the so-called weathered black layer shows definite soil characteristics and it has been formed by the weathering of the habitational deposits of chalcolithic period. This indicates that the place was deserted by the chalcolithic people due to some calamity and considerable period must have elapsed to allow for the growth of vegetation and the formation of soil on the top portion of the habitational deposit. The chemical and physical analyses indicate that the soil formed here belongs to the brown soil group and as such comparatively drier climatic conditions must have prevailed with annual rainfall of about

50 cm accompanied by hot summers and corresponding scrub forest
type of vegetation growing in the locality.

On the basis of further studies, Mujumdar (pers. comm.) thinks that the
rainfall might have decreased to about 300 mm, indicating that the climate
was becoming more and more arid. It is common knowledge that
agriculture in India is a gamble with nature; every third year is a bad year
and every fourth, a famine. Even in our own times, large-scale migrations
take place if there are successive droughts, and it is highly likely that the
same may have happened around 1000 BC. Thus, the early farmers of the
northern Deccan seem to have deserted their settlements at the close of the
2nd millennium BC, at least in the Tapi and the Godavari valleys, but it
appears that the climate in the Bhima valley was slightly more congenial
and, hence, they continued to live there, as the evidence from many sites
in the Upper Bhima valley indicates. It is this culture which has been
labelled as the Late Jorwe and had been identified first at Inamgaon.
    A number of sites was discovered in the Bhima valley, more particularly
in the Bijapur District of Karnatak (Sundara 1968, 1969–70, 1970–71).
All these sites have yielded a typical black-on-red painted pottery,
unmistakably of the Jorwe fabric. But the precise stratigraphical position
of this ware was determined in the course of our large-scale excavations at
Inamgaon (District Poona, Maharashtra) which is located on the right
bank of the Ghod, a tributary of the Bhima (Dhavalikar 1975–76). It is
actually the Late Jorwe which, on the basis of combined evidence of
stratigraphy and antiquities, has been assigned to c. 1000–700 BC. The Late
Jorwe occupation is also present at Songaon (District Poona, Maharashtra),
as is clear from the evidence of pottery which has been described by the
excavator as the 'degenerate' Jorwe ware (Deo 1969, p. 5). It appears from
the available evidence that the Late Jorwe culture was confined to the
Bhima valley and represents the end phase of the Chalcolithic of the
northern Deccan.
    Of all the sites of the Late Jorwe culture, perhaps the most extensive is
that at Inamgaon, where the habitation of this phase was spread over an
area of about 5 ha making it one of the largest Chalcolithic settlements in
the Deccan. Upstream in the Bhima valley the westernmost sites are
Theur and Sashtewadi, both near Poona, which were excavated by S. R.
Rao of the Archaeological Survey of India (IAR 1969–70, pp. 27–9,
1971–72, pp. 35–6). Further southeast, in Karnatak, Sundara (1968) has
brought to light over 20 sites of the Late Jorwe culture which he has
classified separately (Group C). All of them are small settlements,
extending from 1 to 2 ha in extent. It is needless to emphasize, therefore,
that Inamgaon is the largest settlement of this culture.
    The sparse settlement in the Bhima valley in prehistoric times was due
to the fact that practically the whole basin is a dry area with an average
rainfall varying between 400 and 700 mm. Precipitation is high in the
source region, but the early farming communities never occupied the high
altitude areas with more rainfall. Although the low-lying valley terraces
are known for their black cotton soil in the east (in parts of Ahmednagar

and Sholapur districts), the upland areas have a capping of poor soil and their junction with low-lying valleys is marked by a reddish loam. Thus, large tracts of fertile soil were not available and the pioneering colonizers therefore located their settlements in those areas where patches of arable land existed, as at Chandoli, Songaon, Inamgaon, etc. Many of these sites are located in areas where the river takes a sharp meander, as at Inamgaon, or are located on the confluence of rivers, as at Songaon, and are naturally well protected. Thus food, water, and security were all available in these areas.

The excavated sites of the Late Jorwe culture include Songaon, Theur, Sashtewadi, and Inamgaon, all located in the upper Bhima valley. The first three of these have not yielded anything significant, except pottery, but Inamgaon has been systematically excavated for 13 seasons, as a result of which we now have a fairly good idea of the life of the Late Jorwe people. Over 50 houses of this phase have been exposed and they throw a flood of light on the micro-settlement pattern of the period. They are all modest huts, mostly round in plan, and having dwarf mud walls which were probably covered by the wattle-and-daub construction (Fig. 15.2). The roof, probably conical, was thatched. Typha grass (*Typha latifolia*) was used as roofing material, as is done today. It is waterproof and grows abundantly along the banks of streams in the region. The impressions of walls and roofs have been recovered in the course of excavations; they are similar to those of the typha grass. The floor inside the hut was carefully made; it was composed of a layer of gravel over which was rammed yellow silt and black clay. It was frequently plastered with mud and cowdung, as is the case today. It was probably repaired, or rather relaid, after every one or two years, and in one hut we could count 14 successive layers of floors. The courtyard of the house was also similarly well made and plastered with mud. All these huts were rather small in size, their diameter varying from 2 m to 3 m.

Some of the huts contained a set of four flat stones which were meant to support either four-legged storage bins, as at present, or a jar with four legs, the like of which has been recovered in the course of excavations at Inamgaon. In most of these round huts we came across a hearth (*chulah*) which was nothing but a trough-shaped fire pit; very rarely was there a semi-circular clay *chulah* identical with modern hearths. Sometimes, however, the *chulah* was located outside in the courtyard which, as already noted, was well rammed and plastered with mud. The hut was usually so small that with a *chulah* and a storage jar, there was hardly enough space for people to live in and one wonders how a family of six could have been accommodated in it. But it should be stated that in India, the climate being hot and damp, it is the courtyard where much of the life is lived, while the house or the hut is used for keeping valuables, cooking, and storage. However, it appears that each household had more than one hut. This observation is supported by the artifactual evidence and also by the clusters of three or four huts with only one hearth and one set of four flat stones for supporting a storage jar.

The house type marks a noteworthy change from the Early Jorwe to the Late Jorwe, that is from rectangular to circular, and, therefore, an

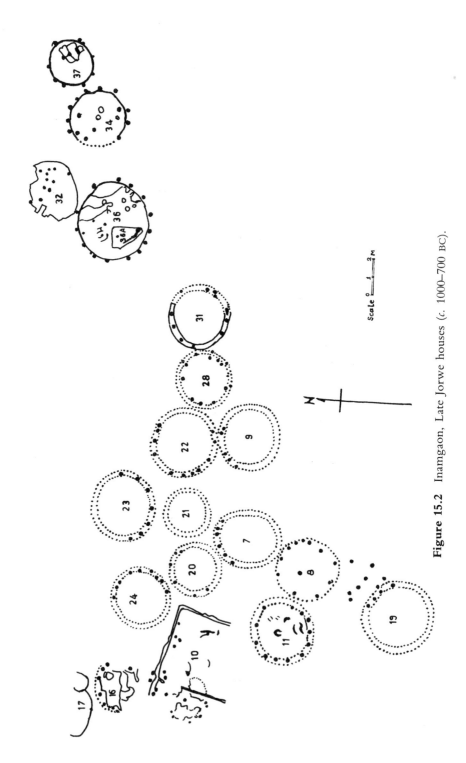

Figure 15.2 Inamgaon, Late Jorwe houses (c. 1000–700 BC).

explanation is in order. Poverty was no doubt the most important cause, and is reflected also in the coarse pottery of Late Jorwe. It was probably the result of the decrease in rainfall, which caused the general economic decline. Dwindling agriculture was also one of the most important consequences. The people obviously could not have afforded the large rectangular houses which their predecessors had built and so they constructed small round huts. Our ethnographic survey in the surrounding region shows that, even today, poor people, especially migrant labour on sugarcane fields, build small, round huts mainly because they are easy to build in a short time and, what is more, they withstand strong winds which are a characteristic feature of arid or semi-arid regions. Inamgaon today is a semi-arid area falling in the rain shadow zone with average precipitation around 450 mm. The ethnographic evidence also suggests that the Late Jorwe people may have lived a semi-nomadic existence (Flannery 1972).

The poverty of the people becomes extremely marked around 800 BC, after which we do not come across well-made round huts, but only patches of flimsy floors and post-holes which do not make a sensible plan. Charred seeds of cultivated grain become scarce, but the number of animal bones increases considerably. It was with great difficulty that we could recover a few houses of the end phase of the Late Jorwe. They are irregular circular or rectangular in plan and have sunken floors. In one hut the owner was buried in a crouching posture and an antler was placed near his feet (Fig. 15.3). All this only goes to show that at the end of the Late Jorwe people were again resorting more and more to hunting and gathering, and living in flimsy huts only seasonally.

Along with round huts, a few rectangular houses, probably belonging to the well-to-do people, have also been recovered. Two such houses were encountered in the eastern periphery of the principal habitation area on the river front. One of these was a multi-roomed structure, part of which was completely burnt by fire, and hence everything inside was found almost intact. There were several storage jars containing lots of charred grain. The wooden posts and the roof had been burnt, trapping a three-year-old child below a wooden post. He died on the spot; his bones and even teeth being charred. In one of the southern rooms of the house was found the burial of the owner, which will be discussed later.

In the courtyard of the house were three small oval huts without any well-made floor. They can better be described as sunken floors. One of these has a small semi-circular verandah in the front in which were found a stone anvil and hundreds of finished and unfinished chalcedony blades, indicating that this was the place where tools were made.

This was the largest house of the Late Jorwe and from the contents it appears that it may have belonged to a very important person, perhaps the ruling chief. The rectangular house-plan is said to be indicative of fully settled life (Flannery 1972), and we may then suggest that, although a majority of the Late Jorwe people led a semi-nomadic existence, at least a few enjoyed the luxury of fully settled life. The same situation exists at present, as will be discussed later.

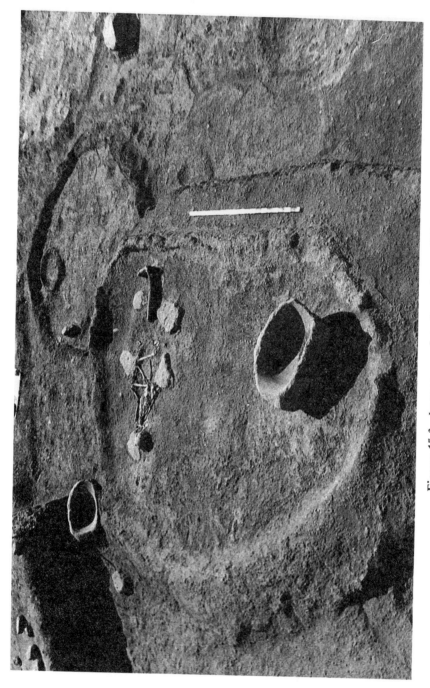

Figure 15.3 Inamgaon, a Late Jorwe hut, c. 800 BC.

We have interesting evidence of the subsistence pattern of the Late Jorwe people from the Inamgaon excavations which have yielded a large number of charred seeds. The prosperity of the Early Jorwe period was a thing of the past in the opening century of the 1st millennium BC. The people nevertheless were farmers, practising subsistence agriculture, cultivating barley as the principal cereal crop. In the initial stage, wheat was grown as was done by their predecessors who, however, could cultivate it because of the artificial irrigation made possible by the diversionary channel through which the flood-water of Ghod was diverted (Dhavalikar 1975–76, pp. 47–8). But soon, it appears, the diversionary channel became silted up and fell into disuse. This also explains the decline in wheat production, which later completely stopped (c. 900 BC). After this, the people subsisted more on barley, and perhaps even more on animal foods, as is suggested by the increase in the quantity of animal bones towards the closing stages of the Late Jorwe. Fruit (Zizyphus) and fish supplemented the diet.

The middle levels of the Late Jorwe period at Inamgaon mark a noteworthy change. From here onwards house plans could be recovered with great difficulty; the quantity of charred grains decreased drastically, suggesting a total decline in agriculture, but correspondingly there is a considerable increase in the yield of animal bones. There is indirect evidence to suggest that the Chalcolithic people subsisted more on animal food than on plant food. The comparative analysis of tooth size and dental indices of the Inamgaon skeletal series gives an impression of a relatively and absolutely large-toothed population indicating that their diet was much coarser (Lucaks 1985, pp. 820–1). Our statistics show that in the early levels of the Late Jorwe, the percentage of bones of cattle was much more than the sheep/goat bones (56 per cent cattle; 25 per cent sheep/goat), but from the middle levels the percentage of the former begins to decrease and that of the latter increases considerably (26 per cent cattle, 52 per cent sheep/goat). It may, however, be stated that the percentage of hunted animals, more particularly gazelle and antelope, also increases, from 5 per cent to 14 per cent. It is, therefore, clear that the people were gradually resorting to sheep/goat pastoralism in the 9th century BC. It is also roughly around this time that the black-and-red ware appears at the site. This may, perhaps, be due to the incursions of the megalith builders of the southern Deccan. Very probably these Iron-age horsemen were to some extent responsible for driving away the Late Jorwe pastoralists from their settlements. This is not unlikely, in the light of the existence of megalithic stone circles near Inamgaon at Pimpalsuti (Ansari & Dhavalikar 1976–77).

The Late Jorwe people worshipped a mother goddess, as did their predecessors. She was probably the goddess of fertility, as one specimen was found in close proximity to pit silos of the Early Jorwe period. Such figures betray crude modelling; their extremities are stumpy and there was no attempt at depicting the facial features. So, also, is the case of a god who was worshipped occasionally, as the discovery of two such unbaked figures near a chulah in the courtyard of a Late Jorwe house suggests. At present, almost identical figures of wheat flour are made and worshipped at the time of community feasts, being immersed in the river or a well after the successful conclusion of the feast (Dhavalikar 1970).

The people believed in life after death. The dead were buried usually within the house floor in a pit specially dug for the purpose. The custom was the same as the Early Jorwe burial practice. Adults were buried in an extended position with head towards the north and legs towards the south, whereas the children were accommodated in two grey ware urns, placed mouth to mouth horizontally in the burial pit. Usually a painted bowl and a spouted jar were also placed in the pit; they probably contained food and water for the dead. Some burials also contained more vessels, which doubtless indicate the economic condition of the family. As in the Early Jorwe period, the portion below the ankle of the dead was, perhaps deliberately, chopped off. This may have been done with a view to preventing the dead from turning into ghosts and running away.

A burial of a different class was found in the only multi-roomed house of this period, which was unearthed at Inamgaon. It was in one of the rooms inside the house, where, in a pit, were recovered two skeletons, one over the other, with their heads towards the north and the legs towards the south (Fig. 15.4). The lower skeleton was that of a male aged about 30 years old, and that over him was of a female aged about 25 years old. The latter has a concussion mark over her forehead which may have been caused by hitting. It is therefore highly likely that she was deliberately stunned and made unconscious and then buried with her consort. Another feature of this double burial is that the lower extremities of the legs of both the skeletons are intact, they were not chopped off as was the general rule. A number of pottery vessels was placed inside the burial pit. All this suggests that he was a very important person of the Late Jorwe community, living in a large multi-roomed house which also had three oval huts in the courtyard, probably for servants. He may, perhaps, have been the ruling chief of the community.

At the end of the Late Jorwe in the 8th century BC the people used coarse painted jars in place of grey urns for burying children. This reflects the general deterioration in the potter's art, which is only a poor survival of the Jorwe painted ware. Another noteworthy feature is that the black-and-red ware, which appears in the end phase of the Late Jorwe, also occurs in burials, but one thing is clear; the people become poor in the latter half of the Late Jorwe, so much so that instead of two urns some of them used only one urn for the child burial.

In the 8th century BC the Late Jorwe people completely vanished from the scene without leaving any trace. The reason for this is not difficult to see. The Inamgaon evidence demonstrates that, in the closing stages, agricultural activity had completely dwindled, so much so that, in spite of our best efforts, very few cultivated grains could be found. The people can therefore be said to have been again slowly reverting to pastoral nomadism. They subsisted more and more on animal foods, as is evident from the sudden increase in remains of animal bones in the middle levels of this period. It is significant that the oldest surviving text (5th–3rd

Figure 15.4 Inamgaon, a double burial, c. 1000 BC.

centuries BC) of the Pali Buddhist canon reports an argument between two ascetics, of whom the more austere censures the other for living as an alms man, begging food cooked by others, when it would be more righteous to subsist by pure (vegetarian) food-gathering in the wilderness (Kosambi 1963). This suggests that food-gathering was an accepted and more respectable mode of subsistence at the beginning of the early historic period, around the 6th–5th centuries BC.

It is highly likely that settlement at Inamgaon was rather seasonal, as very few house plans could be recovered. Their dwellings were flimsy structures of impermanent nature. All this indicates the deterioration of the agricultural economy. This was also the time when the megalith builders from the south were coming into the northern Deccan, as the evidence from Pimpalsuti, which is hardly 5 km from Inamgaon, shows (Ansari & Dhavalikar 1976–77). In all probability, the poor Chalcolithic folk were dominated by the megalith builders, who possessed the horse and effective iron weapons. There was, however, an overlap between the Chalcolithic and the Iron Age and this provides the link between the prehistoric and early historic periods which was thus far missing.

The lifestyle of the Late Jorwe pastoralists was probably not much different from that of the present-day Dhangars in Central Maharashtra, which is a semi-arid region. They are semi-nomadic pastoralists who live in their settlements from May to October, which is the period of the southwestern monsoon. From October and November they leave their settlements (*vadis*) on the plateau and begin their march to the western coast. They move, with all their belongings loaded on horseback, to the Kolaba District on the western coast, for grazing the sheep on the stubble in rice fields where the harvesting is over by October–November. The farmers invite them so that the sheep grazing in their fields fertilize them by their droppings. Again, by April they start to return to their settlements on the plateau, before the onset of the monsoon because the sheep cannot survive in rain.

Some of the Dhangars also own lands, and in their absence the farmers in the villages, who lead a settled life, cultivate their lands. Thus, there exists a symbiotic relationship between the Dhangars and the farmers in a village (Sontheimer 1975, p. 167). It is noteworthy that the Dhangars today occupy roughly those areas where the Late Jorwe settlements have been found. This, however, does not in any way suggest that the present-day Dhangars are the successors of the Late Jorwe people, but what is important to note is that, just as the arid phase in the first half of the 1st millennium BC forced the Chalcolithic farmers to resort gradually to semi-nomadic sheep/goat pastoral nomadism, the same phenomenon is being repeated in the second half of the 2nd millennium AD. Pastoral nomads, the Gavali Dhangars, whose main occupation is rearing water buffalo and whose habitat is the higher rainfall areas of western Maharashtra, have of late started keeping more and more goats because of the deterioration of the vegetation (Gadgil & Malhotra 1979, p. 62)

The culture process discussed in the foregoing pages was, indeed, a remarkable phenomenon in the prehistory of the Deccan. It is, however, important to know that it was not an isolated phenomenon, but was repeated throughout the history of the region. Whenever there were successive droughts and famines, people had no alternative but to resort to pastoral nomadism. (It is significant that some of the dynasties of the early Medieval period, such as the Yadavas of Maharashtra, the Hoysalas of Karnatak, and the Vijayanagar kings of Andhra–Kasnatak, claimed their descent from the Yadavas, who were pastoralists.)There was a general economic decline in the early Medieval period from the 7th to the 10th centuries (Sharma 1986) and again later from AD 1500 to 1900; the effect on human life was the same as during the first half of the 1st millennium BC. It has been observed that from the 13th century until the middle of the 15th century there was no attempt to settle the proper revenue divisions. The land was covered with grass and the main occupation of the people was cattle-breeding (Gune 1953, p. 4 f.n.9). This probably continued until the 19th century when the British conquered Maharashtra from the Peshwas (in 1818). Mountstuart Elphinstone, the representative of the East India Company who was placed in charge of the administration, found that there was no lack of arable land but that there were not enough people to cultivate it, the reason probably being that they were living as cattle-breeders (Sontheimer 1975, p. 149).

The culture change that occurred in the Deccan during prehistoric and historic times was doubtless brought about, in a large measure, by the drastic change in climate. It became more and more arid and, as a result, droughts became of frequent occurrence. The climate change in the first half of the 1st millennium BC has been reflected in the archaeological record, whereas that for the early Medieval (7th–12th centuries) and the Medieval (AD 1500–1900) has been referred to in the literature (Dhavalikar 1987). It is significant that it was roughly during these periods, that is c.1000 BC–500 BC and AD 1500–1900, that Europe is said to have experienced mini Ice Ages (Birks 1986). It appears that whenever Europe has periods of intense cold, the tropical lands suffer droughts. It has been observed that

> Climate is a world-wide integrated system. Significant changes cannot take place in one part of the system without changes occurring in other places. There are dynamic connections that interlink climatic changes in various parts of the globe (Bryson & Ross 1977, p. 509).

References

Ansari, Z. D. & M. K. Dhavalikar 1976–77. Megalithic burials at Pimpalsuti. *Bulletin of Deccan College Research Institute* **36**, 84–8.
Birks, H. J. B. 1986. Late Quaternary biotic changes in terrestrial and lacustrine environments, with particular reference to north-west Europe. In *Handbook of Holocene palaeoecology and palaeohydrology*, B. E. Berglund (ed.), 3–65. New York: Wiley.
Bryson, R. A. & J. E. Ross 1977. Climate variation and implications for world food production. *World Development* **5**, 507–18.

Deo, S. B. 1969. *Sonegaon excavation.* Poona: Deccan College.

Dhavalikar, M. K. 1970. A prehistoric deity of western India. *Man* **5**, 131–2.

Dhavalikar, M. K. 1975–76. Settlement archaeology of Inamgaon. *Puratattva* **8**, 44–54.

Dhavalikar, M. K. 1981–83. Chalcolithic cultures: A socio-economic perspective. *Puratattva* **12–13**, 63–80.

Dhavalikar, M. K. 1984. Toward an ecological model for chalcolithic cultures of central and western India. *Journal of Anthropological Archaeology* **3**, 133–58.

Dhavalikar, M. K. 1987. *Cultural ecology of Maharashtra.* Unpublished paper read at the Second International Seminar on Maharashtra – Culture and Society, University of Poona, Pune, 3–5 Jan.

Flannery, K. V. 1972. The origins of the village as a settlement type in Mesopotamia and the Near East: a comparative study. In *Man, settlement and urbanism,* P. Ucko, R. Tringham & G. W. Dimbleby (eds), 23–54. London: Duckworth.

Gadgil, M. & K. C. Malhotra 1979. *Ecology of a pastoral caste: the Gavali Dhangars of peninsular India.* Calcutta: Indian Statistical Institute.

Gune, V. T. 1953. *The judicial system of the Marathas.* Poona: Deccan College.

IAR 1969–70. *Indian archaeology – a review.* Annual report of the Archaeological Survey of India, New Delhi.

IAR 1971–72. *Indian archaeology – a review.* Annual report of the Archaeological Survey of India, New Delhi.

Khazanov, A. M. 1984. *Nomads and the outside world.* Cambridge: Cambridge University Press.

Kosambi, D. D. 1963. Staple grains in the western Deccan. *Man* **63**, 130–1.

Krishnamurthy, R. V., D. P. Agrawal, V. N. Misra & S. N. Rajguru 1981. Palaeoclimatic influences from the behaviour of radiocarbon dates of carbonates from sand dunes of Rajasthan. *Proceedings of the Indian Academy of Sciences* (Earth Planet. Science) **90**, 155–60.

Lucaks, J. R. 1985. Tooth size variation in prehistoric India. *American Anthropologist* 87, 811–25.

Mujumdar, G. G. & S. N. Rajguru 1965. Comments on soils as environmental and chronological tools. In *Indian Prehistory 1964,* V. N. Misra & M. S. Mate (eds), 248–53. Poona: Deccan College.

Sharma, R. S. 1986. *Historical archaeology and problems of urban history.* Unpublished paper read at the SAARC Archaeology Conference, New Delhi.

Sontheimer, G. D. 1975. The Dhangars – a nomadic pastoral community in a developing agricultural environment. In *Pastoralists and nomads in south Asia,* L. S. Leshnik & G. D. Sontheimer (eds), 139–70. Wiesbaden: Otto Harrasowitz.

Sundara, A. 1968. Protohistoric sites in Bijapur district. *Journal of Karnatak University* (Social Sciences) **4**, 2–23.

Sundara, A. 1969–70. A new type of Neolithic burial in Terdal in Mysore State. *Puratattva* **3**, 23–33.

Sundara, A. 1970–71. Neolithic cultural patterns and movements in north Mysore State, *Journal of Karnatak University* (Social Sciences) **6**, 3–12, and 7, 1–8.

16 *The changing role of reindeer in the life of the Sámi*

PEKKA AIKIO

A personal view from my childhood

In my childhood, the reindeer was the most familiar animal. At home were also cows, a cat, many dogs, and later horses. But reindeer were the most common animals of all. This was quite natural because my childhood home was located far away from other houses in the forest. The nearest neighbour, the post office, the store, and the school were a 10 km trip, and there was no road.

Actually, all the reindeer which I knew as a little boy were domesticated and tame. They were not tied to a tree and were allowed to roam freely in the coniferous forest close by our home.

All of the members of our family had their own reindeer, which were marked with the owner's own mark (Fig. 16.1). This was made by notching the ears of the reindeer with a knife. My mother owned her own semi-domesticated draught reindeer; my grandmother, or in the Sámi language *ahku*, owned several; my younger uncles had their own reindeer too, but their reindeer were half wild and untame. My father owned most of all.

Every draught reindeer was given his own 'personal' name. It was important to know the individual reindeer from the rest of the herd. When Mum or Grandma (Fig. 16.2) went on a trip, they themselves had to go and separate their own draught reindeer from the herd. It would have been bad manners 'to help' them. It was also clear that each person used his own draught reindeer when he made a trip. In Sámi families, the women were always economically independent of the men, and this was especially emphasized by the fact that the women, like everyone else, used only reindeer that they personally owned.

My youngest uncle, Niilo, helped me to list the draught reindeer we owned in the 1950s. The list included 66 names (see Table 16.1) of which only five or six had Sámi language names. All the rest were Finnish.

My earliest memories concerning the 30 semi-domesticated draught reindeer which we had at home come from the time when I was four or five years old (I was born in 1944). My father was extremely overjoyed when a son (myself) was born to him, and he notched my personal reindeer ownership mark on the biggest and strongest male reindeer calf that he owned. Later this reindeer was castrated. When I was five years

Figure 16.1 The author aged 12 (1956) with a draught reindeer called *Pehkossuivakko*.

old, this reindeer, called *Sloagga*, was the master of the herd. My special job was to prepare a treat every day out of rye flour and water which was only served to *Sloagga*. This leader of the herd with his big antlers always remembered this, and every day during winter, he came with his herd to pay us a visit in our yard. While I served *Sloagga* his special food, all the other reindeer were given horsetails, *Equisetum* sp., which they ate with great relish. The reindeer passed a couple of hours in our yard and then went back to feeding on lichen in the nearby forest.

Even in my childhood in the 1950s, the reindeer was the companion of the Sámi. The reindeer were, as a general rule, domesticated. Of course, there were wild reindeer in the herd. But the herd as a whole was not nearly as wild or undomesticated as it is at present.

Figure 16.2   The author's grandmother wearing a reindeer coat and carrying the author's little sister, springtime, 1956.

There are many examples of how an individual reindeer showed its companionship for a man or a woman or for another reindeer. My Uncle Jouni (Fig. 16.3) told the story of one female reindeer who got up from her sleeping place in the dead of winter, walked some tens of metres to a draught reindeer with an injured leg, and began to lick and care for the other reindeer's wounds. My uncle watched this and observed that the draught reindeer's injury got better.

When the wolf pack howled nearby, the whole herd would press together near the *kota* (the conical Sámi tent) and its fire in order to gain some protection. The draught reindeer always stuck close to the Sámi's home, and if the reindeer travelled away from this home, it was with reluctance. However, when they were headed back again home, their hooves really flew!

Table 16.1 List of the names of the draught reindeer of the author's family in the 1950s.

Akselin lainakko	Musikki
Aplikki	Myrre
Hiljanen	Mäenpään nulppo
Hiljanen nulppo	Nalhta
Hirrosarvi	Njaiti
Hirvensarvi	Nulppo
Isohärkä	Nylhtä
Isokelosarvi	Näppärä
Isopälli	Oma pikkuhärkä
Itsellinen	Onkisarvi
Jantikka	Paha nulppo
Jekke	Palonojan merkkinen
Julle	Pehkossuivakko
Juovoja	Pekan härkä
Jyty	Pekan pikkuhärkä
Kartanon Jaakko	Pikkukello
Keampa	Pikkumusikki
Kelosarvi	Pikkusuivakko
Kilkuttaja	Pälli
Kumppari	Rymysuivakko
Kusilukkari	Silmäpuoli
Källi	Sloagga
Körri	Stuokki
Laiska	Suivakko I
Laiskanulppo	Suivakko II
Lankakorva	Tassukka
Laukki	Toinen pikkuhärkä
Laukunkantaja	Tonnari
Loikkari	Tulikäpälä
Luosto	Valkko
Mitätön härkä	Vanha suivakko
Musta härkä	Viikari
Musta nulppo	Äidin pikkuhärkä

When a draught reindeer grew old, it was slaughtered and eaten. There were never any tears. If we didn't eat it, the wolves surely would, and why feed the wolves?

A big herd of reindeer always meant more to the Sámi than an individual reindeer. The Sámi's close relationship with the big reindeer herd was always brought forth in the oral tradition. A big reindeer herd gave people a sense of security because a large herd buffered the family from any catastrophe which might befall the herd (Fig. 16.4). There was a wonderful balance between the lives of the people and their herds. A large reindeer herd meant that there would most likely be food on the table and clothes to wear, even if, in a catastrophic situation, some of the reindeer died. The emphasis on the economic aspect of large-scale reindeer-herding in ethnographic research tends to give a rather one-sided picture of the

Figure 16.3 The author's Uncle Jouni, with his dog, at a reindeer-calf marking
site, in front of a turf hut, in the 1950s.

relationship between the Sámi and their reindeer.

Even at the end of the 1950s, the reindeers' position as companions to
the Sámi was unshakable. Then in our fifty-family reindeer-herding
association, there were hundreds of semi-domesticated draught reindeer.
Traditionally, the Sámi herders have owned many draught reindeer.
According to statistics, of all the draught reindeer that were not
slaughtered, 55–60 per cent belonged to Sámi owners. A little boy like
myself didn't think twice about it; it was the way it had always been and
would always be. But in the 1960s dramatic changes took place, of which
the so-called 'snowmobile revolution' was not the least. In the 1970s, there
were about 2300 snowmobiles in Finland, Sweden, and Norway (NKK
1981). In other words, one snowmobile per 300 reindeer. In addition, a lot
of other modern technical aids began to be used in reindeer-herding,
including walkie-talkies, cars, tractors, aeroplanes, and, in Sweden, even
helicopters. Today, our 13-year-old son has only one semi-domesticated
draught reindeer, and in the whole reindeer-herding association to which I

Figure 16.4 Reindeer in an autumn round-up, 1977.

belong, there are only ten draught reindeer – not for the Sámi's own use
but only for the tourists. During the past ten years, the racing of draught
reindeer has started again. Reindeer are used for driving tourists, but, first
of all, the reindeer races have become very popular. It looks as though this
enthusiasm for racing will save the tradition of racing draught reindeer and
will preserve this very valuable skill. Times have changed.

Of course, a reindeer-herding family owned more reindeer other than
merely draught reindeer. In general, the entire herd consisted of 10–20 per
cent draught reindeer.

The origins of reindeer-herding

Reindeer-herding originated in Eurasia. Traditionally in Eurasia, reindeer-
herding has taken place in the entire region north of the 0° isotherm. Of
the total 4.5 million reindeer in the world, 3 million are domesticated
(Andrejev 1977). Of these 3 million, 77 per cent are in the Soviet Union
and 21 per cent in Finland, Sweden, and Norway. A smaller number of
reindeer live in North America, Scotland, Greenland, Iceland, and on
South Georgia Island in the Southern Hemisphere near the Falkland
Islands.

There is hardly any information on reindeer-herding or on reindeer
themselves from ancient times. The small amount of written data and their
direct absence tempt one to assume that reindeer-herding, at least in some

form, is a quite new phenomenon. However, cultural researchers (e.g. Birket–Smith 1972) warn us not to base too much on written sources or, at least, the lack of such sources. Reindeer-herding arose as the product of a widespread Eurasian Arctic culture. The Arctic peoples did not generally leave copious literary traces. The oldest historical information on reindeer-herding dates from the year AD 499. The annals of the Liang dynasty of China tell of the mythical land of Fu-sang where the inhabitants used moose-like animals (in other words, reindeer) as draught animals and for milk (Laufer 1917). The Italian explorer, Marco Polo, who visited Mongolia at the end of the 1200s, also tells of a people who rode moose-like animals. In 1302, the Persian historian, Rashideddi, further tells of a reindeer-herding, nomadic people living in the area around Lake Baikal. This was a people who Wiklund (1947) believed were one and the same as the Soyot or Tungus people, relatives of the Samoyed.

Quite well known is the story of Ohthere's report to the English King, Alfred the Great, in the year 892. Ohthere himself owned 600 reindeer, of which six were decoy animals. Ohthere's report is considered to be the first preserved written reference to reindeer-herding in Scandinavia. But Wiklund estimates (1947), based on old linguistic features, that the Sámi people had already engaged in reindeer-herding even before this – perhaps a long time before this.

Many linguistic and ethnologic factors have pointed to the conclusion that reindeer-herding originally began in one place and then spread elsewhere. Birket–Smith (1972) notes Kai Donner's opinion that reindeer-herding could have begun with the Samoyed people at the beginning of the Bronze Age and was fully developed long before our written history. Some common features regarding reindeer-herding exist throughout the whole of Eurasia, such as the use of the lasso, notching the ears of the reindeer to make ownership marks, the use of skis, and the castration of reindeer by biting.

Recently, support has been given to Wiklund's proposal (1947) that reindeer-herding developed in several locations independently of one another.

Man's earliest contact with reindeer, and in its original form of wild reindeer, occurred in the form of hunting which took place by means of traps, pits, and trenches, as well as by fences or corrals. A decoy animal was used to capture wild reindeer.

Everywhere in Finland are found signs of a wild-reindeer-hunting culture reaching far back into the distant past. For example, there are systems of trenches, complete with different types of fences, that were used in hunting. There are also examples of isolated fences used for the same purpose.

It is not possible to say with certainty how long ago these types of hunting were used. But it is, in any case, certain that long before our system of reckoning time began, people were hunting reindeer in these ways. As an example of this, Helskog (1977) mentions one rock drawing located at Jiebmaluok'ta near Alta, Norway (Fig. 16.5). The drawing shows a corral area with the shape of a four-leaf clover. Inside the corral

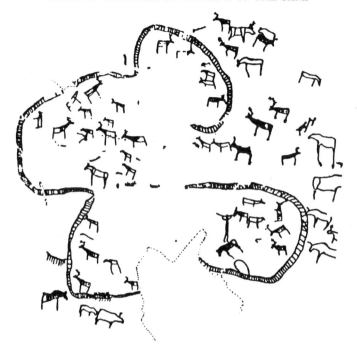

Figure 16.5   The rock drawing at Jiebmaluok'ta, Alta (Helskog 1977).

are drawn reindeer, elk, and perhaps also boats. Many reindeer are coming inside the corral through an opening. The hunter, with spear in hand, is pursuing the reindeer into one corner. These rock drawings are situated 23.5–24.5 m above sea level, and from this elevation, it is possible to calculate that the age of the drawings is between 5500 and 6000 years. In other words, the ancient inhabitants of this area had already developed, about 3500 years BC, a hunting and entrapping culture based on making use of indigenous reindeer. With the possibility of boats in the drawing, fishing can also be postulated.

In the 1970s, in the northwest corner of Finland, near the city of Tornio, a piece of a reindeer horn was found which was dated as being 34 000 years old (Fig. 16.6).

The graphic terminology used by the Sámi people to describe reindeer and its way of life is also clearly very old. Already in the time of wild-reindeer hunting, it was possible to use words to describe exactly the structure of the wild-reindeer herd and the different types of reindeer. Different researchers have collected close to 1000 terms related to the reindeer (for example, Itkonen 1948).

Quite often it is emphasized that the completely nomadic way of life based on reindeer-herding and practised by the Scandinavian reindeer Sámi people did not come about until the end of the Middle Ages. The definition between total and semi-nomadism should not lead, by means of

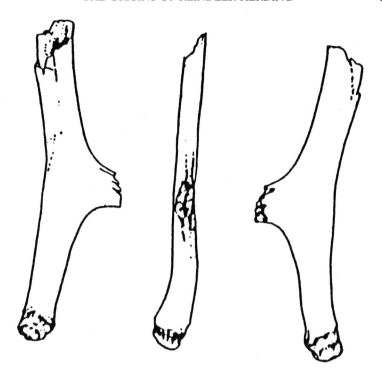

Figure 16.6   Three views of the Tornio antler (Siivonen 1975).

conceptual obscurity, to uncertainty because long before the appearance of the so-called large-herd reindeer-herding, there already existed an advanced multiplicity of ways to herd reindeer. In the polar regions of Eurasia, the wild reindeer was already an important source of game for the inhabitants, and, subsequently, the practice of herding tame reindeer appeared. Reindeer-herding, in so far as the anthropologists are concerned, began in a state of confusion and inconsistency. Samuli Aikio (1977) has also remarked on the reasons for the mystery which surrounds research on reindeer-herding with obscurity, and states that even the history of large-herd reindeer-herding is, in part, difficult to interpret. Nevertheless, it is certain that even from the beginning, reindeer-herding has been character-ized by many variations in customs and in the many different ways of herding reindeer. This diversity can be seen in modern Finnish herding as well.

The first written evidence of the keeping of large herds is to be found in the tax rolls of the 1500–1600s. On the other hand, it is known that archers (in other words, those who paid the bow tax) in Pohjois–Pohjanmaa in the north of Finland also paid church tithes according to the tithe statute of 1335. The tithe was every tenth reindeer calf, or one tax unit for each calf (Tengström 1820–22).

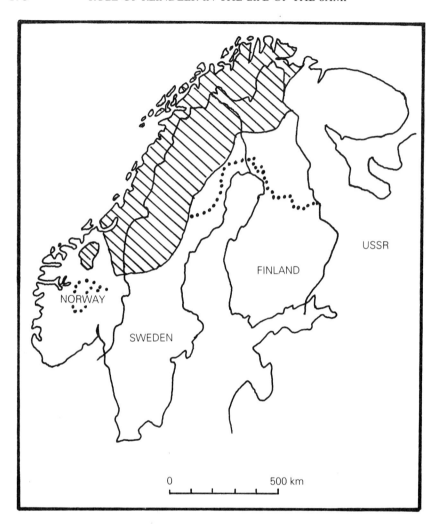

Figure 16.7 Present-day reindeer-herding areas in Norway, Sweden, and Finland. The shaded area denotes Sámi settlement. The dotted line represents the southern boundaries of the reindeer-herding areas.

Nickul (1970) displayed a more uncertain attitude than Wiklund toward the ability of the tax rolls to give information on reindeer-herding. He found that conclusions about the abundance of reindeer could not be drawn from taxes paid in reindeer hides, because the Sámis needed reindeer hides for their own use and it was cheaper to pay taxes from the hides of other hunted animals. Fish could also be used to pay taxes. Nickul continued by saying that if there was no other way to pay the tax, then people resorted to paying with reindeer.

Reindeer-herding under the pressure of change

Figure 16.7 shows the reindeer-herding areas in Finland, Sweden, and Norway. These areas are almost the same as those where the Sámi people live. Only in Finland is the Sámi area smaller than the reindeer-herding area.

During the past 30 years, great technological and economic changes have affected reindeer-herding. From the beginning of the 1960s, perhaps the most dramatic change in reindeer-herding occurred – the appearance of the snowmobile.

The Sámi language has lost its position as the reindeer-herding language. Reindeer-herders are becoming Finnicized. Formerly, reindeer-herders used almost 1000 Sámi terms describing reindeer. The present-day young herder masters hardly ten Sámi terms. At the southern border of the Sámi home territory in Finland, the Sámi language in the 1950s was almost completely replaced by Finnish, even in reindeer-herding (see Table 16.1). The surprising fact is that reindeer-herding does not preserve the Sámi language. This is in contradiction to the commonly held view (Aikio, M. 1984).

Draught reindeer are no longer needed for transportation. Fortunately, the skill of how to raise and train a draught reindeer has been preserved because these reindeer are used by tourists and for racing. Without these modern uses for the draught reindeer, they would have totally vanished, along with the skill to train them. Also, the use of the special reindeer-herding dog has dramatically decreased.

The present-day reindeer-herding legislation is based on the point of view of cattle-raising and agriculture. Reindeer-herding is viewed as a part of agriculture, and all plans for the development of reindeer-herding are modelled after cattle-raising. The primary purpose of reindeer-herding, in this view, is for meat production. Here lies an obvious contradiction: the reindeer is a small animal; its carcass weighs under 25 kg (Aikio, P. 1978).

Reindeer-herding is perhaps the most efficient way to utilize the scant natural resources of the northern terrestrial ecosystem. This can be seen when comparing the amount of reindeer meat with other meat produced in three different zones (as defined by the author) of the Finnish reindeer-herding area (Fig. 16.8). In the northernmost zone (i.e. the Sámi home territory), the total amount of reindeer meat produced equals 90 per cent of total meat production. In the southernmost zone (i.e. the agricultural and forestry area), this is reversed. When thinking of the entire Finnish reindeer-herding areas, the reindeer thrives equally well in the northern and southern areas. However, agriculture and forestry fares best in the south.

During the past 25 years, considerable changes have occurred in the structure of the reindeer herd. Because the reindeer has become a part of the meat-producing machinery of the market economy, man has tended to maximize production. As a result, the proportion of adult females in the Finnish reindeer herd has increased substantially, as has the proportion of slaughtered calves (Fig. 16.9). This did not happen in traditional Sámi reindeer-herding.

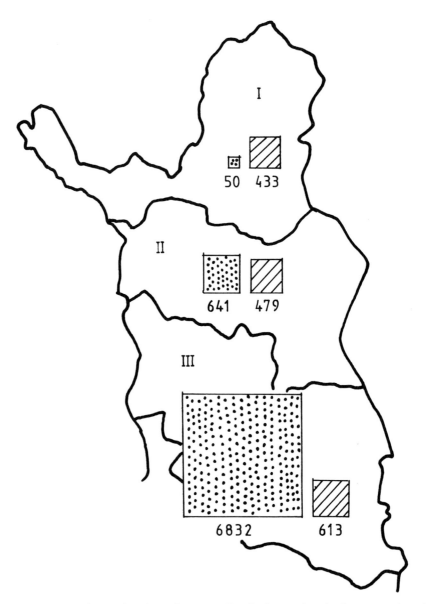

Figure 16.8 The produced reindeer meat (hatched square) and other meat (dotted square) in 1980 in three different zones of the Finnish reindeer-herding area. Zone I indicates the Sámi home territory in the north, which consists of fells and protected forests. Zone II indicates the area where meat production from reindeer-herding and agriculture is equal. Zone III indicates the agricultural and forestry area. The figures under the squares give meat production in metric tons.

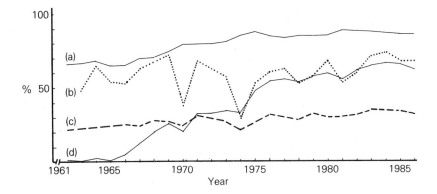

Figure 16.9 The curves symbolize the structure of the Finnish reindeer herd from 1961–86. (a) The percentage of females in the adult reindeer herd; (b) the percentage of calves born; (c) the percentage of calves in the entire reindeer population; (d) the percentage of slaughtered calves out of the total number of reindeer slaughtered.

The reindeer owned by the Sámi graze on natural lichen pastures. We have not, to date, developed an extra source of food for the reindeer which would be economical and useful in the fell areas, and which could be distributed to large herds. As a result, we cannot increase the number of reindeer above the present figure in Finland and in the other Nordic countries. This means that the reindeer is an insignificant source of meat in terms of national production. The authorities in charge of development actions view the reindeer only as a meat-producing domesticated animal. This view is not completely valid. In reality, the reindeer is a half-wild close relative of the North American caribou. In an agrarian society in the coniferous forest zone, the reindeer could have a totally different role than in a sub-arctic fell area or at the forest limit, where the climate is a limiting factor to other human activities.

Historically, reindeer-herding is part of the Lapp livelihood, along with hunting and fishing. Reindeer-herding is successfully combined with other occupations in the sub-arctic environment, which has small primary production. This can be seen in the research carried out by Eino Siuruainen (1976). Based on this research, I have made a diagram which plots the number of reindeer owned against different occupational groups (Fig. 16.10). In Finland at the beginning of the 1970s, according to this diagram, farmers, fishermen, forestry workers, and those working in the service industries owned a small percentage of the total number of reindeer. However, by far the largest share of the reindeer herd was owned by reindeer-herders.

In the Nordic countries in the 1970s, and especially in Finland, reindeer-herding developed in the direction of modern deer-hunting. The reindeer

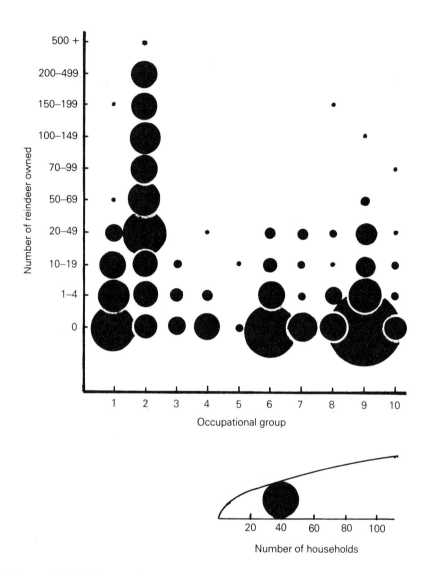

Figure 16.10 The number of reindeer owned plotted against different occupational groups. Professional group of the head of the household: (1) farmer; (2) reindeer-herder; (3) fisherman; (4) forestry worker; (5) contractor; (6) labourer; (7) service-industry worker; (8) businessman; (9) other occupations; (10) unknown.

as an animal has become wilder and has significance mainly as an animal to be slaughtered. In North America, reindeer-herding was introduced as a means to supplement the diet of the native people. This attempt has encountered many difficulties because there they do not have the thousands of years of traditional reindeer-herding experience behind them as is to be found in Eurasia. In North America, the role of the reindeer was to be an animal for hunting.

Concluding remarks

The traditional Sámi reindeer-herding culture, including all its characteristics, has been developed over thousands of years to be highly specialized in its know-how and technology. The modern urban society has never understood the real nature and significance of this type of reindeer-herding. Perhaps it is for this reason that the outside world has misunderstood and misinterpreted the history and development of reindeer-herding.

Could reindeer-herding be as young as it often is presented in the scientific literature? Can scientific research be so limited that it only leans on written sources? There are no written documents on the piece of reindeer antler which was found near the city of Tornio (Fig. 16.6) and which is 34 000 years old. However, a Sámi reindeer-herder who saw a drawing of this antler concluded that it had belonged to a castrated male reindeer. If this is true, then, this reindeer belonged to a reindeer-herding system and was not wild. Researchers have almost always seen spears and other weapons in old rock drawings. What would happen if the human figures in the rock drawings in the city of Alta (Fig. 16.5) really represent people with instruments for capturing animals and not for killing them? Why couldn't the reindeer, like the caribou, have migrated thousands of kilometres from the Mediterranean to the north of Europe and back every year during the latter part of the Ice Age? Perhaps the rock drawings and cave paintings of middle and northern Europe, together with unprejudiced archaeological research, can provide new and surprising results in the historical study of reindeer and the Sámi people.

Acknowledgement

I wish to thank Nick Gardner for his help in editing and translating this text.

References

Aikio, M. 1984. The position and use of the Sámi language: historical, contemporary, and future perspectives. *Journal of Multilingual and Multicultural Development* 5(3&4), 277–91.

Aikio, P. 1978. Reindeer herding in Finland 1975–76: production and labor. *Nordia* 12, 273–82. Oulu.

Aikio, S. 1977. Some minor remarks on the economic history of the Sámi [in Finnish]. *Suomen antropologi* **2**, 91–4.

Andrejev, V. N. 1977. World reindeer herding: its structure and areal distribution. Summary by Elis Pålsson [in Finnish]. *Ottar* **101**, 12–18. The University of Tromsø.

Birket–Smith, K. 1972. *Kulttuurin tiet* (Cultural roads). Porvoo, Helsinki: WSOY.

Helskog, K. 1977. Reindeer round-up in the stone age [in Finnish]. *Ottar* **101**, 25–9. The University of Tromsø.

Itkonen, T. I. 1948. *Suomen lappalaiset vuoteen 1945* (Finnish Lapps to 1945). Helsinki: WSOY.

Laufer, B. 1917. The reindeer and its domestication. *Memoirs of the American Anthropological Association* **4** (2). New York: Kraus Reprint Corp., 1964.

Nickul, K. 1970. *Saamelaiset kansana ja kansalaisina* (The Sámi as a people and citizens). Helsinki: Suomalaisen kirjallisuuden seuran toimituksia (Publication of the Finnish literature society) 297.

NKK 1981. *Reindriftsnaeringen på Nordkalotten*, Nordkalottkommittens publikationsserie (Reindeer economy in the North Calotte, A North Calotte Committee publication) no. 12.

Siivonen, L. 1975. *New results on the history and taxonomy of the mountain, forest, and domestic reindeer in northern Europe*. Proceedings of the First International Reindeer and Caribou Symposium, Fairbanks, Alaska, 33–40.

Siuruainen, E. 1976. The population in the Sámi area of Finnish Lapland. A regional study with special emphasis on rates and sources of income. *Acta Universitatis Ouluensis* Series A **40**, Oulu, *Geographica* **2**, 1–138.

Tengström, J. 1820–22. *Afhandling om presterlinga tjenstgöringen och aflöningen i Åbo Erke–Stift, Utgifven af Domkapitlet i Åbo, delen 1, Åbo* (A study of the tasks of the priests and their remuneration in the Bishopric of Turku, part 1. Turku: The Judicial Church District of Turku.

Wiklund, K. B. 1947. *Lapparna* (The Lapps). Stockholm: Bonniers.

17 *The geographical distribution and function of sheep flock leaders: a cultural aspect of the man–domesticated animal relationship in southwestern Eurasia*

YUTAKA TANI

Introduction

On a medallion of the 17th century by Zincgreff Julius Wilhelm (Henckel & Schöne 1967, pp. 537–8), there is a scene of a flock of sheep leaping down into a river from a bridge. Four sheep are already in the water. Two are jumping and the others stand on the bridge waiting their turn. The first one, with large strongly curved horns and a bell around his neck, is in the forefront in the stream (Fig. 17.1).

The medallion illustrates the ideal relationship between the leader and his followers with the allegory of the flock of sheep, with the ewes following the male. From the comments accompanying the medallion we know that this leading male is a wether: a castrated male.

Sheep have a natural disposition to follow and the flock's direction of movement is consequently determined in general by the leading group. Sheep are timid when they face a gap, or a stream, or something dangerous. When the leading group stops, the followers jostling with each other begin to flow sideways like a stream of water intercepted. On such occasions, the flock sometimes divides into two, or it changes direction and does not follow the course planned by the shepherd, who may adopt several techniques of control during his herding activities in order to make the flock travel and graze along the route he has planned, without dividing or losing members.

The most common and simplest action is to menace a flock by running at it. By shouting and running at the sheep, however, all that a shepherd can accomplish is to chase several sheep close to him or make the stray sheep return to the main group. A sheepdog can chase stray sheep back to the main group, and it can change the direction of a flock, but unless it is

SIC DUCIS URGET AMOR.

Figure 17.1 Medallion by Zincgreff Julius Wilhelm, 17th century (Henckel & Schöne 1967).

intensively trained, its effectiveness remains relative. Goats mixed in the flock tend to run in front, due to their recklessness and rashness, and stimulate the sheep that follow, but they are too capricious to keep pace with the sheep. Another way to control a flock on the move is to utilize a natural leader in the flock itself, if such an individual exists, and with training the shepherd can control the movements of the whole flock by means of vocal commands to the flock leader. If the leader is female, there is no problem, but if the leader is male, sexual activity becomes a problem. Sexually active rams may not obey the shepherd, and so he separates the breeding rams from the ewe group during the summer to prevent them from mounting at will. In order to utilize the male flock leader, it is preferable to castrate him. The medallion mentioned above shows that the western European shepherd has used a castrated male as the flock leader for at least the last 300 years.

It is, however, not only in western Europe but also in the Mediterranean area and in the Middle East that shepherds utilize castrated males as flock leaders. In Abruzzo, central Italy, where I carried out field research, this kind of guide wether (*guidarello* in Italian) is called *manziero*. Every year, the shepherd selects a few breeding males from the male yearlings, most of whom are slaughtered within a year. From among these selected young rams, when they reach two years of age, the shepherd picks one as a candidate for *manziero* and castrates him.

The castrated male holds a unique position in terms of the man–domestic animal relationship. So far as the shepherd views the castrated male leader as a member of the flock, he belongs to the category of the dominated group. However, being castrated, he is alienated from being an essential element of the flock's reproductive process. Moreover, being instructed to respond to the shepherd's vocal commands, he works as an agent of the shepherd, and in this context he belongs to the dominating shepherd's side. By sacrificing his sexual capability and learning the shepherd's vocal orders, he obtains the role of the leader of the flock (dominated) and at the same time the role of the agent for the shepherd (dominator). The sacrifice of his sexual power by castration allows the wether to take the position of mediator between the dominator and the dominated. If most of the flock is female, it means that he takes the role of the guardian of the female group. We can see that the role of the guide wether is functionally analogous to the eunuch.

The eunuch can be defined as the mediator between the emperor and the people, and transmitter of the emperor's commands and will to them. The eunuch was also the guardian of the harem in many imperial courts. The eunuch took the function of mediator between the dominator and the dominated.

We do not know where and when the idea of the eunuch originated. However, the position of the eunuch in human political institutions corresponded to that of the castrated male flock leader in the shepherd–flock relationship. Behind the idea of eunuch that had been adopted in the imperial courts in ancient times, can we suppose an analogy in the techniques of the guide wether to political institutions?

As shown later, however, the geographical distribution of the utilization of the guide wether is not world-wide. In some areas, shepherds adopt different methods to control the flock. They utilize the female sheep instead of the guide wether (from now on these leader sheep used for herd control, regardless of sex and other conditions, will be termed 'flock leader' or 'guide sheep'). Any herd control technique can be regarded as a pastoralist people's cultural expression of how they look upon and treat their domestic animals. Moreover, if some of their cognitive or technical features were analogously applied to the management of man or vice versa, we can discover their hidden way of thinking on the inter-relationship between man and his domestic animals. Has the eunuch, as a political institution, a certain relevance to the guide wether technique from the cultural and historical point of view?

In my field survey on the man–animal relationships of pastoralists it was one of the main purposes to make clear the geographical distribution of the guide wether techniques. But, during the field survey, I found that other kinds of flock leader are utilized in certain areas. Moreover, it was known that some pastoralists place multiple female guide sheep in a flock, and others utilize both types of flock leader, male and female, in the same flock. This multiple leadership raises questions of why they put multiple leaders in one flock and how can the multiple leaders guide the flock harmoniously in one direction. Questions to be asked are:

(a) What types of flock leader are utilized for herd control?
(b) What is the function of these flock leaders in herd control?
(c) How can we interpret the meaning of the multiple leadership?
(d) What is the position of the guide wether type of herd control, historically and technically, in comparison with the other types of flock leader?
(e) In which areas and what periods did each type of herd control come into existence?

Flock leader types and their geographical distribution

The use of a guide wether for flock control, as mentioned above, was seen in western Europe as well as in central Italy. According to the shepherds in Abruzzo in central Italy, they train the guide wether (*manziero*) as follows. After castration, the shepherd puts a short woollen rope on the wether's neck and walks along with him, like walking a dog, and teaches the wether to respond to his touch. When the wether becomes habituated to him, the shepherd begins to instruct the wether on how to respond to the vocal commands: stop (*ferma*), advance (*avanti*), come (*vieni*). To teach the wether what each vocal command means, the shepherd shouts the commands and pulls the rope by hand in a way to make the wether learn how he should react. The shepherd gently pats him or gives him something to eat whenever he behaves properly. In the next stage, after the castrated ram has mastered how to respond to vocal commands, the shepherd puts a long woollen rope around his neck and puts him in the midst of the flock in the grazing-field, and continues to train the wether until this novice *manziero* behaves obediently even in the flock. Having finished this training the shepherd takes off the rope. Now, without any physical control, the trained wether (true *manziero*) goes ahead or stops or returns upon receiving the shepherd's vocal commands, even from a long distance.

The shepherds consider the *manziero* to be the leader of the flock. They say that the *manziero* goes in advance and dares to dash against difficulties, following the shepherd's vocal order. In the winter time, when the sheep must move through the snow, he makes a path in advance of the flock (for more details, see Tani 1976, 1977).

The shepherd gives the *manziero* a special personified name, for example *Generale*, *Capitano*, *Mussolini*, etc. On the other hand, the ewes do not have any proper names. Each ewe is nominally referred to according to the classificatory term based on her body colour patterns. The shepherds set great value in their guide wether *manziero* and the theft of a guide wether leads to a vendetta.

Regarding the pastoralist activities of central Italy, two types of organization can be found (Tani 1977).

(a) Non-professional – each family in the mountain village usually keeps 20–30 sheep for the daily supply of milk. The owners of these sheep take it in turns to herd them every few days.
(b) Professional – the professional shepherd keeps in trust other owners' sheep under contract and takes care of them, forming a joint flock with his own sheep. Usually the herder has a family partner; his brother or son, who always attends to the flock which includes his own sheep.

The *manziero* is usually found in the latter type of herding organization.

In Greece, the Sarakatchani shepherds in Ipiros and Thessalia use a guide wether for herd control. They call him *kriari ghisemia*. A ram of more than two years of age is chosen and castrated. They interestingly adopt the verb ευνυχλεω (to make eunuch) for the act of castration. The shepherd gives him a proper name and he is trained to react to the shepherd's vocal orders. Interestingly, in each flock of 300–500 head, two or three *kriari ghisemia* are mixed in. Moreover, as mentioned later, the Sarakatchani use other kinds of sheep for guiding the flock. Why do they have other kinds of guide sheep besides guide wethers? Do each of these two categories of guide sheep carry out different functions corresponding to the distinct herd-control situation? Apart from these questions, it is certain that the Sarakatchani utilize the guide wether. All Sarakatchanis visited by the author were professional pastoralists. They keep their own sheep and migrate with the family.

In Crete as well, the shepherd raises a castrated sheep as a guide wether. A castrated goat is sometimes used instead of a sheep. Every year, the shepherd castrates four or five males aged 2–3 months. They do not have any systematic method for training the leader as observed in central Italy. They believe that shouting, beating, and throwing stones accompanied by vocal commands are sufficient to make wethers learn to lead the flock. They also greatly value their own guide wether, and the theft of a guide wether can be the cause of vendettas.

Among the nomadic Yöluk of Anatolia (who are of central Asian origin) it is reported that there is no evidence for the utilization of the guide wether (Matsubara 1983). On the other hand, the nomadic Kurds near Hakkari in southeastern Turkey use a wether as flock leader. Usually a two-year-old ram is castrated and put in the flock after training by the same method as in central Italy. Sometimes a castrated he-goat (*teke*) is also used. They said it was usual to place only one guide wether in a flock. In Iran, according to Digard, the shepherd of nomadic Baxtyari also raise the guide wether (Digard 1981, p. 57).

Going further east towards Afghanistan, data were collected on several pastoralists of different ethnic groups; the Kandahari of Durrani Pashtun, the Uzbecki, the Arabi, and Shaghni of mountain Tajik, who ascend every summer to the Shewa high plateau (*Dasht Shewa*) in Badakhshan (see details in Tani & Matsui 1980). The first three groups are nomadic and the last, the Shaghni, are sedentary farmers in the villages along the Pamir River, whose shepherds come there seasonally for transhumance. The Shaghni do not use the guide wether, but the other three pastoralist groups do. The only difference from the Mediterranean pastoralists is that they castrate not the ram but the he-goat. The Arabi select and castrate one-year-old he-goats. They said that such he-goats can be used for about ten years as leaders. The Arabi and the Uzbecki call the he-goat leader *sarkka* and the Kandahari of Pashtun call it *mukhi*. In the summer quarter, the Kandahari shepherds do the night-time herding, not individually but collectively. Looking at the formation of the flock and its direction of movement, two shepherds, one in the front and another at the back, approach the flock with vocal calling and whistling. This guiding method reminds us of that described by Baskin as a primordial herding-control method (Baskin 1974, p. 530). The castrated he-goat is usually bought by the owner of the flock (*tsessten*) in the market. The shepherd (*shpun*) is in charge of the flock and is employed by *tsessten* with an annual contract. His position as a shepherd is not stable. The he-goat is not given any proper name. Between the shepherd and the he-goat, there is not any individual command-to-obey relationship.

Further east, in the northeastern territory of India (Kashmir), the pastoralists of Gujar Bakkalwala (goat-herders) and the Kashmir shepherds (sheep-herders) did not know about the use of the guide wether, and they do not care for the flock with as much attention as observed among the shepherds in the areas mentioned above. It was seldom that their flocks were followed by any shepherds in the daytime. We have no information about the Tibetan pastoralists in the Ladakh and the Chinese territory, except for the northern Zanskar area of the Ladakh which I visited, and they did not use a guide wether. Most goat-owners of the Bakkalwala, even though they migrate in the summer quarter with their families, do not attend to the flock, but hire a servant-shepherd by contract. In the Zanskar of the Ladakhi, on the other hand, the village women attend, in turn, to the flock collected from the villagers, as well as having their own sheep amongst the flock.

Though more evidence should be collected from different places, to fill the gaps between the areas visited, it can be said that the use of the guide wether or castrated goat for herd control is found along a southern belt from West Europe to Afghanistan, with the exception of areas where pastoralists in Turkey have migrated from the north.

Now, turning north from the Mediterranean area, the data on the Romanian pastoralists will be presented. I visited the following three places: the southeastern plains of Dobrogea near the Black Sea (around Topalu), the southern Carpathian mountains in the province of Sibiu (around Tilişca), and the northern Carpathian mountains in the province of Bistriţa and Nasaud (Ardan, Şebis, and Sieuţ: see details of my field studies in Tani 1982, 1987).

In Dobrogea, the guide wether used for herd control is called *batal*. The shepherd often sleeps using as a pillow a lamb chosen to be a *batal* in the future, to form a familiar relationship. In the spring when the lamb is one or two years old, the shepherd castrates him. A special proper name is given to him and he is always called by name. By often feeding *mamaliga* (boiled maize paste) to the lamb, it is trained to follow the shepherd and respond to commands. To enhance the appearance of the guide wether, the shepherds even try to correct the curve of the horns using the hot *mamaliga*. They said that only one guide wether is usually placed in a flock. But in the flocks observed, the shepherds often pointed out three or four guide wethers.

In southern Carpathia (province of Sibiu) also, the use of a *batal* was observed, although it was not found in all the flocks. The shepherds said the *batal* is only placed in the large flocks, and usually only one in each flock. The proper names given to the *batal* are often taken from human names. According to the etymological dictionary by Hasedeu, the term for the guide wether *batal* is derived from the Greek 'βαταλλος', meaning 'castrated berber'. The connotation 'guide wether' must be added to this term as an extension of the original meaning. The shepherds at Tilişca, who seasonally migrated to Dobrogea until the socialist revolution, said that the use of *batal* was introduced from Dobrogea. In Dobrogea, trade with Greek merchants for milk products had been frequent. Taking into account the Greek origin of the term *batal*, it is highly probable that the utilization of the guide wether was introduced from the southern Balkan peninsula.

The shepherd is a professional who joins up the sheep of the village farmers in his flock, as well as having a large number of his own sheep. Generally speaking, the herding team is composed of eight shepherds. Half of them take in turn the charge of herding each week.

Further north in Romania, the utilization of the guide wether *batal* completely disappears. In the province of Bistriţa and Nasaud, the shepherds know of the *batal* but never use it. Instead of the castrated male, they have another flock leader that is, interestingly, female. This guide ewe is called *fruntaşa* which literally means 'the female leader that goes in front'.

There are always some young ewes that approach the shepherds without fear. The shepherds mark these and begin to tame them by calling their names and feeding *mamaliga* to them every morning. After the marked females have learned their own names, the shepherds teach the meaning of vocal callings and whistles. The names of address are not any special proper name but simple classificatory terms corresponding to the colour pattern of each sheep. During the summer the flock is usually composed of the sheep owned by several different shepherds, who take care of the flock in turns for one week. If you ask them to point out the *fruntaşa*, they indicate about ten *fruntaşas* in one flock. Why are so many leaders put in one flock? How do they use them to manipulate the movement of the flock? Apart from these questions, which will be discussed in the following section, the main point is that in northern Carpathia (Romania) they use the female as a flock leader instead of the castrated male.

The labour organization is just the same as that of southern Carpathia. When I first asked the shepherds if they utilized castrated males as flock leaders, some denied their use with a disparaging tone, even though some of them knew that southern Carpathian shepherds use wethers. It is not clear if this repugnance is derived from a cultural base, such as, for example, the importance of the feminine or low evaluation of the status of the 'castrated'. Or, are these two types of flock leaders, male and female, incompatible for a technical reason? Evidence against this is found among the southern Carpathian and Dobrogean shepherds, who use the guide wether, but, at the same time, raise the *fruntaşa* type of leader and use her in parallel with the *batal*. They call her, instead of *fruntaşa*, *cîrmace* which means 'the one to be followed'. The training method for the female guide sheep is the same as that in northern Carpathia. The coexistence of two kinds of leadership here, however, raises the following question. If both guide wether and guide ewe carry out the same leadership role, which do the other flock members follow? If both lead the flock, does not the flock divide? It seems better to have only one leader.

Multiple leadership, however, is not confined to southern Carpathia. Even the pastoralists of Sarakatchani in Greece, who use the guide wether, also train a kind of guide ewe and ram. Besides multiple *kriari ghisemia* (guide wethers), two kinds of special sheep, *manari* (male) and *manara* (female) are used for herd control by the shepherd. Every morning just before departure and in the evening after returning back to the camp, *manari* and *manara* are give a handful of barley by the shepherd. When the flock starts from the camp site for grazing, the shepherd leads the flock with whistles. These *manari* and *manara* follow the shepherd and the rest follow smoothly. To train these *manari*, they select a certain male or female lamb from among the yearlings and keep him or her in the shepherd's hut in order to be tamed. Instead of allowing the lamb to suck from its mother, they feed it milk from a nursing bottle by hand. They treat it like a pet, often holding it in their arms. After weaning, they feed the lamb barley by hand and call it by name. After establishing a familiar relationship between shepherd and sheep, the shepherd begins to teach it the meanings of vocal orders.

Even if the details of the training method of the Sarakatchani are different from those of Romanian *fruntaşa*, the main principles of training by feeding are common to both. The peculiarity of the former, however, is in the use of the non-castrated male as well as the female. In northern Carpathia, the shepherd makes use of the guide ewe only. In southern Carpathia, the guide wether is used with the guide ewe. Even though *manari* and *manara* are categorized in one class by a common word root, we must say that the Sarakatchani utilize for herd control three kinds of flock leaders: guide wether, guide ram, and guide ewe. The same kinds of guide sheep are used by Cretan shepherds.

Further east, in northwestern India, the Ladakhi transhumance shepherds give milk by cup to establish a familiar relationship with certain lambs. The Bakkalwala of Kashmir also intervene between the ewe and her offspring, just after the delivery, and give milk to the lamb. They say that it is possible to have a close relationship by this feeding, but neither the Ladakhi nor the Bakkalwala train flock leaders.

The collected data are too scattered to draw a complete distributional map of the different types of flock leaders. Moreover, the historical evidence is insufficient to identify the place where these techniques originated and the process of how they diffused to the regions where they are in use at present. However, the method for raising the guide wether through castration and training can be considered more elaborate than that of raising the non-castrated leader, since the idea of forming a familiar relationship by feeding and taming sheep like an obedient pet seems to be more natural. Even though the present geographical distribution of herd control by non-castrated leaders is very limited in area, this type of leader may have been the first in the early stage of herding. Subsequently, herd control using the guide wether may have been initiated in the eastern part of the Mediterranean area and diffused to the eastern part of the Middle East, and westwards to western Europe. These are my tentative inferences regarding the historical origins of herd control techniques by the trained flock leader.

Behaviour and function of flock leaders in the herding situation

In the Bistriţa area of Romania, I asked the shepherds the meaning of *fruntaşa*. Many of them answered that in contrast to the *codaşa* which moves always in the back of the flock, the *fruntaşa* is the sheep that runs in the front of the flock, and is the conductor of the flock. To the question: 'Is the offspring of the *fruntaşa* also likely to become *fruntaşa*?', not all, but many shepherds said 'yes'. If we take these answers at face value, the *fruntaşa* may be a natural leader with an innate tendency to lead the flock. But some of the shepherds were reluctant to confirm this. They said that the *fruntaşa* does not always lead the flock and denied the existence of any tendency for the *fruntaşa* offspring to inherit the leadership from her mother.

Leaving aside the possibility of inheritance, multiple *fruntaşas* were found in all the flocks. If the *fruntaşas* behave similarly to the supposed natural leader and if the *fruntaşas* move independently, how could the flock remain united? Why don't the shepherds keep only one leader in a flock? These questions led me to study the actual behaviour of *fruntaşas* in the herding situation.

The observations were carried out in the summer on a flock of 500 sheep owned by eight shepherds at Mt. Câlimani in northern Carpathia.

As a rule, four shepherds, organizing a team, took charge of the herd on alternate weeks.

It was realized at an early stage that the shepherds gave different answers on the number of *fruntaşas* in the flock. At first, to avoid complication, I asked only one of these shepherds (from now on referred to as A) to point out all the *fruntaşas* in the flock. Surprisingly, 20 sheep were pointed out. I put a strip of differently patterned cloth on each of them to make their position in the flock obvious.

The position of each *fruntaşa* was counted every 15 minutes over three hours (12 times). The movement of the flock as a whole is as smooth as flowing water. The positional order of each sheep incessantly changes because of pausing, grazing, and hurrying up. Though it was very difficult to count the exact order of the 20 *fruntaşas*, their approximate position could be estimated. Although there should have theoretically been 240 scores in the 12 times of the observation (20 *fruntaşas* × 12 times = 240), it was actually possible to count the positional scores for only 197 cases because I lost the chance to witness 43 cases in the crowd of sheep.

It was quickly noticed that the so-called leader *fruntaşas* did not tend to march in the front of the flock. One was around 15th, and another was about 250th. Some of them were even about 10th and 20th from the last and it often happened that one at 25th at a certain time was found at 90th, 15 minutes later. According to the observations, a *fruntaşa* was in an advanced position, within the first ten, in only nine scores among the total number (197).

If we suppose that these 20 *fruntaşas* share the same behavioural disposition with the common sheep, the theoretical probability that at least one of the 20 *fruntaşas* is in the first ten should be 0.34. If we take into consideration the missed cases and make a revision, the actual possibility that at least one *fruntaşa* was in the first ten becomes 0.05. The theoretical probability for at least one of these 20 *fruntaşas* to be in the first 50, should be 0.99, while the actual possibility was 0.28. In both of them, the actual frequency that the *fruntaşa* was in the advanced position was very low in comparison to the theoretical values. The data imply that these *fruntaşas* might have been clustered together, or the sheep were not randomly distributed in the flock. Those 20 *fruntaşas* belong to the same subgroup owned by one shepherd. Did the tendency to be clustered in the joint flock depend on it? One day a clue to the function of the *fruntaşa* was discovered, when I noticed that another shepherd on herding duty (from now on referred to as B) took a piece of *mamaliga* from under his belt while he shouted a vocal command to urge the flock to start from the camping site in the morning. He showed it to the sheep who were in front of the flock coming out from the opened gate. He gave it selectively only to certain sheep who did not bear any coloured strips. He explained that these sheep were his (B's) *fruntaşas*. Responding to my request to point out his *fruntaşas* among the sheep rushing ahead to obtain a small piece of *mamaliga* from his hand, he indicated five sheep as his *fruntaşas* among the 15 sheep in the first group, while none of shepherd A's *fruntaşas* were found within the first 15 heads rushing ahead towards shepherd B. At this moment 18 of the 20 *fruntaşas* marked by shepherd A (those that I could identify immediately) were in the following positional order: 24, 25, 28, 55, 75, 85, 207, 245, 250, 290, 340, 345, 370, 380, 381, 390, 400, and 405 (average: 232nd among 500 sheep). This meant that when shepherd B was in charge of the flock, his *fruntaşas* promptly approached him and consequently led the flock, while the *fruntaşas* of shepherd A were far behind. Urged by the invitation of his clapping and whistling, his *fruntaşas* dashed ahead and then the others smoothly followed after them. To sum up, the *fruntaşas* are the sheep who are tamed by each shepherd through feeding in order to respond to his vocal orders and are his favourites, and do not respond to the other shepherds.

Even though these *fruntaşas* do not naturally tend to lead the flock during grazing, they follow their master without delay when attracted by the shepherd's offer of *mamaliga* and calls, and their movements in turn cause the rest of the flock to follow. In the latter sense, they can be defined as the inducers of flock movement. Their function depends on the order-to-obey relationship established by training through feeding, and on the disposition of the sheep in general to follow. As regards the problem of multiple leaders, we can say that the small initial inducement by only one *fruntaşa* does not have sufficient effect. The more numerous the inducers, the more effective the first is in causing the others to move. Since all the *fruntaşas* make their common initial move by simultaneously responding to the shepherd's order, the plurality of *fruntaşas* does not bring about any problem in herd control.

As further evidence for the utility of multiple *fruntaşas*, I often took note in the field that the shepherd called the name of another *fruntaşa* when one *fruntaşa* reacted slowly and her initial move was not sufficient to urge the rest of the flock to follow. When the shepherd wanted the flock to turn back or stop, he whistled with his fingers, called the name of a *fruntaşa*, and directed it to come back or stop. But, if the initial move of that sheep did not cause the others to follow, he often repeated the same order, and called other *fruntaşas* by name to invoke a more intensified inducing effect.

The disagreement on the number of *fruntaşas* between the shepherds who take care of the same flock is not strange, now that we know that each shepherd raises his favourite *fruntaşas* from his own sheep in the flock. Moreover, if we consider the shepherd's preferential feeding of his *fruntaşa*, we can see the offspring of a *fruntaşa* sometimes become *fruntaşas*, as some shepherds recognized. The lamb, being together with her mother *fruntaşa*, may have more chances to have close contact with the shepherd. If the shepherd is likely to feed the sheep that does not fear him, such offspring would be more likely to be chosen as *fruntaşas* in future.

I have no quantitative observations on the non-castrated male and female (*manari* and *manara*) of the Sarakatchani shepherd. However, they are tamed like pets from their infancy and are continuously fed. In the morning, before departure, they are given a handful of barley and then follow after the shepherd who walks in front of the flock. In the evening, they walk in the forefront expecting the offer of barley, even if the shepherd does not accompany them. The shepherd treats them mostly the same as *fruntaşas*.

To sum up, the *fruntaşa* (or *cîrmace*) of Romania, as well as the *manari* and *manara* of Greece, work as mediators between the shepherd and the flock. Through feeding and training, they become obedient agents reacting to the shepherd's vocal order. The plurality of these so-called leaders in a flock intensifies the effect of leadership.

The actual function of the guide wether will now be discussed. At first, the data on the guide wether, *batal*, in Dobrogea will be presented. Three guide wethers were placed in the flock observed. To observe the positional order of these *batals* in the flock, coloured strips were fixed on each. On the departure from the wooden fence in the morning, these *batals* tended to march in the rear half of the group, and it was rare to find them in the front. Only the goats mixed in with the sheep tended to dash ahead and induced the others to follow. Later on, while the flock was in the field grazing it was widespread and slowly moved as a whole in a certain direction, sometimes being led by several sheep in the lead group. Even in such a situation *batals* were seldom found within the first group. The positional order of the *batal* was really random.

However, there were occasions when the *batal* led the flock. For example, one day the flock was proceeding along the bank of a river, and it happened to come against a gully forming a dry stream bed. The lead group halted there without any attempt to jump over. The shepherd immediately called a *batal* by his name and approached him. This caused the *batal* to jump over the gap. The others followed his initiative and jumped one by one. To establish this order-to-obey relationship, the shepherd gives salt to the guide wether while calling his special name. In this sense the *batal* has the same role as the *fruntaşa*. Nevertheless, the shepherds of southern Romania keep both in the flock. The shepherd emphasizes the *batal*'s effectiveness in leading the way in deep snow. His bravery and toughness is the reason why they utilized the *batal* with the female guide sheep not only southern Romania but also in Greece.

Now, the last type of guide sheep found in Greece, the *manari* (non-castrated male), has not yet been discussed. Among all pastoralists, it is common to separate yearlings from the ewe group after they are weaned. As long as they keep the *manari* in the yearling group as a guide sheep, there is no problem. But in the mating season when all categories of sheep, including the ewes, are joined together into a flock, the *manari* might be less eager in his duty. In this respect, the *manari* cannot be as effective a guide in comparison to the guide wether.

For the same reason, the *fruntaşa* and *manara* as well may be less effective because in the mating season she will be the objective of the breeding ram. The guide wether, with the sacrifice of his sexual ability, should be the best as flock leader.

Discussion

We have described the flock leader as artificially trained and with a close relationship with the shepherd, even though there are various grades of elaboration in the training.

If a shepherd does not have his own sheep but is under annual contract, like the Kandahari nomads, it is impossible for him to foster such a familiar relationship between himself and the flock leader. Among the Kandahari, the use of the castrated he-goat was most commonly seen. They did not give any proper names to these goats. Even though they tame the he-goats by feeding, they might rely more on the goats' recklessness and rashness than on control through training. In central Italy the non-professionals who take care of the flock of the villagers' sheep do not use the guide wether. We know that the he-goat is utilized not only in Afghanistan but also widely in the Mediterranean and Near and Middle East, where the professional shepherd often works for the owner of large flocks.

Now if we classify the guide sheep into two categories: the non-castrated (female and male) and the castrated (male), direct observation shows that neither type had any natural inclination to lead the flock. They take the role of leader only when the shepherd takes the lead in front of the flock or sends vocal orders to them. Being conditioned by training and instruction, the guide sheep can understand his own name and understand the vocal order to stop, or return back, or go ahead, etc. Through this communication they make the initial move, responding to the shepherd's orders, and consequently induce the others to follow. Both types of guide sheep achieve basically the same function. The difference between them is at first found in the wether's toughness and bravery. The guide wether works effectively even in the mating season and in the winter snow. Nevertheless, in northern Carpathia where the snowfall is heavy, the guide wether is not used.

In Greece, at least among the Sarakatchani and the Cretan shepherds, both types of flock leader, the castrated and the non-castrated, were utilized. The utilization of the guide wether seems to have been recently introduced into southern Carpathia from Greece.

We do not have any clue which of the types, the non-castrated or the castrated, was developed first. With respect to the effectiveness as flock leader, the guide wether is, however, more elaborate than the non-castrated types, as discussed above.

Now, it is opportune to raise a question. On which domestic animal was the castration-training technique at first applied: on the ram or on the bull? It is more probable to have been applied to the bull. We know from a Sumerologist, Maekawa, that systematic castration and training of the bull had been applied from the 3rd millennium BC under the Ur III dynasty in Mesopotamia, to produce plough-animals and to lend them to the allotted farmers (Maekawa 1979, p. 99).

Is the guide wether the analogical extension of the idea of a trained castrated bull? I can imagine another way for the beginnings of the guide wether. With the supposition that the *manari* and the *manara* types of flock leader had been practised in an earlier stage, if shepherds began to castrate the *manari*, the guide wether would have become established. The technique to produce an obedient male by castration and training was perhaps developed when man first began to use horses and oxen for draught and riding. But we can also suppose that the eastern Mediterranean shepherd who already had the non-castrated leader (*manari* and *manara* type) learned this castration-training technique from the civilized world of the ancient Near East and applied it to their flock control.

Leaving aside the concern about the historical origin of such techniques, it is opportune to return to the problem of the eunuch. Maekawa (1979, 1980, 1982) has shown from the Sumerian cuneiform tablets of Lagash, of the Ur III dynasty, that many sons of captured or slave women, who worked in the specialized weaving camps were used for physical labour and categorized by the special term for the castrated ploughing bull; *amar-KUD*. While the female offspring of the slave women were recruited again as weavers in the maternal group, their male offspring were castrated and used for physical labour and separated from the maternal group. This is an interesting example of the application of cattle management techniques to the management of slaves.

We do not have any historical data on how and where the institution of the eunuch was originally developed. Moreover, at present there is no evidence to suggest that the guide wether was the model of the eunuch. But it seems possible that this kind of social control, using the castrated male, prepared the cultural background for the idea of the eunuch. This is, however, just a sideline. The main purpose of this chapter has been to present information on the variety of flock leaders, to give the main distribution of each type, and to define the nature and the function of flock leaders on the basis of field data. More detailed field data and further discussion of the shepherd's intervention into the sheep-flock relationships are published in Tani (1987).

References

Baskin, L. M. 1974. Management of the ungulate herds in relation to domestication. In *The behaviour of ungulates and its relation to management*, V. Geist & F. Walther (eds), 530–42. Morges: IUCN Publications, N.S. 242.

Digard, J.–P. 1981. *Techniques des nomades baxtyâri d'Iran*. Cambridge: Cambridge University Press.

Henckel, A. & A. Schöne 1967. *Emblemata: Handbuch zur Sinnbild Kunst des XVI. und XVII., Jahrhunderts*. Stuttgart: J. B. Metzlersche Verlagsbuchhandlung.

Maekawa, K. 1979. Animal and human castration in Sumer, I. *Zinbun* **15**, 95–140. Kyoto: Research Institute for Humanistic Studies, Kyoto University.

Maekawa, K. 1980. Animal and human castration in Sumer, II. *Zinbun* **17**, 1–56. Kyoto: Research Institute for Humanistic Studies, Kyoto University.

Maekawa, K. 1982. Animal and human castration in Sumer, III. *Zinbun* **18**, 95–122. Kyoto: Research Institute for Humanistic Studies, Kyoto University.

Matsubara, M. 1983. *Yuboku no Sekai* (Ethnography of the nomadic Yöluk people in Anatolia). Tokyo: Chuōkoronsha.

Tani, Y. 1976. Bokuchiku Bunkakō (On the pastoralistic culture: interaction between man and domestic animals and analogical application of the pastoralistic managemental techniques into the social control of man) [in Japanese]. *Bulletin of the Research Institute for Humanistic Studies* **42**, 1–58. Kyoto: Kyoto University.

Tani, Y. 1977. *Italia chūba sanson, iboku hitsujino kanri ni tsuite* (On the management techniques of the sheep flock among the transhumance shepherds in central Italy). *Society and culture of Europe–field research reports*, 117–67. Kyoto: Research Institute for Humanistic Studies, Kyoto University, 1976.

Tani, Y. 1980. Man–sheep relationship in the flock management techniques among north Carpathian shepherds. In *Preliminary report of comparative studies on the agro-pastoral peoples in southwestern Eurasia*, Y. Tani (ed.), 67–86. Kyoto: Research Institute for Humanistic Studies, Kyoto University.

Tani, Y. 1982. Implications of the shepherd's social and communicational interventions in the flock – from the field observation among the shepherds in Romania. In *Preliminary report of comparative studies on the agro-pastoral peoples in southwestern Eurasia*, Y. Tani (ed.), 1–18. Kyoto: Research Institute for Humanistic Studies, Kyoto University.

Tani Y. 1987. Two types of human interventions into the sheep flock: intervention into mother–offspring relationship, and raising the flock leader. In *Domesticated plants and animals of the southwest Eurasian agro-pastoral cultural complex*. Vol. 2, Y. Tani (ed.), 1–42. Kyoto: The Research Institute for Humanistic Studies, Kyoto University.

Tani, Y. & T. Matsui 1980. The pastoral life of the Durrani pashtun nomads in Northeastern Afghanistan. In *preliminary report of comparative studies on the agro-pastoral peoples in southwestern Eurasia*, Y. Tani (ed.), 1–31. Kyoto: Research Institute for Humanistic Studies, Kyoto University.

18 *Cattle in ancient North Africa*

JULIET CLUTTON–BROCK

Introduction

The purpose of this chapter is to enter the discussion on the origins of domestic cattle in North Africa, their phenotypes, and the dates of their first appearance in the archaeological record. There has been contention about this subject since the first descriptions of the ancient remains of wild cattle from Algeria in the late 19th century, but more especially since the discovery of the wonderful rock art of the Tassili n'Ajjer mountains, which depicts a diversity of domestic cattle and other bovids (see Fig. 18.1). Today, the comprehensive reviews of Smith (1980, 1986), Banks (1984), Wendorf *et al.*(1984), and Gautier (1987) provide up-to-date accounts of the archaeology of cattle-keeping in the Sahara and West Africa.

Theories on the origins of cattle in North Africa

The traditional view on the origins of domestic cattle in Africa is that the animals were brought to the continent across the region of Suez, approximately 6000 years ago, which is about 2000 years later than cattle were first domesticated in western Asia and Greece. These long-horned humpless cattle are said to have spread by diffusion westwards and then south into North and West Africa and the central Sahara, where they are represented in the Tassili rock art (Lhote 1959, Epstein & Mason 1984). Epstein believes that these first long-horned cattle were replaced by a new wave of humpless short-horned cattle around 4000 years ago, and that these followed the same diffusion route except that they did not reach the central Sahara.

Humped cattle, according to Epstein, were introduced considerably later (c. 3500 BP) across the Horn of Africa. These were cattle with cervico-thoracic humps and they were the precursors of the true zebu or Indian humped cattle, which have a hump only over the thoracic vertebrae. The neck-humped cattle spread across Africa through Sudan and reached West Africa where they interbred with the primitive longhorns to produce the Fulani breed. Meanwhile, cattle proliferated in Ethiopia where the neck-humped cattle were crossed with long-horned animals, around 3000 years ago, to produce the Sanga breeds. It was from the Sanga that the remarkable Ankole cattle of Uganda, and the Afrikander cattle of South Africa are derived.

Later still the true zebu was introduced to Africa, also through Somalia, and these thoracic-humped cattle rapidly displaced the older Sanga breeds because they were more rinderpest-resistant and gave a higher milk yield.

These theories have been put forward by Epstein & Mason in their review of 1984, and are based on deductions from the distributions of present-day breeds in Africa and on evidence from pictorial representations in the Sahara, Ethiopia, Sudan, and ancient Egypt. The archaeological evidence, however, presents a different story.

The differentiation and dating of cattle remains from North Africa

There is now general agreement amongst archaeozoologists that the progenitor of all domestic cattle was the extinct aurochs, *Bos primigenius*, although the zebu (Indian humped cattle) could be descended from the subspecies, *Bos primigenius namadicus* (Clutton–Brock 1981, Epstein & Mason 1984, Grigson 1985). Formerly, it was believed that in Europe there were two species of ancient wild cattle, *Bos primigenius* and a smaller short-horned form, *Bos longifrons* or *brachyceros*. Better dating of archaeological sites and more rigorous examination of the material has shown, however, that examples of this small 'species' were, for the chronologically older specimens, the females of the aurochs, whereas the more recent specimens are the remains of small prehistoric domestic cattle. Similarly, in North Africa two wild species of cattle were described from early finds, these being *Bos primigenius* and a small form, *Bos ibiricus* or *africanus*. Today, it seems clear that the small form was the female of *Bos primigenius* (Banks 1984).

Remains of cattle that are believed to have been domestic from their rather small size have been retrieved from sites of around 8000 BP in Greece, Anatolia, and western Asia, while, from Mehrgarh in Baluchistan, Meadow (1984) has reported the recent finding of a clay model of a humped ox dating to around 6000 BP. The theoretical view that descendants of these early cattle first entered Africa through Egypt and that they were long-horned is not borne out by the subfossil finds. The bases of the horn cores on a skull from Badari, now in the British Museum (Natural History), which should be at least 5000 BP, indicate that the animal had rather short, upstanding horns. In any case, this skull should be directly dated by radiocarbon before its context is accepted, because the skull of an ass from the same level at Badari has produced a date of 700 ± 120 BP (OxA–564). There are also two horn sheaths from a very long-horned ox or cow from the Fayum in the British Museum (Natural History) collections, but these, too, should be dated before they can be accepted as predynastic.

The earliest securely dated finds of cattle in a cultural context in Africa at present come from Capéletti in Algeria, where dates were obtained of 6530 ± 250 BP for the lowest levels (where the cattle were not certainly domesticated), and 5100 ± 150 BP for the upper level, where domestic cattle predominated (Roubet 1978). Other sites from further south, in the

Table 18.1   Sites in the Sahara and Sahel from which remains of domestic cattle have been identified (from Smith 1980, p. 495).

Site	Absolute date (BP)			
Uan Muhuggiag	5952	5405		
Adrar Bous	5780	5740	5130	4910
Karkarichinkat Sud	3960	3640	3310	
Karkarichinkat Nord	3950	3710	3620	
Dhar Tichitt	3465	3350	3620	
Kintampo	3560–3220 (suggested dates)			
Daima	2400			

Sahara and Sahel, where remains of domestic cattle have been identified, are given in Table 18.1. These sites are also shown in Figure 18.1, together with those localities cited for pictorial representations. From none of these sites do the cattle remains show any signs of having been long-horned, and, indeed, the bones from Kintampo are so small that they resemble those of the modern dwarf cattle of West Africa, the Ndama breed (Carter & Flight 1972). These short-horned subfossil remains conflict with the appearance of the cattle from the rock art of the Tassili mountains, for which Lhote (1959) proposed four periods. The earliest phase he claimed had pictures of the extinct North African buffalo, *Homoiceras singae*. Then came Lhote's Bovidian phase from which there are a multitude of paintings showing both long and short horns and polled cattle. The people who created these paintings were cattle pastoralists who appear to have had no agriculture, and who may have lived about 5000 years ago. Later, there was Lhote's horse phase which was protohistoric, and finally a camel phase dated to around 2000 BP.

From the indications of a savanna environment in which a diversity of wild ungulates could survive as well as humans and their cattle, the artists of the Tassili mountains must have been living there before the climate began to be really arid, around 4000 years ago. It could be a matter of chance that no osteological material from these long-horned cattle has survived, or it could be that these were rare and valued animals that were more often painted than seen.

Long-horned cattle were seen to the north of Tassili by Herodotus who described them in a famous passage, written about 2450 BP: 'In the Garamantian country are found oxen which, as they graze, walk backwards. This they do because their horns curve outwards in front of their heads, so that it is not possible for them when grazing to move forwards, since in that case their horns would become fixed to the ground.'

Smith (1980, p. 492) quotes from this passage but infers that the cattle described by Herodotus were wild *Bos primigenius*. I disagree with him over this, for I am sure that these were domestic long-horned cattle that must have looked very similar to the present-day English Longhorn breed.

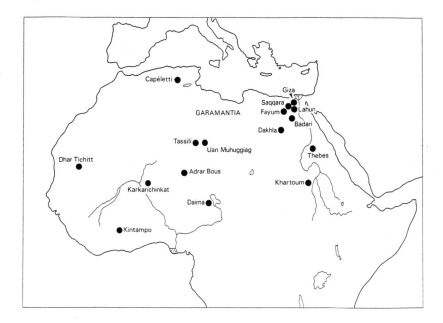

Figure 18.1 Sites and regions in North Africa where either subfossil remains or pictorial representations of ancient domestic cattle have been found.

These cattle can have exactly the same problem with their long down-curved horns from which the tips have often to be removed.

Cattle in the Nile valley

The theoretical view claims that cattle first entered Africa through Egypt. There is, however, no well-dated osteological or pictorial evidence to substantiate this claim as yet. The cattle remains from the predynastic sites of Badari and Fayum, as mentioned previously, could be earlier than 5000 BP but they have not been directly dated, and those from Neolithic sites such as Dakhleh Oasis and the Khartoum sites appear to be from wild *Bos primigenius* (Churcher 1982, Smith 1984). The only radiocarbon date obtained from cattle remains directly (on a skull of a sacrificial ox from Lahun which had black hide and medium length, upstanding horns) is very late, being 3420 ± 80 BP (BM–1420). All other information on domestic cattle comes from the later sites, such as Saqqara, which can be well-dated historically, of course, but by this time the animals were fully domesticated and yield nothing new about their early history.

Many of the paintings of cattle from ancient Egypt are very well known. From Giza there are pictures of spotted and polled cattle (4420–4270 BP) being carried in boats, which shows that by this period they were already highly bred, placid animals, and there are many pictures of cattle from Thebes, dating from around 3411–3070 BP. These are long-horned, lyre-horned, and some polled. But perhaps the most interesting painting is at Thebes, dated to c.3400 BP, showing humped cattle together with a herd of long-horned cattle. This is the earliest known date for humped cattle in Africa.

Discussion

It is widely believed that the ass, the cat, and the Guinea fowl were first domesticated in Africa and that these were the only taxa which this continent contributed to the world's assemblage of domestic animals (see, for example, Shaw 1977, p. 110). The Guinea fowl can be allowed as an original African domesticate, but the cases for the ass and the cat are more dubious as both could equally well have been first domesticated in western Asia and only later in the Nile valley. The same may be true for cattle as there is a definite possibility that these livestock may have been domesticated locally within the northern Saharan region, albeit about 1000 years later than in western Asia. A number of archaeologists have discussed this proposition (Shaw 1977, Smith 1980, Banks 1984) and there does not seem any good reason to gainsay it, just as Meadow (1984) believes that cattle may have been locally domesticated on the eastern margin of the Middle East at Mehrgarh, although here the animals would probably have been humped. There is no evidence to suggest that the earliest domestic cattle in Africa were humped, or indeed that they were long-horned.

Gautier (in Wendorf et al. 1984) has postulated that the remains of cattle from early Neolithic sites in the Bir Kiseiba region of the eastern Sahara could be from domesticated animals derived from wild Bos primigenius in the Nile valley. The bovid elements are, however, very fragmentary and their specific identification and domestic status are questionable, as discussed by Smith (1986). The provisional date for this material is around 9000 BP but, in addition, the specimens should be directly dated before their context can be substantiated.

After looking at the question of when cattle were first domesticated in Africa, either locally or arriving there by diffusion, it is then pertinent to ask why it was necessary to go to the trouble of husbanding livestock, rather than continuing to hunt wild animals for meat. Perhaps the answer lies in the additional resource of milk which could be life-saving in a region that was becoming ever more arid. Shaw (1977) has discussed the movement of cattle-keeping southwards from the Sahara to the Sahel and West Africa over a period of 3000 years, and he has linked this diffusion with the gradual drying of the climate since 4000 BP (see the isochronic diagram in Shaw 1977, p. 108).

More recently, the changing climate of the Sahara and the Sahel throughout the early Holocene, together with the early history of pastoralism in Africa, have been discussed in detail by Banks (1984) and in many of the papers in Clark & Brandt (1984). David (1982, p. 54) has briefly mentioned the roles that different breeds of cattle may play in the changing pattern of pastoralism over time, and it should not be forgotten that there are many sides to this, with the intermingling of sheep, goats, and camels with cattle, and, indeed, the herding of sheep and goats may have occurred earlier than that of cattle in Africa, as it did in western Asia.

Before 8000 BP there was a hunting and fishing economy in the central and southern Sahara that flourished around numerous large lakes inhabited by hippopotamus, crocodiles, and fish. This apparently ideal environment began to dry up shortly after this period, but there was a renewed wet phase around 7000 BP, and it was during this period that the innovation of herding began in Africa. By 6000 BP pastoralism was widespread throughout the Sahara and extended to the Nile valley. Desertification began around 4500 BP, putting pressure on the prehistoric peoples who were forced to follow the river systems that drained southwards.

Shaw (1977) and Smith (1980, 1984) have suggested that the wetter conditions of the Sahara in the early Holocene enabled the tsetse-fly belt to stretch further north into the Sahara, and this formed a barrier beyond which, to the south, cattle could not survive. With the increasing aridity humans, cattle, and tsetse flies moved southwards.

By 3300 BP the pastoralists had moved into the savanna regions, and it was around this time that it seems that humped cattle made their appearance in pictorial representations. They were gradually interbred with the local cattle, or replaced them, and this may have been because humped cattle were better suited to the long-distance migrations that were essential for survival in arid regions. This pattern of transhumance, centred on herds of cattle, camels, sheep, and goats, continued for 5000 years, undoubtedly with recurrent famines but never on the scale of the present time. In an important review by Sinclair & Fryxell (1985) the modern disruption to the carrying capacity of the Sahel has been succinctly put in their abstract:

Migratory pastoralists have traditionally lived with their cattle in balance with the vegetation. This balance was disrupted in the 1950's and 1960's by (i) the settlement of pastoralists around wells, and (ii) the cash crops coincident with increasing human and cattle populations. This has resulted in continuous famine in various parts of the Sahel since 1968. In addition, widespread soil denudation may be causing climatic changes towards aridity. Long-term climatic trends in the past 3000 years point to human interference rather than climatic change as the cause of famine. The evidence suggests that the Sahel problem is a man-made famine caused by overgrazing and not by lack of rain.

If we ignore the ancient, adaptive strategies of the pastoralists, deserts and famines will spread, so that the fragile semi-arid regions of the world will become totally uninhabitable. What better proof could we have of the need to study the past?

References

Banks, K. M. 1984. *Climates, cultures, and cattle: the Holocene archaeology of the eastern Sahara.* Dallas: Southern Methodist University.

Carter, P. L. & C. Flight 1972. A report on the fauna from the sites of Ntereso and Kintampo rock shelter six in Ghana; with evidence for the practice of animal husbandry during the second millennium B.C. *Man* **7**(2), 277–82.

Churcher, C. S. 1982. Dakhleh Oasis project of geology and palaeontology: interim report of the 1981 field season. *Journal of the Society for the Study of Egyptian Antiquities* **12**(3), 103–14.

Clark, J. D. & S. A. Brandt (eds) 1984. *From hunters to farmers: the causes and consequences of food production in Africa.* Los Angeles and London: University of California Press.

Clutton–Brock, J. 1981. *Domesticated animals from early times.* London: British Museum (Natural History) & Heinemann.

David, N. 1982. The BIEA southern Sudan expedition of 1979: interpretation of the archaeological data. In *Culture history in the southern Sudan,* J. Mack & P. Robertshaw (eds), 49–57. Nairobi: Memoir 8 of the British Institute in East Africa.

Epstein, H. & I. L. Mason 1984. Cattle. In *Evolution of domesticated animals,* I. L. Mason (ed.), 6–27. London: Longman.

Gautier, A. 1987. Prehistoric men and cattle in North Africa: a dearth of data and a surfeit of models. In *Arid North Africa. Essays in honor of Fred Wendorf,* A. E. Close (ed.), 163–87. Dallas: Southern Methodist University Press.

Grigson, C. 1985. *Bos indicus* and *Bos namadicus* and the problem of autochthonous domestication in India. In *Recent advances in Indo–Pacific prehistory,* V. N. Misra & P. Bellwood (eds), 425–8. New Delhi: Oxford & IBH Publishing.

Lhote, H. 1959. *The search for the Tassili frescoes.* London: Hutchinson.

Meadow, R. H. 1984. Animal domestication in the Middle East: a view from the eastern margin. In *Animals and archaeology.* Vol 3: *early herders and their flocks,* J. Clutton–Brock & C. Grigson (eds), 309–38. Oxford: BAR International Series 202.

Roubet, C. 1978. Une économie pastorale, pré-agricole en Algérie orientale: le Néolithique de tradition capsienne. *Anthropologie* **82** (4), 583–6.

Shaw, T. 1977. Hunters, gatherers and first farmers in West Africa. In *Hunters, gatherers and first farmers beyond Europe,* J. V. S. Megaw (ed.), 69–126. Leicester: Leicester University Press.

Sinclair, A. R. E. & J. M. Fryxell 1985. The Sahel of Africa: ecology of a disaster. *Canadian Journal of Zoology* **63**, 987–94.

Smith, A. B. 1980. Domesticated cattle in the Sahara and their introduction into West Africa. In *The Sahara and the Nile,* M. A. J. Williams & H. Faure (eds), 489–503. Rotterdam: A. A. Balkema.

Smith, A. B. 1984. Origins of the Neolithic in the Sahara. In *From hunters to farmers: the causes and consequences of food production in Africa,* 84–92. Los Angeles & London: University of California Press.

Smith, A. B. 1986. Review article: cattle domestication in North Africa. *The African Archaeological Review* **4**, 197–203.

Wendorf, F., R. Schild & A. E. Close (eds) 1984. *Cattle-keepers of the eastern Sahara: the Neolithic of bir Kiseiba.* Dallas: Southern Methodist University Press.

19 *The development of pastoralism in East Africa*

PETER ROBERTSHAW

Mparan amu iyata suami (You have friends because you have animals).
Samburu proverb

Little is known about the prehistory of the pastoral peoples of East Africa and the development of the pastoral mode of production in the region. This sad state of affairs is rendered more depressing by the lack of historical awareness among pastoral development planners. So often the research of these planners involves 'reconstructions' of 'traditional' systems of livestock management practices, which amount, in effect, to the projection of present systems of animal husbandry on to what can be gleaned from ethnographers' accounts of early patterns of land tenure and mobility. Similarly, with the emotive issue of whether pastoralists cause desertification, the archaeological and palaeoenvironmental evidence that might answer the question is very rarely invoked.

The first half of this chapter aims to dispel some of this gloom by outlining what is known about the introduction and development of pastoralism in East Africa. The second half examines in some detail the implications of several archaeological faunal assemblages that seem to document hunting-cum-pastoral subsistence. This combination of subsistence pursuits is extremely rare in East African ethnography and, furthermore, would appear to involve an internal contradiction within the social relations of production between the principles of collective access to hunted resources and of divided access to domestic livestock. Several ways are explored in which this contradiction may be resolved.

Setting aside extravagant claims for the presence of domestic animals in eastern Africa at an exceptionally early date, the archaeological evidence indicates that pastoralism was first practised in East Africa late in the 3rd millennium BC. The sites that document this are located in the Lake Turkana basin of northern Kenya (Barthelme 1985). It seems to have taken approximately 1000 years more for domestic animals to reach the Rift Valley and Highlands of central and southern Kenya, and Tanzania (Ambrose 1984). However, ceramic studies (Collett & Robertshaw 1983) suggest that further fieldwork may lead to a narrowing of this chronological gap. Cattle, sheep and goats, and probably donkeys, were the animals involved. While donkeys, and perhaps cattle, were domesticated in North Africa, sheep and goats are of Near Eastern origin; thus their introduction to East Africa must be seen as the result of a process of diffusion or human population movements. Most authorities infer relatively small-scale immigrations of pastoralists from the north, perhaps Cushitic-speakers from Ethiopia (Ambrose 1984). Clarification of the subsequent development of the early pastoral communities of East Africa is rendered obscure by a complex culture-history, which to the struggling archaeologist appears at times to verge on the chaotic (Bower *et al.* 1977, Collett & Robertshaw 1983). Several further immigrations into East Africa by pastoral/agricultural groups are indicated from historical linguistic evidence (Ehret 1971, 1974) and receive support, though not unequivocal, from archaeological data (Ambrose 1982). These pastoralists were stone-tool, rather than metal-users, and obsidian was the preferred raw material, being found in substantial quantities at sites up to 100 km or more from the major sources in the central Rift Valley (Merrick & Brown 1984).

While we know, obviously, that the early pastoral peoples of East Africa kept livestock, we have as yet no direct archaeological evidence for grain cultivation. However, 'indirect evidence' (Shaw 1976) for the growing of cereal crops has been obtained from studies of site locations, ground stone tools, and historical linguistics. Thus, from the perspective of diet these peoples were mixed farmers. However, consideration of regional settlement patterns suggests that pastoralism rather than cultivation was the preferred mode of subsistence, for areas such as the eastern Highlands of Kenya, possessing rich, fertile soils, do not seem to have been settled; instead people inhabited the savanna grasslands which, on the whole, are less suitable for cereal cultivation, due to thinner soils and less rainfall. Thus, it seems likely that domestic livestock were of prime importance to the early food-producing communities of East Africa, not necessarily in the sense that these animals provided the major portion of the diet, but rather in the sense that there was a cultural preoccupation with livestock to the extent, perhaps, that people who lost their stock would have chosen to remain in the grasslands, rather than opt for an agricultural way of life in the adjacent highlands (Robertshaw & Collett 1983). This preoccupation – some might call it infatuation – with livestock is prevalent among several modern East African groups, the classic example being the Nuer: the Nuer social idiom is said to be a bovine idiom, yet much of their diet consists of millet, the cultivation of which is regarded as an unfortunate necessity (Evans–Pritchard 1940).

The cultivation of cereal crops may have been an 'unfortunate necessity' too for the pastoralists of the 1st millennium BC, as there were no agricultural peoples from whom they could obtain grain by barter. Survival entirely from the produce of their herds is a most unlikely hypothesis, since the ethnographic data make it quite clear that no pastoral society could survive indefinitely without access to agricultural produce (Monod 1975, p. 134). The need, then, for the early pastoralists to grow crops, as well as tend their herds, facilitates the construction of a model of settlement patterns. This model, for which some support can be derived from existing archaeological data, has been set out in detail elsewhere (Robertshaw & Collett 1983) and need not be repeated here.

Around AD 200 early Iron-age farmers settled the then sparsely populated highlands adjacent to the savanna grasslands. Thus, from this time on pastoral peoples would have been freed from the necessity of cultivating crops themselves; presumably they could now exchange small stock for grain with their agricultural neighbours, as has been documented, for example, for Maasai–Kikuyu relations at the beginning of this century (Waller 1976, Berntsen 1979). Given the cultural emphasis placed upon livestock by the pastoralists, we can conjecture that an economic shift took place around the mid-1st millennium AD, from mixed farming with some hunting to specialized herd management with grain obtained through trade (Robertshaw & Collett 1983).

The model outlined above is complicated by historical linguistic evidence of a migration of eastern Nilotic-speakers into the savanna regions in the 1st millennium AD, who displaced or absorbed the previous inhabitants. An archaeological correlate for this event has not been established with any certainty (Robertshaw 1984 *contra* Ambrose 1982). Even if population immigration in this period is eventually proven, it must be emphasized that it is the coming of Iron-age farmers to neighbouring regions, not the identity of the pastoralists, which is crucial to the development of specialized herd management strategies in the savanna.

The role of hunting among the early pastoral communities of East Africa will now be considered. The great majority of the faunal assemblages from archaeological sites of this period are dominated by bones of domestic livestock with very few, if any, remains of wild species (Gifford–Gonzalez & Kimengich 1984). This pattern holds true even for areas which support large numbers of wild ungulates. However, we now know of two, possibly more, exceptions where sites have yielded significant quantities of both domestic stock and wild animals, primarily medium to large ungulates. These sites are Prolonged Drift, located in the Central Rift Valley and dated to the mid-1st millennium BC (Gifford *et al.* 1980), and the Elmenteitan levels at Gogo Falls, situated about 20 km east of Lake Victoria and dated to about the 1st century AD (Robertshaw 1985). Both are extensive open-air occurrences (Fig. 19.1).

Ingold (1980) has argued that the change from hunting to pastoralism is not simply a shift in ecological relations from herd pursuit to herd management (*contra* Foley 1982), but it involves a major transformation in the social relations of production and the ideology of prestige. Pastoralism is based on the principle of divided access to living animal resources, i.e. the ownership of animals, while hunting rests on the principle of collective access where an animal becomes someone's property only with its death.[1] A pastoralist gains prestige by accumulating wealth on the hoof, but for the hunter it is the demonstration of his skill and his generosity in distributing the meat from his kills which win him respect. Thus, archaeological sites which document a combination of pastoralism and hunting appear from this theoretical perspective as surprising phenomena where the implied contradictions in the social relations of production and prestige require explanation. This impression is reinforced by a perusal of

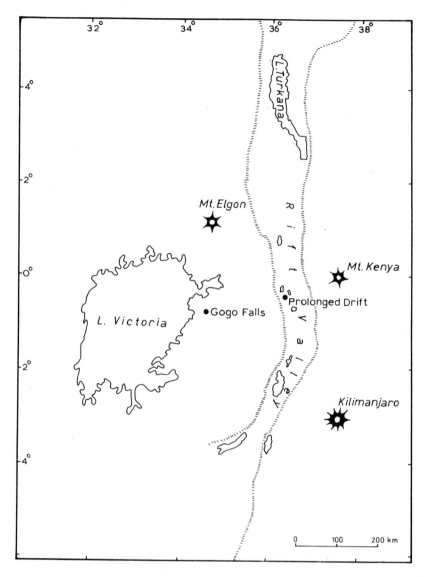

Figure 19.1 Location of sites yielding bones of both domestic stock and wild animals.

the ethnographic literature on the pastoral peoples of East Africa, in which hunting is, as a rule, very rarely mentioned. Could it be that the archaeological sites document not only a system of subsistence no longer extant in East Africa (Gifford–Gonzalez 1984) but also, perhaps, different cultural attitudes towards animals? To answer these questions we need to look more closely at the ethnographic evidence.

Many ethnographies of pastoral peoples either completely fail to mention hunting or remark simply that it is a rare activity, perhaps resorted to more often in times of severe drought (e.g. Carr 1977, p. 200). For some groups, wild animals are only of interest in their relationship to domestic livestock: for the Nandi 'wild animals are of two kinds: those that injure livestock and those that do not' (Huntingford 1953, p. 127). Not only is hunting a rare activity for many pastoralists, it is also regarded with some contempt: the Nuer insist that only the absence of cattle makes a man engage in hunting other than casually (Evans–Pritchard 1940, p. 73). Among the Turkana the word 'hunter' (*egulokit*) is almost synonymous with that for a poor man (*erkeboton*). 'Gazelle are the herds of the poor man' (Gulliver 1963, p. 33). Similarly, 'since Borana Gutu equate wealth with livestock hunters are by definition poor people' (Dahl 1979, p. 178). Furthermore, for Borana at least, among whom reside an endogamous 'caste' of Wata hunters, these people are not only poor but they are also 'unworthy' since they lack any concern for the capital expansion of their resource base (Dahl 1979).

Only one or two East African pastoral peoples appear to hunt regularly under normal climatic and environmental conditions. These are the Murle and possibly the Datoga (Barabaig), but no figures are available for the amount of labour invested in hunting, or the yields achieved. Murle hunting involves the interception of migrating animals at river crossings; it is interesting to note that elaborate rules exist among the Murle for the distribution of all meat, both game killed during the hunt and cattle or sheep slaughtered at ceremonies (Lewis 1972). Among the Datoga a distinction is made between the 'common pastime' of hunting with bow and arrow and the pursuit with a spear of dangerous prey, such as lions and Maasai, in order to demonstrate bravery (Umesao 1969).

Although regular hunting by pastoral peoples seems a rare phenomenon, we should be aware of the possibility that the pastoralists' contempt of hunting may sometimes have rubbed off onto the ethnographer; for example Spencer (1965) in his monograph on the Samburu makes no mention of hunting, yet recent fieldwork has revealed that not only do Samburu warriors regularly hunt to feed themselves at outlying grazing camps, but also there are Samburu resident in mountainous areas who might best be described as herder-foragers (Louise Sperling pers. comm.).

If we overlook the suspicion that some ethnographers may have under-reported the occurrence of hunting, it seems fair to generalize, on the basis of the ethnographic literature on East African pastoralists, that hunting is an activity of 'poor' people – either those whose herds have been seriously depleted by drought or disease, or 'castes' of hunters such as the Wata, who live in symbiotic fashion among pastoralists. Thus, we may speculate that the archaeological sites referred to earlier which contain considerable numbers of bones of both domestic stock and wild ungulates may be the settlements of 'poor' pastoralists attempting to re-establish their herds after some sort of calamity (cf. Robertshaw & Collett 1983, p. 74). This, and the alternative suggestion of the existence of 'castes' of hunters among the prehistoric pastoralists, may be tested by further archaeological research, in which, perhaps, the identification of particular items of material culture as being symbols of wealth and prestige will be an important goal. The possibility that the settlements are those of hunters occasionally bartering or raiding livestock from neighbouring pastoralists may be discounted, at least for Gogo Falls, because the faunal remains at this site accumulated within a very large mound of prehistoric animal dung. The diversity of the faunal assemblage accruing from hunting may indicate whether the hunters were 'poor' pastoralists or 'true' hunters, for pastoralists, as a general rule, will eat only those wild species that can be culturally linked to domestic animals, e.g. eland may be eaten because they are 'like' cows but animals with claws and fur may not be eaten. For the moment at least, it would seem sensible to set aside the possibility that prehistoric subsistence systems were so removed from anything known ethnographically that they might even have embodied different cultural attitudes towards animals.

Finally, we may note that accounts exist which demonstrate how a 'worthy' man may re-establish himself as an independent herd-owner through hunting:

> Ware [a Borana Gutu] had lost all his stock through misfortune, and his clansmen decided to give him ten cattle. Ware was grateful but decided to return all the cows and keep only one ox. This ox he bartered for two doublesized spears (*bode*). Then he went out to hunt with his spears and killed two big elephants. With the profit from the sale of two sets of tusks he was able to invest in a herd of 60 cattle . . . Ware thus re-established himself as a herdowner, but he continued to kill game and to increase his herds until he had hundreds of cattle. (Dahl 1979, p. 179)

Although ivory may not perhaps have been such a precious commodity in the prehistoric era, the exchange of meat, skins, and other products of the chase for domestic livestock may have made it feasible for a herd-owner to replenish his stock, and in so doing create a bone midden akin to those of Prolonged Drift and Gogo Falls. Thus, these faunal assemblages, containing bones of both wild and domestic animals, should be seen as the refuse heaps of pastoralists for whom hunting was the means to a pastoral end.

Acknowledgements

I thank Cory Kratz, Louise Sperling, and Donna Klump for sharing information and ideas.

Note

1 However, pastoralists do have collective access to grazing, while hunters may have divided access to particular resources, e.g. honey among the Okiek (Kratz pers. comm.).

References

Ambrose, S. H. 1982. Archaeological and linguistic reconstructions of history in East Africa. In *The archaeological and linguistic reconstruction of African history*, C. Ehret & M. Posnansky (eds), 104–57. Berkeley: University of California Press.

Ambrose, S. H. 1984. The introduction of pastoral adaptations to the highlands of East Africa. In *From hunters to farmers: the causes and consequences of food production in Africa*, J. D. Clark & S. A. Brandt (eds), 212–39. Berkeley: University of California Press.

Barthelme, J. W. 1985. *Fisher-hunters and Neolithic pastoralists in East Turkana, Kenya*. Oxford: BAR International Series 254.

Berntsen, J. L. 1979. Economic variations among Maa-speaking peoples. *Hadith* 7, 108–27.

Bower, J. R. F., C. M. Nelson, A. F. Waibel & S. Wandibba 1977. The University of Massachusetts' Later Stone Age/Pastoral 'Neolithic' comparative study in Central Kenya: an overview. *Azania* 12, 119–46.

Carr, C. J. 1977. *Pastoralism in crisis. The Dasanetch and their Ethiopian lands.* University of Chicago, Department of Geography, Research Paper no. 180.

Collett, D. P. & P. T. Robertshaw 1983. Pottery traditions of early pastoral communities in Kenya. *Azania* 18, 107–25.

Dahl, G. 1979. *Suffering grass: subsistence and society of Waso Borana*. Stockholm: Stockholm Studies in Social Anthropology.

Ehret, C. 1971. *Southern Nilotic history*. Evanston: Northwestern University Press.

Ehret, C. 1974. *Ethiopians and East Africans; the problem of contacts*. Nairobi: East African Publishing House.

Evans–Pritchard, E. E. 1940. *The Nuer*. Oxford: Clarendon Press.

Foley, R. A. 1982. A reconsideration of the role of predation on large mammals in tropical hunter-gatherer adaptation. *Man* (NS) 17, 393–402.

Gifford, D. P., G. L. Isaac & C. M. Nelson 1980. Evidence for predation and pastoralism at Prolonged Drift: a Pastoral Neolithic site in Kenya. *Azania* 15, 57–108.

Gifford–Gonzalez, D. P. 1984. Implications of a faunal assemblage from a Pastoral Neolithic site in Kenya: findings and a perspective on research. In *From hunters to farmers: the causes and consequences of food production in Africa*, J. D. Clark & S. A. Brandt (eds), 240–51. Berkeley: University of California Press.

Gifford–Gonzalez, D. P. & J. Kimengich 1984. Faunal evidence for early stock-keeping in the Central Rift of Kenya: preliminary findings. In *Origin and early development of food-producing cultures in north-eastern Africa*, L. Krzyzaniak & M. Kobusiewicz (eds), 457–71. Poznan: Polish Academy of Sciences.

Gulliver, P. H. 1963. *A preliminary survey of the Turkana*. University of Cape Town, School of African Studies, New Series no. 26.

Huntingford, G. W. B. 1953. *The Nandi of Kenya: tribal control in a pastoral society*. London: Routledge & Kegan Paul.

Ingold, T. 1980. *Hunters, pastoralists and ranchers*. Cambridge: Cambridge University Press.

Lewis, B. A. 1972. *The Murle: red chiefs and black commoners*. Oxford: Clarendon Press.

Merrick, H. V. & F. H. Brown 1984. Obsidian sources and patterns of source utilization in Kenya and northern Tanzania: some initial findings. *The African Archaeological Review* **2**, 129–52.

Monod, T. 1975. Introduction. In *Pastoralism in tropical Africa*, T. Monod (ed.), 1–183. London: International African Institute.

Robertshaw, P. 1984. *The prehistory of pastoralism in Kenya*. Unpublished seminar paper, School of Oriental and African Studies, University of London.

Robertshaw, P. 1985. Preliminary report on excavations at Gogo Falls, South Nyanza. *Nyame Akuma* **26**, 25–6.

Robertshaw, P. T. & D. P. Collett 1983. The identification of pastoral peoples in the archaeological record: an example from East Africa. *World Archaeology* **15**, 67–78.

Shaw, T. 1976. Early crops in Africa: a review of the evidence. In *Origins of African plant domestication*, J. R. Harlan, J. M. J. de Wet & A. B. Stemler (eds), 107–53. The Hague: Mouton.

Spencer, P. 1965. *The Samburu*. London: Routledge & Kegan Paul.

Umesao, T. 1969. Hunting culture of the pastoral Datoga. *Kyoto University African Studies* **3**, 77–92.

Waller, R. 1976. The Maasai and the British 1895–1905. *Journal of African History* **17**, 529–53.

20 *Cattle and cognition: aspects of Maasai practical reasoning*

JOHN G. GALATY

Introduction

In anthropology, the question of pastoralist rationality and irrationality has long served as an apt case with which to address the more general issue of how ideas and belief systems influence economic life. Herskovits' (1926) notion of a 'cattle culture' has been associated with the point of view that pastoralism was directed and informed by essentially religious rather than economic precepts, and thus was essentially 'irrational' from the economic point of view. An abundance of more recent research on the economic lives of pastoralists has made quite clear the forms of rationality underlying herding systems (Evans–Pritchard 1940), among more specialized pastoralists for milk-based subsistence, among many agro-pastoral communities, both for food and for other ends such as traction, manure, and to provide stores of value (Galaty in press). However, economic studies have tended to infer forms of rationality from economic practice, rather than from systems of thought and the assumptions, beliefs, and aims which underlie pastoralist discourse about their productive activities (Galaty 1984). In this respect, 'rationality' implies the logic rather than the 'practical reasoning' (Sahlins 1976) underlying pastoralism.

There are, however, cognitive concomitants of herding systems, apparent in the quasi-formal cultural classification of domestic animals, of the arid environment, and of forms of labour, and also in the more informal reasoning implicated in pastoral experience as a dynamic process. These cognitive concomitants of pastoralism do not simply replicate in mind the objective properties of livestock, the rangelands and the labour process, but represent a cultural selection and ordering of the terms by which pastoralism is carried out. Such cognitive systems maintain a certain cultural autonomy *vis-à-vis* the pastoral process *per se*, the former being both internally structured and the subject of reflection and discourse. However, these cognitive capacities, which are in part realized both in the form of discrete symbolic systems and, somewhat less coherently, in the form of implicit knowledge, do underpin pastoral practice, which in turn provides the context within which these systems are acquired (Cole & Scribner 1974). To what extent do young pastoralists, removed – most often by schooling – from the routine practice of herding, attain the cognitive skills and the command of knowledge normally associated with pastoralism? If they do not, this may imply either that the cognitive concomitants of pastoralism are embedded in concrete practical experience, without which this competence is not acquired, or that such individuals are also likely to be divorced from the more general settings in which they might gain exposure to more autonomous systems of knowledge (Scribner 1984, Street 1984).

With such questions in mind, this chapter examines certain aspects of the practical reasoning underpinning Maasai pastoralism: the logic underlying extensive semi-nomadic specialized animal husbandry, the informal reasoning associated with the pastoral experience, the more formal symbolic classification systems through which cognitive apprehension of the pastoral field of activity takes place, and the processes of identifying and remembering individual animals. By concentrating on 'practical' reasoning, I do not mean to diminish the significance of more abstract aspects of 'cultural' reasoning (Sahlins 1976), such as the religious assumptions made about the relation between animals and people, or the social values attributed to livestock, in exchange or bridewealth. The case can be made that Maasai pastoral ideology reflects certain higher order symbolic notions about the social, economic, and religious value of cattle. This essay focuses, however, on what might be construed as the 'first-order' level of pastoral rationality, the cognitive apprehension, perception, and classification of domestic animals, the point at which mental and material categories join and culture constitutes practical life. This domain of primary cognition is especially pertinent at a time when what Braudel once called 'material life' (1973), the daily routines of local and domestic economy, is being transformed by the spread of commercial production oriented to demand far from the rangelands and the intimate tending of herds, and when schooling is progressively altering the nature of the cognitive experiences of young Maasai, thus transforming the basis of knowledge on which pastoral practice has for so long rested.

The logic and experience of pastoralism

The logic of specialized pastoralism, as practised in several semi-arid land communities in eastern Africa, can be distinguished from that of ranching in several respects pertinent to the cognitive concomitants of animal husbandry. The most obvious difference is that ranching is a commercial activity, in which forms of instrumental rationality enter into production decisions; thus primary production and the market are closely linked through judgements and assessments of commercial value. While domestic-level animal husbandry may also involve commercial sale, livestock may not be produced as commodities *per se*, but as objects of multiple use and personal ties. Thus, market calculation is only one, and an imperfectly developed, mode of valuation of livestock. In contrast to ranching, pastoralism involves relatively high labour intensity, which in less technical terms implies that there are relatively more people active in and dependent on the pastoral process than is true in ranching. This factor of labour also makes possible the development of personalized ties between people and animals. Such ties may lie at the heart of the complex symbolic and emotional aspects of pastoralist ideas and attitudes regarding animals, but also have implications for production, since ease of herd management, animal handling, and efficient production of milk depends on a sort of 'taming' process enhanced by close personal ties and physical contact between people and their domestic animals (Ingold 1980).

Cognitive processes, such as classification, description, pattern recognition, and problem solving, associated with the use of cattle and other livestock are not simply individual responses to concrete situations of environmental use and animal production. Such cognitive operations reflect capabilities which are institutionally shaped, both by the patterns of activity proper to pastoralism and by the cultural strategies and formal classificatory systems collectively encoded for carrying out pastoral tasks. This is an important observation, since 'psychological' perspectives are often theoretically distinguished from the 'cultural' as individual to collective forms of behaviour. And, as previously suggested, 'cognition' can be assessed both in terms of the cognitive practices of individuals and the collective representations or institutionalized forms of symbolic classification of a group. But, however spontaneous or individually generative cognitive activity may be, it reflects the content of social experience and learned principles of cognitive organization. Pastoralists reveal an impressive ability to recognize and identify their animals, a skill which led early observers to attribute to them phenomenal powers of memory. Contributing to this ability to recognize hundreds of individual animals is the development over time of personalized knowledge of a herd.

In one interview, it was observed that:

A man gets his cattle when they are few. So he actually comes to know all the calves that are born into the herd, until they become many. By the time he has a very large herd, he also knows them by their colours and their cattle 'houses'. So you find that the man is in no sense a stranger to his own herd.

But given the gradual acquaintance of herders with members of the herd, there are those who are known to be skilled in recognition and identification, the *Abarani*, a person with a very good memory, who can look at a cow very quickly and know which one it is, observing colours and idiosyncratic features. Describing the able herdsboy, it was suggested that:

One can bring such a young boy to look after a large herd and the boy can master the cattle very quickly after very few instructions.

The able *Abarani* is contrasted to the inept *Mankalioni*, who cannot 'master' the herd, who cannot tell the difference between two similar cows.

Young pastoralists who attend school are perforce removed from the setting in which the cognitive capacities of herding are learned, both through experience and tutorial. Drawing schoolchildren outside the context of direct learning and indirect experience of the domain of animals and their husbandry inevitably influences both the content and organization of cognitive reasoning. Yet, other skills, such as literacy and arithmetical calculation, learned through school may also be pertinent to animal husbandry (Goody 1977). This may be especially true when such skills are combined with the transformation of needs, values, and social and economic orientations, often associated in rural schools with the new social and cultural experience which perforce accompanies the pedagogical experience. Taste for a diversity of foods, store-bought commodities, access to an economy of money, and class-inflicted identities correlated with educational status, indirectly shape new attitudes towards animals and their production, which are seen more in terms of impersonal commerce than personal and domestic livelihood. But such a shift also implies an increasing lack of experience in or declining motivation regarding pastoralism, which can result in a decline in cognitive capacities, including the content and organization of the skills and knowledge, entailed by careful, personal, and productive husbandry. For many educated pastoralists, the shift in orientation is away from pastoralism altogether, but for many others the shift is in the logic of animal husbandry from pastoralism towards commercial ranching (Galaty 1981).

It is clear that intimate and extensive experience with livestock inculcates the capacity to identify and remember animals in *general*, but that personalized acquaintance with the individuals of a specific herd leads to knowledge and recognition of animals in *particular*. Such perception of animals is in part holistic, depending on the recognition of *patterns* of visual and behavioural features which represent the uniqueness of each individual. The personal recognition and identification of up to 1000 individual animals may be remarkable, but when understood as resulting from years of intimate contact with a herd, and with each individual animal from birth, no more remarkable than the common capacity to recognize up to 1000 human faces of people known over a span of years. However, the holistic nature of personalized pattern recognition, so important to pastoralism, is in practice underpinned by analytic processes, in part subject to conscious reflection and deliberation. These conscious and unconscious processes involve the perceptual encoding of features, not simply based on visual form but drawn from more formal cultural classification systems. These classification systems, representing a multi-levelled formal structure of cognition based on language, provide herders with the means to comprehend and master the social and natural worlds of pastoralism by ordering the complex, intimate, and nuanced experience of domestic animals, productive resources, and labour in terms of a much smaller number of symbolic attributes and features.

Cattle and classification

In addition to patterns of identification built up through experience, livestock identification also occurs through systems of symbolic classification of a cultural order, in particular through earmarks and brands which signify the owners' family and descent group, livestock names, and descriptives. In the case of names, the Maasai endow all cattle in a particular 'lineage' with a single appellation, proper not to the apical bull but to the bovine 'matriarch'. Table 20.1 is a list of conventional cattle names, representing some of the more commonly used. As can be seen, names tend to be representational rather than arbitrary, being slight deformations or transformations of other terms, with appellative prefixes denoting the gender of the animal. For example, *Nkeyi* is a slight phonological variant of *Keri*, meaning 'speckled' or 'splotched', with a feminine gender prefix (*na*). The masculine form would be *Lukeyi*. The names thus serve not only as indices of particular cattle lineages or individuals, but also signify in a representational manner, 'connoting' as well as 'denoting'. There tend to be five distinct modes of signification in cattle naming, three utilizing symbols of transactions, two forms of visual description.

In the Transaction Mode proper, names depict the type of transaction by which the apical animal was acquired: the return of a debt, an exchange, for bloodwealth, bridewealth, as a 'pure' gift to a friend, from a raid, etc. In the Donor Mode, the name signifies the giver of the apical animal. In the Reciprocal Mode, the name signifies the reciprocal presentation for which the apical cow was exchanged: 'reminders' of oxen, donkeys, or other animals exchanged, the 'eye' for which a gift was compensation, the 'girls' given in marriage, or the 'meat' of the stolen animal, for which the return was compensation. There are two Descriptive Modes, which use 'descriptives' by colour, form, or status to signify either the animal given in exchange (Reciprocal Description Mode), or the animal received (Reflexive Description Mode). It is especially interesting that only one of the naming modes actually signifies the apical animal received and owned, while all the others signify the transaction itself, the animals given in exchange, and thus no longer in possession, or the exchange partner. Cattle names, then, are less about bovine possessions than about social ties and transactions. In most cases, names which denote animals actually connote – or 'signify', in the form of the images or descriptions which constitute names – not the animal but the social and economic transactions by which it, or its apical ancestor, was acquired.

In contrast, 'descriptives', being based on attributes, by necessity connote, but can also be used to denote. In this sense, descriptives serve many of the functions of names, and, indeed, constitute names in the Descriptive Modes noted above. In such cases, names presumably began as descriptives, and thus describe as well as denote. But Maasai livestock descriptives refer to individuals in a way conventional names cannot, and

Table 20.1 Maasai cattle names.

Name	Meaning	Comment	Signifies
Kusaka	debt returned	payment, overdue debt	TRANS
Nawolli	exchanged one	received in exchange (*a-wol*)	TRANS
Meretoi	helper	gained after a struggle	TRANS
Kirror	bloodwealth	from killer of kinsman	TRANS
Torosi	bloodwealth	from killer of kinsman	TRANS
Kiripa	for care (*a-ilip*)	given for care	TRANS
Reyio	by raid	stolen in raid	TRANS
Noosilan	bridewealth	bridewealth payment	TRANS
Kerreti	a blessing	given to family member	TRANS
Entotua	in friendship	gift from dear friend	TRANS
Nyorra	our love	from beloved friend	DONOR
Nemwala	of Wala	received from Wala	DONOR
Nenkong'u	of the eye	payment by bride's father	RECIP
Noo-Ntoyie	for girls	bridewealth	RECIP
Nor-Mong'i	reminder of an ox	exchanged for ox	RECIP
Nyamu	for meat	for animal stolen and eaten	RECIP

Mon'go Olkutai	reminder of an ox	exchanged for ox	RECIP
Noosirkon	for donkey	exchanged for donkey	RECIP
Noonkeya	for children	received for children	RECIP
Mong'o kinyi	reminder of a small ox	exchanged for a small ox	RECIP
Mong'o Sampu	for striped ox	exchanged for striped ox	DE/RC
Noldere	for dusty (*enderit*) one	exchanged for whitish animal	DE/RC
Ngiyo	of greyish-brown one	exchanged for greyish-brown	DE/RC
Nado Yukunya	of red-headed one	exchanged for red-headed cow	DE/RC
Nayasha	of white-patched one	exchanged for *N-arasha* coloured	DE/RC
Leyaai	yellowish (*Olerai*)	named after *Acacia seyal*	DE/RF
Nkeyi	speckled (*Keri*)	white variegation	DE/RF
Kerenket	in a hole	born in valley or hole	DE/RF
Nawoyoyi	suckler	suckled when mother dies	DE/RF
Kutiti nki	with small teats	describes the animal	DE/RF
Sampu	brown-striped	describes cow	DE/RF
Kurtolet	slightly-blind	describes cow	DE/RF
Nenkong'u nabo	one-eyed	describes cow	DE/RF

TRANS, signifies the transaction; DONOR, signifies the giver of the animal; RECIP, signifies the reciprocal presentation exchanged for the animal; DE/RC, description of the reciprocal animal; DE/RF, reflexive description of the animal itself.

Table 20.2 Maasai classification of livestock by status.

	Males	Females
cattle:		
calf	*olashe*	*enkashe*
immature	*olbung'ai*	*entawuo*
mature	*oloing'oni*	*enkiteng'*
castrate	*olkiteng'*	
goats:		
kid	*olkuo*	*enkuo*
immature	*ositima*	*esupen*
mature	*oloror*	*enkine*
castrate	*olkine*	
sheep:		
lamb	*olkuo*	*enkuo*
immature	*ositima*	*esupen*
mature (ram/ewe)	*olmeregesh*	*enker*
castrate	*olker*	

most cattle names do not, by evoking the distinctive features of specific animals rather than serving simply as demonstratives. While names are used to *call* cows, especially to their calves, descriptives are used to *refer* to them, for the benefit of those who do not share personalized acquaintance with the herd (Galaty 1982).

Cattle descriptives tend to utilize five distinct modes of attribution: status, colour, pattern, horns, and distinctive characteristics. In the status mode (see Table 20.2), an animal is denoted by its sex (male/female), maturity (calf/immature/mature), and reproductive status (castrate/whole).

It is worth noting that the principles underlying lexemes in the status mode make discriminations relevant to animal reproduction, between offspring, pre-productive yet whole animals, productive animals, and castrates. Yet, lexical symmetry occurs not between the productive males and females but between castrates and productive females, with productive bulls, he-goats, and rams representing the 'marked' category. In fact, very few productive males are kept for breeding purposes, both as a strategy of selection and as a mechanism of herd control, since whole males are notoriously ill-behaved and disruptive of the herd. The quintessential animals, which constitute the bulk of the herd, are thus lexically paired: castrates and productive females. Symbolically, the ideal male pastoralist analogue is not the bull but the ox, which exemplifies virtues of size, strength, beauty, and sociability, as opposed to the wiry, scrawny, and ill-behaved bull (Beidelman 1966). The lexical parallels between goats and sheep reflect as well the joint management of the two species in a single herd (*in-tare*), in contrast to the separate and more prominent cattle herd (*em-boo*).

The second and third principles are not really distinguished by Maasai, and should be grouped together as a single principle of 'colour-pattern' or

Table 20.3 Maasai elementary colour classifiers.

Narok	'black or bluish'
Naibor	'white'
Sirua	'tannish' (i.e. 'colour of eland')
Sero	'dark brown' (i.e. 'wilderness')
Elerai	'cream-like or yellowish' (*Acacia seyal*)
Barrikoi	'reddish'
Nanyokie	'brownish red'
Nado	'red'

Table 20.4 Maasai elementary pattern classifiers.

Mugie	'dark, blackish variegation' (splotched)
Keri	'white streaked or lined' (cf. *Oldoinyo Keri*, Mount Kenya)
Tara	'brown spots on the body'
Sintet	'spotted' (black)
Arus	'spotted' (black and white on stomach)
Owuaru	'many brown splotches' (lit. 'beast-like')
Sampu	'striped' (brown on the body)
Olmotonyi	'white with a brownish head' (lit. 'vulture-like')
Dere	'whitish with black around the neck' (lit. 'dusty')
Ngencherii	'brownish and reddish'
Narasha	'black, red, or brown with white sides'
Nairimo	'multiple small spots'
Otialei	'white eye'

'visual form'. It is only to heighten the distinctiveness of the pattern mode that I treat it here separately from the colour mode (Table 20.3). Patterns are more complex, being evoked by both single lexemes and phrases, referring to the distribution of bodily colours (Table 20.4). Patterns may also be evoked by compound phrases, composed of nouns and qualifiers, which signify colour combinations or colours localized on the animal's body (Table 20.5).

In the horn mode, the presence or absence of horns, their shape, and their combination with other features provide principles for the genesis of descriptives (Table 20.6). Other distinctive features regarding physical state can be used to describe animals, apart from colour, pattern, and horns (Table 20.7).

In actual discourse, elements derived from the five principles are combined to form descriptives, as in *Enkiteng' Odo Lukunya*, 'the cow with a red head'. Although colour, pattern, horns, and other characteristics are often mutually exclusive, they can be combined with the basic status of the animal in an additive way, such as in *Enkiteng' Narok Oomswuarak*, 'black cow with long horns', and in cases of ambiguity a further feature can always be added to distinguish the animal in question from similar ones. Thus, it is possible to identify cattle not only through personal knowledge and individual names, but also through descriptives by which they can be referred to in such a way as to render each animal in a herd or

Table 20.5 Maasai compound colour-pattern classifiers.

Narok Sopia	'black with some white hair'
Narok Lukunya	'black head, white body'
Narok Abori	'black stomach'
Naibor Oshoke	'black with a white stomach' (black animal)
Nado Lukunya	'red head, white body'
Sampu Sintet	'brownish streaks and black spots'
Sampu Mugie	'blackish in colour with brownish streaks'
Sampu Kumpau	'with reddish stripes' (tobacco colour)
Sampu Lerai	'cream colour with brown streaks'
Narasha Sampu	'white sides with brownish stripes'
Narasha Sintet	'white sides and black spots'
Keri Nasiantet	'white streak with small black spots'
Keri Ng'iro	'brown with a white streak on the back'
Keri Barrikoi	'reddish with white on the back'
Keri Abori	'white stomach' (white-streaked below)
Keri Enyokie	'red body with white line across the body'
Enkeri Sarioni	'short, incomplete stripe on back'
Keri Ai	'long white streak on body with bluish spots'
Keri Sampu	'brownish stripes with white line'
Ol tara Mugie	'blackish with large brown spots'
Owuaru Leleshwa	'whitish with brown splotches'
Nairimo Enyokie	'with many red spots on the body'
Nairimo Erok	'with many black spots on the body'

Table 20.6 Maasai horn configuration classifiers.

Elementary terms	
Ruma	'without horns'
Arro	'horns facing down'
Ng'elesh	'one horn up and one down'
Compound terms	
Arro Oirrag Imowuarak	'horns facing down and close to head'
Ng'elesh Napir	'one horn up, the other down, and fat'
Narok Ruma	'black with no horns'
Narok Oomswuarak	'black with long horns'
Mugie o Mowuarak	'dark brown with horns'
Ong'ow Imowuarak Siadi	'horns facing the back'
Mugie o Mowuarak	'dark brownish with horns'
Imowuarak Siadi	'horns to the side'
Ogama Imowuarak	'with horns that almost interlock at front'
Olua Lukunya	'horns facing the sides'

Table 20.7 Maasai somatic livestock classifiers.

Shompole	'blind in one eye'
Ng'ojine	'lame'
Erruk	'humped' (i.e. large hump)

region fairly distinctive. As can be seen, the actual number of lexemes are relatively limited, but in combination with other lexemes great specificity can be quickly achieved. To the unfamiliar, cattle or other animal species may appear homogeneous, but by using only a few of the five descriptive principles just mentioned they can be easily distinguished in verbal form, facilitating actual description and identification.

The system underlying Maasai cattle descriptives is mastered by most people involved with the pastoral process, and involves familiarity with basic terms, the rules of their combination, and the conventional associations between visual patterns and these linguistic descriptives. Colour patterning also conveys symbolic meaning and bears cultural value, but that topic is beyond the scope of this chapter (Fukui 1979).

As an indirect test of the mastery of descriptives by schoolchildren and those not attending school (home-people), several boys were requested simply to enumerate as many cattle descriptives as possible. It could be assumed either that the responses of schoolboys would reflect less command of livestock vocabulary and descriptives than those of 'home' boys, or that the facility with which boys are able to verbally respond would increase with schooling, offsetting any differences in knowledge or cognitive competence in the pastoral domain.

The schoolboys generated only 57 per cent of the number of descriptives generated by the home boys, the latter appearing able to list descriptives virtually indefinitely. There was also a strong tendency for schoolboys to list more elementary terms and fewer complex compound terms than home boys. Schoolboys listed 79 per cent elementary and 21 per cent compound terms, while home boys listed 56 per cent elementary and 44 per cent compound terms. The compound terms listed by home boys also tended to evoke a greater range of principles and types.

It seems accurate to conclude from this sample that all boys tend to know some of the basic elements and rules of generating descriptives, but that the non-schooled have a much larger repertoire of colours and patterns at their disposal, and are able to generate multi-lexemic combinations with greater facility than are schoolboys. Girls also demonstrate mastery of the principles of livestock descriptives, and tend to generate, spontaneously, more descriptives for small-stock than cattle, probably due to their greater involvement in sheep and goat-herding. Those removed from direct and daily involvement in animal husbandry through schooling tend not to master terms, rules, or associations to the same extent that the non-schooled do, presumably due to differences in exposure and motivation.

Cognitive processes: identifying the missing animal

There are three quite different cognitive tasks entailed by the identification or recognition of livestock:

(a) identifying one's own animal among those of others, a problem of pattern recognition;

(b) identifying another's animal among one's own, a problem of matching patterns and verbal descriptions; and
(c) recognizing when an animal is absent from one's own herd.

The third task is quite different from the first two, since what must be identified is absent rather than present; however, the task also involves pattern recognition, in that an entire herd forms a complex configuration which is altered when an individual is missing. This cognitive ability, of recognizing the 'missing figure', is quite important in pastoralism, where a relatively small number of herders care for a relatively large number of animals, and where theft and loss to predators is always a threat. Personal familiarity with one's herd surely underlies the ability to recognize which animal out of several hundred is *not* present, and this holistic 'sense' of a slightly incomplete or imperfect pattern plays a role in this modality of perception. However, underlying any holistic and synthetic assessment, this process is essentially 'analytical', and depends not on one principle but on the interaction of several in what proves to be a systematic process of deliberation and cognitive 'search' rather than one of global holistic apperception.

A herd-owner often meets his herd a distance away from the main kraal, as it is brought home by the young herders, and helps to drive the herd along the final stretch to the main gate where the calves await their mothers. The individual casually examines the animals and looks for signs of sickness or for disruption of the usual herd pattern. The scrutiny continues as the herd enters the gate and settles throughout the great central kraal, in the midst of the circular boma. As evening comes, the herd-owner continues to 'ruminate' on the herd, reflecting on whether all the livestock have returned. This examination, scrutiny, and reflection involves a fairly self-conscious evocation of a number of quite different dimensions of classification of animals; in effect, a 'search' can occur down one line of classification, followed by another cross-cutting 'search' down another. Following the principle of redundancy, the more dimensions of classification that exist, the greater the likelihood of coming upon the missing cow.

Maasai elders proved able to discuss in great detail the procedures used in scrutinizing their herds and checking for missing animals. One elder reported:

Normally, we elders and younger men who have livestock stand next to the gate when the animals are returning. You give them a very thorough look to see which one has gone into the *Enkang'* (kraal) and which one hasn't. You also thoroughly observe their colours. So by the time they have all gone into the home, you know that such-and-such a cow has disappeared and such-and-such a cow is present.

Then you walk around the cattle yard in order to identify which cow you haven't seen. The owner of the herd always knows which cattle are generally left behind by the rest, and which are always leading the herd. You can also observe carefully to know which ones always graze at the sides of the herd, refusing to stay with the others. So when the cattle come in you look for all those cattle first before the rest.

The dimensions of classification are essentially spatial and genealogical, but also involve visual signs which distinguish individuals from each other, and set out especially vulnerable categories. The moving herd has a definite structure, with certain individuals and cohorts walking in the front, the sides, and the back of the herd; some lead while others commonly spread the herd, drawing by example a few neighbours away from the major body to more peripheral areas of grazing. Often the same spatial configuration will be maintained as the herd moves towards home, and the herder and the manager will initially look for the more ambitious animals, who might tend to get lost or to stray. With their arrival at the main gate, animals often enter in a roughly similar order, and at that time men and women will search for individual animals who need special care: those in milk who must be paired with calves, the diseased or injured, or those which are pregnant. Thus, the herd is implicitly segmented into special categories, each having a few individuals. Then the herd settles into the great enclosure, the *Boo*, and assumes a spatial pattern which also has stability and regularity over time. Some individuals stay near the fence, others near their owners' homes, while others prefer the centre of the enclosure. As people move through the herd, checking their animals, they use the location of animals within the kraal as another dimension of organization, whereby the missing animals can be identified.

The structures of animal classification described in the previous section are explicitly drawn upon in the process of herd assessment. Finally, the pastoral family sits inside the home, eating supper or talking. In addition to the topics of the day, a good herd-manager will silently consider the well-being of the herd, and will deliberate on his animals. A common technique often mentioned is to think about each genealogical livestock 'house', the descendants of a given female cow classified together by being called by a single name. A person can quickly review an entire herd through its genealogical structure, by which it is broken down into a series of cognitively manageable units. One informant relates:

Another trick we use is that we owners know our cattle 'houses' [lineages] and colours very well. If we are doubtful of a given animal, we can start counting one house of a given cow to the end, and then start on another one. We are also very keen about the colours. By doing this, nothing will be overlooked, as you might know that a particular cow has a certain number of calves and you also know its colour very well. Also, every woman knows the number of her cattle and their colours. Also the husband knows which cows belong to which woman. In the process of counting, he will actually identify that a particular cow which may belong to one of his wives is not present. The women can quickly tell if any of their cows are missing.

With somewhat less elaboration, a herdboy mentioned many of the same principles of scrutinizing a herd:

> I count the cattle according to their houses. If a cow has one calf, I count that. If it has more I count them until I have accounted for them all. I also note the cattle that are always behind and those that always lead the rest. And also those that graze at the sides of the herd, and the sick ones. I also look for those cows that have young calves. During the rainy season, these cows are quick to sneak back home. When I discover that one is missing, I will look for it in other herds. I will also describe it [to other herders] while I look for it, according to its colour, its earmarks and firemarks [brands] it may have.

Then, it is said, one sleeps and during sleep thinks and dreams about cows, and in the middle of the night or in the morning the herder or owner will know if such-and-such an animal is missing. The elder describes:

> Another method we use is that when one goes to sleep one counts one's herd in one's sleep. In most cases, you may find out that there is a cow or an ox you haven't seen while counting the herd. In the morning, one checks the herd once more to see if the cow or ox one had not seen in one's sleep is in or not. Sometimes it's likely that it may be in but was not noticed. It is also true that a person finds the cattle that he hadn't seen in his sleep are missing in the morning. Then he will ask the person who actually tends the herd to check and find out the last time he had seen it.

The association between sleep, dreaming, and cognitive deliberation on one's cows was drawn not by one but by every informant consulted. It was suggested separately by many that by dreaming one could often review one's herd and come up with a missing animal more effectively than during conscious waking review of the herd.

Some concluding remarks

The perceptual process described is not holistic, nor is it iterative, by which a missing animal would be identified if the count was off. Rather, the herd is symbolically differentiated by a series of analytical dimensions, some of which are permanent, such as genealogy, some of which are quite ephemeral, such as states of well-being. A quite large herd is conceptually manageable when broken down into a much smaller number of categories, through a process similar to what has been called 'chunking' by cognitive psychologists. Genealogy represents the most well-defined and unambiguous principle of classification. One herd of 54 animals described was subdivided into 13 'houses' or matrilineages, composed of an average of four animals each, with a range from two or seven members. Of the herd 80 per cent were female, 20 per cent male, and of the females 16 were identified as having already given birth, 27 presumably being heifers. Extrapolating, a herd of 150 might be subdivided into around 37 houses, though such a large herd might actually tend to be composed of fewer houses with more members. Since these houses are also allocated to specific matricentric households in the polygynous family, the place of any given animal within the genealogical structure is quickly narrowed.

The herd's use of space outside and inside the kraal represents perhaps the most direct and accessible system of scanning by the herder to determine whether the herd is in order. But genealogy (through naming) probably represents the most reliable mode of cognitive assessment of the wholeness of the herd. But taken together, the modes of cognitive classification and processing constitute a structure, by which the herd is symbolically subdivided in several cross-cutting ways, providing *redundancy* in the ordering of animals and for the implicit 'search' for missing members. Any individual animal is classified spatially, genealogically, and individually; the herder 'searches' each dimension, and if in doubt 'searches' another, based on the state of the herd at the critical moment. The impressive memories of pastoralists are so not because of their greater potential for perceptual memory, holistic pattern recognition, or cognitive capacities for 'storage' and 'review', but because of the nature of their symbolic organization of domestic animals. Multiple dimensions of cultural classification provide for cognitive organization and redundancy, allowing them to efficiently process smaller 'chunks' of the herd and to cross-check one result with another.

These forms of 'practical reasoning' represent technical achievements in pastoral culture, underpinning careful husbandry and facilitating higher overall productivity, both in terms of labour and aggregate output. Knowledge of individual animals in a herd, systems of naming and describing them, and mastery of processes of classification and identification, are not evenly or randomly distributed throughout pastoral communities, although there are always those known for their cognitive skills or lack of them. Rather, pastoral cognitive competence – regarding both content and process – is gained both through tutorial and experience, and depends not just on passive participation but also on the sort of motivation and high valuation of pastoralism that was once taken for granted. The cultural systems of symbolic classification and cognitive processing described here represent a critical concomitant of pastoralism viewed as a highly productive and viable way of life in the semi-arid rangelands of East Africa. The knowledge and cognitive competence so represented may well decrease as pastoralist children turn towards formal education, and the subject matter of the school curriculum, and away from the pastoral experience through which competence is acquired. It should not be assumed that pastoral economic life merely represents a set of work routines, or a simple strategy of subsistence production, which can only be improved through education or the more elusive process of 'modernization'. This chapter has attempted to demonstrate that there are cognitive concomitants of the resilient arid-land adaptation represented by specialized Maasai pastoralism on which successful animal husbandry depends, and the loss of this basis of knowledge, motivation, and strategy will influence the viability of this arid-land economy and culture.

Acknowledgements

This research was supported by the joint McGill/Kenyatta University College project on Cognition, Education and Work, through a Cooperative Grant from the International Development Research Centre of Canada. For additional research support, I am also indebted to the Graduate Faculty of McGill University, the Social Sciences and Humanities Research Council of Canada, and the Fonds F.C.A.R. of Quebec. The assistance of the Bureau of Educational Research of KUC and its Director, Professor George Eshiwani, is greatly appreciated, as is collegial comment by other members of the project in Kenya and at McGill. The competent research assistance of Mosinko Ole Tumanka in gathering much material used here is acknowledged with appreciation.

References

Beidelman, T. O. 1966. The ox and Nuer sacrifice: some Freudian hypotheses about Nuer symbolism. *Man* **1**(4), 453–67.

Braudel, F. 1973. *Capitalism and material life 1400–1800*. New York: Random House.

Cole, M. & S. Scribner 1974. *Culture and thought*. New York: Wiley.

Evans–Pritchard, E. E. 1940. *The Nuer*. Oxford: Clarendon Press.

Fukui, K. 1979. Cattle colour symbolism and inter-tribal homicide among the Bodi. In *Warfare among East African herders*, K. Fukui & D. Turton (eds). Osaka: Senri Ethnological Studies.

Galaty, J. 1981. Land and livestock among Kenyan Maasai: symbolic perspectives on pastoral exchange, social change and inequality. In *Change and development in pastoral and nomadic societies*, J. Galaty & P. Salzman (eds), 68–88. Leiden: Brill.

Galaty, J. 1982. Being 'Maasai', being 'people-of-cattle': ethnic shifters in East Africa. *American Ethnologist* **9**(1), 1–20.

Galaty, J. 1984. Cultural perspectives on nomadic pastoral societies. *Nomadic Peoples* **16** (October), 15–29.

Galaty, J. in press. Pastoral and agro-pastoral migration in Tanzania: factors of economy, ecology and demography in cultural perspective. In *Production and authority*, J. Bennett & J. Bowen (eds), Society for Economic Anthropology, Washington DC: University of Americas Press.

Goody, J. 1977. *The domestication of the savage mind*. Cambridge: Cambridge University Press.

Herskovits, M. 1926. The cattle complex of East Africa. *American Anthropologist* **28**, 230–72, 361–88, 494–528, 633–64.

Ingold, T. 1980. *Hunters, pastoralists and ranchers*. Cambridge: Cambridge University Press.

Sahlins, M. 1976. *Culture and practical reason*. Chicago: University of Chicago Press.

Scribner, S. (ed.) 1984. Cognitive studies of work. *The Quarterly Newsletter of the Laboratory of Comparative Cognition* **6** (1&2), 1–46. San Diego: Center for Human Information Processing.

Street, B. 1984. *Literacy in theory and practice*. Cambridge: Cambridge University Press.

21 *Prehispanic pastoralism in northern Peru*

TOM McGREEVY

Introduction

In the summer of 1982 I conducted an archaeological survey on the high grasslands around the town of Huamachuco in northern Peru (see Figs 21.1 & 21.2) under the auspices of the Huamachuco Archaeological Project. Evidence for prehispanic pure pastoralism was sought in this area. Unexpected results forced a study of Peruvian camelid pastoralism so as to allow interpretation of the observed patterns from the north highlands in prehispanic times (McGreevy 1984). This chapter summarizes that study.

The South American camelids

The animals herded in Peru were llamas and alpacas, members of the family Camelidae, genus *Lama*. There is debate about whether they should be separated at a species or a subspecies level (Miller 1979, pp. 1,4) but there is no doubt that to Andean natives they are functionally distinct (Miller 1979, p. 7). Furthermore, conditioned by their functions, the range of alpacas is narrower than that of llamas.

Llamas are used today primarily as beasts of burden. Average loads vary between 20–40 kg, which can be carried up to 30 km per day (Stouse 1970, p. 138). Secondary uses include wool and dung production. Llama wool generally is coarser than alpaca wool and is used for the production of household goods such as blankets, rugs, and rope (Franklin 1982, pp. 465, 468). It can be used for the manufacture of clothing, but alpaca wool is preferred. Dung is used both as fuel and as a fertilizer (Flores Ochoa 1975, p. 307). Llamas are also used for meat and bones; fat, sinews, and pelts are all used in household products (e.g. Stouse 1970, p. 138). They play an important role in ritual during individual rites of passage and at times of collective stress (Stouse 1970, p. 138, Flores Ochoa 1975, p. 307).

Alpacas are primarily wool producers. Their wool is very fine and is used for the production of textiles (Webster 1973, pp. 121–2). Other uses are similar to those of llamas, save that alpacas do not carry cargo and are less important in ritual. Alpacas are the most important animals in Peru today because their wool can be easily traded or sold, and will become increasingly so as it penetrates further into international markets (Franklin 1982, pp. 468–9). Llamas become less important as mules and trucks replace them.

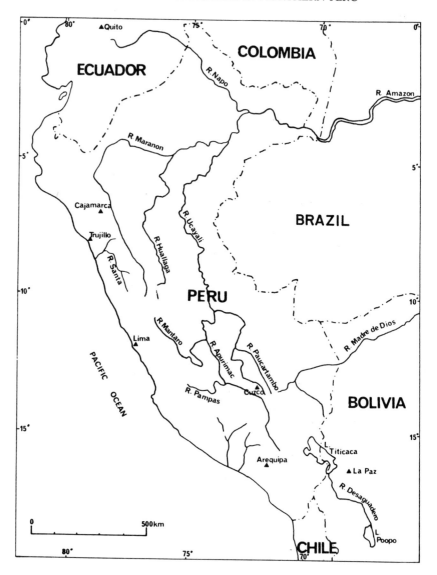

Figure 21.1   General map of Peru.

In Inca times the relative importance of the animals appears to have been reversed. The llama's functions were much as they are today; in addition, llama wool had greater use and llamas played a larger role in ritual. Their wool was used for the cloth of the common people where alpaca wool was absent. However, the Inca state is known to have redistributed alpaca wool into areas where alpacas were absent, to be woven by the people both for their own and the state's use (Murra 1965, pp. 102, 124–5). Thus

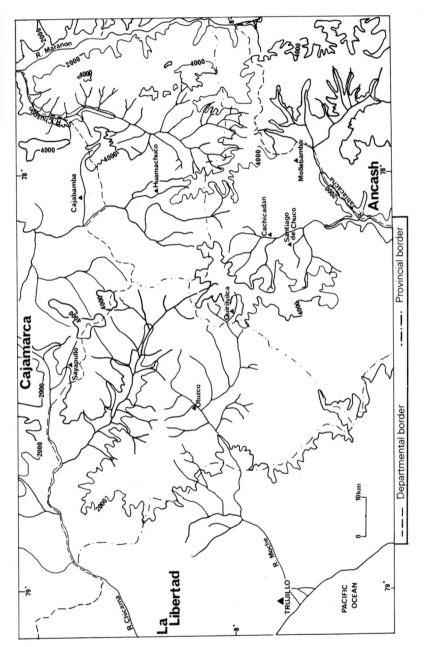

Figure 21.2 Map of north central Peru.

the degree of increased importance for llama wool in textile production is complex to determine. In ritual, llamas played a role in state religion, in addition they were used in individual and collective ceremonies (Murra 1965, p. 195, Tschudi 1969 (1885), p. 131). The great importance of llamas to the Inca can be seen in two facts. First, gifts of llamas by the state were sufficiently special to help ensure loyalty of subjects (Murra 1965, p. 205). Secondly, the Inca strongly encouraged the expansion of llamas throughout the empire (Murra 1962, p. 711).

There is less information from the Inca period about alpacas. Certainly they were kept primarily for wool production. Almost all the fine cloth used by the élite was woven from their wool. These textiles were among the most highly valued commodities in Peru (Murra 1962). Thus, the importance of alpaca wool must not be underestimated. Other alpaca uses were similar to today. In addition, they had an increased role in ritual in Inca times, although the role is unclear. What is clear is that the Inca did not try to expand the range of alpacas as they did llamas. This, as will be argued, is due to the specialized environment alpacas require to produce good wool.

The question of the relative importance of the animals in earlier times remains unanswered. However, I suggest that alpaca wool has always been important since domestication of the animal. The acquisition of alpaca wool by groups without direct access to alpacas is an interesting research problem (Topic *et al.* 1987). Llamas became important in northern highland Peru by the Early Intermediate Period, as they were present in most areas by this time (see Table 21.1 and summary in McGreevy 1984, pp. 117–19). The importance of llamas derives only from their day-to-day use.

Alpacas range from central Peru to extreme northern Argentina, rarely occurring below 4300 m above sea-level (ASL) (Gilmore 1948, p. 441, Gade 1969, p. 341). They are highly dependent upon the succulent pasturage found in high wet areas, known as *bofedales*. When raised in these areas they are generally healthy, but when raised elsewhere they are prone to diseases, the most serious of which is mange (Webster 1973, p. 121, Orlove 1977, p. 207, and summary in McGreevy 1984, pp. 45–8). When affected by any disease, alpacas suffer a decrease in wool quality and quantity (Orlove 1977, p. 207). Since they are kept for wool production, their range is restricted to the zone where they produce the best wool,

Table 21.1 Peruvian chronology (prehispanic periods).

Late Horizon	AD 1476–1532
Late Intermediate Period	AD 1000–1476
Middle Horizon	AD 600–1000
Early Intermediate Period	200 BC–AD 600
Early Horizon	900–200 BC
Initial Period	1800–900 BC
Preceramic periods	?–1800 BC

which is not to say that they cannot survive in other zones. It has been argued that in prehispanic times alpacas were found down to 3500 m ASL (Flores Ochoa 1982, p. 69). However, it has not been conclusively proven that herds living at this altitude were functioning as wool producers. Rather, several of the sites appear to be collection points for slaughter.

Llamas are found today scattered throughout Peru, Bolivia, Chile, Argentina, and Ecuador, rarely below 2700 m ASL (Gade 1969, p. 341). They survive well upon drier forage than alpacas and are less susceptible to disease (Flores Ochoa 1975, p. 303). In prehispanic times llamas were more widespread, living on the coast as well as in the highlands (Gilmore 1948, p. 437).

While it is likely that only llamas would have been found in northern Peru in prehispanic times, the type of pastoralism that would have been practised in the area is less clear.

Types of pastoralism

There are major physiographic differences between highland northern and southern Peru. In southern Peru there are large expanses of high grasslands suitable for herding. Most of the land is over 4000 m ASL (see Fig. 21.3). Agriculture has limited potential even with crops adapted to high altitude, due to early frosts, hail, snow, or excessive dryness (Isbell 1978, p. 307); thus crop failure is not uncommon. In the north, the land is more broken and lower as well as wetter, and agriculture has much greater potential.

Pure pastoralism and agropastoralism are found in Peru today. Pure pastoralism is found in the south on the high grasslands as an adaptive response to an environment largely unsuited for agriculture. Both llama and alpaca herds are raised with a preference for the more valuable alpacas. Access to other goods is through trade or barter (Flores Ochoa 1975, pp. 309–11, Custred 1977, p. 119).

Camelid agropastoralism is found today only on the eastern slope of the Andes and in deep valleys in highland southern Peru. Agriculture is the main subsistence source with pastoralism complementary to it. Llamas are the most important herd animal as they provide labour for transportation (Webster 1973, p. 121, Yamamoto 1981, pp. 127–8) and dung for fertilizer, which can be crucial for maintaining soil fertility (Guillet 1983, p. 563).

The antiquity of pure pastoralism is a matter of debate. This adaptation conflicts with 'verticality', a model of land use which presupposes that lands from different production zones are controlled by a single community which thereby has direct access to the products of each (Murra 1975). This model might be relevant only for the last few centuries of prehistory.

In northern Peru in prehispanic times, it is believed that agropastoralism was the dominant high-altitude land-use strategy. As in the eastern Andes, high-altitude herding lands are interspersed with lower agricultural lands and not isolated from them.

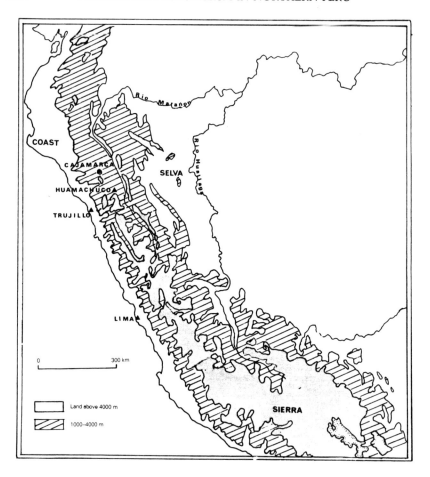

Figure 21.3   General physiography of Peru.

North Peruvian Survey data

Roughly 1200 hectares of land between 3700–4250 m were surveyed for evidence of prehispanic pastoralism (McGreevy & Shaughnessy 1983). Of the 29 sites found, only four gave any indication of a pastoralist orientation. Two of these were undated while the other two contained both Late Intermediate Period and modern sherds. None of them was a long-term habitation site and only one had substantial corral space, with seven large corrals. Thus, the evidence suggests that in prehispanic times this zone was not used as a long-term habitation area by pastoralists, and that short-term use occurred only in late prehispanic periods.

An area of 275 hectares between 3300–3600 m was surveyed as a comparative sample. In this elevation 16 sites were found, all dating from the Early Intermediate Period (McGreevy 1984, pp. 148–54). The sites were predominantly agropastoralist, with the layout of some suggesting that specialist herders may have cared for the animals. This suggestion is tentative as no excavation was carried out. Sites in this area were situated along a ridge-top with ready access to higher grazing lands and lower agricultural lands. Daily herding on the higher lands would leave few traces. This type of settlement pattern is not unique in this area (see for other examples McGreevy 1984, p. 166, Shaughnessy 1984, DeHetre 1979, pp. 32, 143–4). Further survey obviously needs to be conducted in the north highlands to determine the general high-altitude land-use strategy; however from the present evidence it appears the strategy was agro-pastoralism.

Conclusions

It is suggested that pure pastoralism was not practised in prehispanic highland northern Peru. Rather agropastoralism was the norm on high-altitude lands. Llamas were the herd animals providing labour power, dung, wool, and meat. Alpacas were absent due to their requirements for good wool production. Animal numbers increased over time, especially under the Inca who encouraged breeding as state policy. In terms of settlement pattern this adaptation suggests that most settlements would be located near the junction of herding and agricultural lands rather than in the herding lands proper. Sites found in the herding lands were unlikely to have been pastoralist, unless they dated from Inca times when the large expansion in herd numbers occurred. Models to help understand high-altitude land use in northern Peru should not be derived from southern or central Peru where the possibility of pure pastoralism was much stronger, but rather from the eastern slopes where agropastoralism is the dominant subsistence adaptation.

Acknowledgements

I wish to thank the Huamachuco Archaeological Project directed by Drs John and Theresa Topic for having me do the survey. It was funded by the Social Sciences and Humanities Research Council of Canada. Permission for the survey was given by the *Instituto Nacional de Cultura* of Peru. Luis Franco Quezo acted as my field assistant. Both Dr Theresa Topic and Roxane Shaughnessy have read and commented on this chapter and I thank them for their assistance.

References

Custred, G. 1977. Peasant kinship, subsistence and economics in a high altitude Andean environment. In *Andean kinship and marriage*, R. Bolton & E. Mayer (eds), 117–35. Washington: Special Publication of the American Anthropological Association 7.

DeHetre, D. A. 1979. *Prehistoric settlement and fortification patterns of La Libertad, Peru: an aerial photographic analysis*. Unpublished Master's thesis, Department of Anthropology, Trent University, Peterborough, Canada.

Flores Ochoa, J. A. 1975. Sociedad y cultura en la puna alta de los Andes. *America Indigena* **35**(2), 297–319.

Flores Ochoa, J. A. 1982. Causas que originaron la actual distribucion espacial de las Alpacas y Llamas. In *El hombre y su ambiente en los Andes centrales*, L. Millones & H. Timoeda (eds), 63–92. Senri Ethnological Studies 10. Osaka: National Museum of Ethnology.

Franklin, W. L. 1982. Biology, ecology, and relationship to man of the South American camelids. In *Mammalian biology in South America*, M. A. Mares & H. H. Genoways (eds), 457–89. University of Pittsburgh Pymatuning Laboratory of Ecology Special Publication Series Vl 6.

Gade, D. W. 1969. The Llama, alpaca and vicuna: fact vs fiction. *Journal of Geography* **68**, 339–43.

Gilmore, R. M. 1948. Fauna and ethnozoology of South America. In *The handbook of South American Indians*, Vol. 6, J. H. Steward (ed.), 345–464. Bureau of American Ethnology, Bulletin 143. Washington: Smithsonian Institution.

Guillet, D. A. 1983. Toward a cultural ecology of mountains: the central Andes and the Himalayas compared. *Current Anthropology* **24** (5), 561–74.

Isbell, W. H. 1978. Environmental perturbations and the origin of the Andean state. In *Social archaeology: beyond subsistence and dating*, C. L. Redman, M. J. Berman, E. V. Curtin, W. T. Langhorne Jr., N. M. Versaggi & J. C. Wanser (eds), 303–13. New York: Academic Press.

McGreevy, T. H. 1984. *The role of pastoralism in prehispanic and modern Huamachuco*. Unpublished Master's thesis, Department of Anthropology, Trent University, Peterborough, Canada.

McGreevy, T. H. & R. E. Shaughnessy 1983. High altitude land use in the Huamachuco area. In *Investigations of the Andean past*, D. H. Sandweiss (ed.), 226–42. Ithaca: Cornell University Latin American Studies Program.

Miller, G. R. 1979. *An introduction to the ethnoarchaeology of the Andean Camelids*. Unpublished PhD dissertation, University of California (Berkeley). Ann Arbor: University Microfilms.

Murra, J. V. 1956. *The economic organization of the Inca state*. Chicago: University of Chicago.

Murra, J. V. 1962. Cloth and its functions in the Inca state. *American Anthropologist* **64**(4), 710–28.

Murra, J. V. 1965. Herds and herders in the Inca state. In *Man, culture and animals*, A. Leeds & A. P. Vayda (eds), 185–215. Washington: AAAS Publication 78.

Murra, J. V. 1975. El control vertical de un maximo de pisos ecologicos en la economia de las sociedades Andinas. In *Formaciones economicas y politicas del mundo Andino*, J. V. Murra (ed.), 59–115. Lima: Instituto de Estudios Peruanos.

Orlove, B. S. 1977. *Alpacas, sheep and men: the wool export economy and regional society in southern Peru*. New York: Academic Press.

Shaughnessy, R. E. 1984. *High altitude land use, past and present, in the Huamachuco area, north highlands, Peru*. Unpublished Master's thesis, Department of Anthropology, Trent University, Peterborough, Canada.

Stouse, P. A. D. 1970. The distribution of llamas in Bolivia. *Proceedings of the Association of American Geographers* **2**, 136–40.

Topic, T. L., T. H. McGreevy & J. R. Topic 1987. A comment on the breeding and herding of llamas and alpacas on the north coast of Peru. *American Antiquity* **52** (4), 832–5.

Tschudi, J. J. von 1969. (1885) La Llama (translated by G. Custred). In *Mesa redonda de ciencas prehistoricas y antropologicas*, Vol. I, 123–38. Lima: Pontifica Universidad Catolica del Peru.

Webster, S. S. 1973. Native pastoralism in the south Andes. *Ethnology* **12**(2), 115–33.

Yamamoto, N. 1981. Investigacion preliminar sobre las actividades agropastoriles en la Distrito de Marcapata, Departmento de Cuzco, Peru. In *Estudios etnograficos del Peru meridonal*, S. Masuda (ed.), 85–137. Tokyo: University of Tokyo.

22 Andean pastoralism and Inca ideology

GORDON BROTHERSTON

In the long history of man's relationship with other animal species a critical difference occurred with his domestication of certain of them, in order to make of them pastoral creatures. This is the truer the closer we hold to a definition of the pastoral process which respects Ducos's emphasis: 'domestication can be said to exist when living animals are integrated as objects into the socio-economic organisation of the human group, in the sense that, while living, those animals are objects for ownership, inheritance, exchange, trade, etc., as are the other objects (or persons) with which human groups have something to do' (see Bökönyi, ch. 2, Ducos, ch. 3, this volume). Historically, most evidence on the matter has been gathered in the Old World rather than the New, and has given rise to major debates on the role of pastoralism within the 'Neolithic Revolution', in particular its relation to hunting economy on the one hand and to agriculture on the other.

In this Old World tradition one thing, however, emerges with clarity. And that is how thoroughly pastoralism has been inscribed in the twin ideological supports of western culture: the Graeco–Roman classics, and the Bible. Through the former, the pastoral has contributed first premises to political science and whole modes to art and literature, while the latter strenuously privileges flock-keeping morality through such key examples as Cain, the demeaned agriculturalist, and Abel, the preferred shepherd; Abraham and the ram that substitutes for Isaac; the commandment that lists wife after ox and ass; the parable of the sheep and goats, and of the lost sheep; and the role of Christ himself as both pascal lamb and almighty shepherd-lord. So pervasive, indeed, has this ideology been within the western tradition that it has proved most hard to isolate and define. A signal case is that of Jean–Jacques Rousseau, whose highly influential essay 'On the Origins of Inequality' provides the hinge between the classical and Biblical traditions and the new era of social enquiry typified by Marx. Asking why it should be that man, born free, is everywhere in chains, Rousseau repeatedly compares the exploitation undergone by human masses with that of animal herds and draws an explicit parallel between the fall of 'natural' man and the domestication of such animals as horses. Yet such was the power of the pro-pastoral prejudice that he found himself holding other culprits responsible for this degeneration: metallurgy (in a reworking of the world-age story found in Hesiod), and agriculture, Cain's calling.

Something of the prejudice that afflicted Rousseau can perhaps be felt

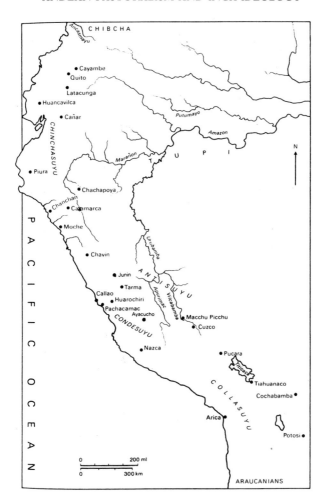

Figure 22.1 Map of the ancient empire of Tahuantinsuyu.

still in approaches to pastoralism in western scholarship even today; in any case, the links between the economy and the ideology of pastoralism have hardly been over-explored, not least when the New as well as the Old World is taken into account. Indeed, the New World offers in this respect a welcome term of comparison since its ideological patterns, being less familiar to western eyes, are likely to be more readily detectable. Moreover, the single focus of pastoralism within the New World, the Andes, and the former Inca empire Tahuantinsuyu ('Four Districts', Fig. 22.1), holds a special and distinctive place in that continent, making yet more purposeful comparisons possible; for while Tahuantinsuyu exhibits the usual native American traits in such practices as weaving, curing, agriculture, metallurgy, and so on, in other respects, not least its social and political organization, it emerges as no less unique than it does on account of its animal herding.

Fundamental to the present enquiry into pastoralism and the New World is the existence of evidence of not just an archaeological or anthropological order but of primary texts, classics of the native American tradition. From Tahuantinsuyu itself come two, both dating from c. AD 1600, which are of especial value. One is Guaman Poma's *Nueva corónica y buen gobierno*, an encyclopaedic survey of Peru sent as a letter to Philip III of Spain (edited by Murra & Adorno 1980, Pease 1980). The first part of this work or *Nueva corónica*, with its tabular arrangement of chapters and illustration-sets, bears out its author's claim (p. 367)[1] that it was in part transcribed from *quipu* sources: this knotted-string device, whose first use appears to have been pastoral, served as the means of official communication and memory in Tahuantinsuyu (see Ascher & Ascher 1981). The other work, composed entirely in Quechua, the language of Tahuantinsuyu and of millions of Andeans today, stems from Huarochiri, on the road between the capital Cuzco and the Pacific coast (Trimborn 1939–41, Arguedas 1975). It affords many insights into the role of animals in Andean cosmogony and religion while offering an internal critique of metropolitan Inca practice (Spalding 1984). As editions and translations have become available over recent decades, the *Nueva corónica* and the *Huarochirí Narrative* (see Trimborn 1939–41, Arguedas 1975) have been increasingly drawn on by scholars, notably Murra (1980), who devotes a whole chapter to 'Herds and herders'. Yet there is much that these sources still have to offer on this subject. This is especially true when they are related to their *quipu* precedents (which even Murra derogates with the telltale epithet 'pre-literate'); when they are put together with other examples of Tahuantinsuyu literature, like its liturgy, poetry, and drama; and when this corpus is placed in turn in its larger native American context.

Tahuantinsuyu and its antecedents

On the high island of the Andes, four types of camelid have their natural home: the llama and alpaca, and the guanaco and vicuna (or the short-necked llama, see Shimadu & Shimadu 1985). Having in common with the Old World camels a remote and long-vanished ancestor in the North American Ohio Valley, these animals differ somewhat in size, the llama being the largest; in quality and colour of wool, the vicuna having the finest and the alpaca the most; and in range, the guanaco extending down on to the Argentine pampa. Most important, only the first two have been successfully domesticated, the llama possibly deriving from the guanaco.

Using bone samples, Browman (ch. 23, this volume) has put the beginnings of domestication back to about 4000 BC and has identified centres at Junin northwest of Cuzco, and at Lake Titicaca to the southeast, suggesting a third possible site at Ayacucho in between. He further detected an important difference in emphasis between Junin and northern

Tahuantinsuyu generally, where camelids were used mainly for meat, and Titicaca, where they served mainly as suppliers of wool and for transport, uses which appear to have spread north as late as AD 500, that is at dates comparable with the spread of other more generally recognized traits of Tiahuanaco culture, which likewise had its base on the shores of Lake Titicaca.

At any event, long before the rise of the Inca, Tiahuanaco and the cold, high plateau around Lake Titicaca, where agriculture is restricted, was reputed to be the richest in llama herds, and caravans plied regularly from there to the Pacific coast. Moreover, this area, known as the Collao or Colla country, is consistently referred to as the one from which the Inca drew support when they came to establish their own centre of power at Cuzco, a century or two before the European invasion. One account, stemming from the Aymara-speakers of the Collao, says simply that the Inca made off with Colla animals in order to build up herds of their own (Matienzo 1567, quoted by Murra 1980, p. 52). From the Inca side it was more a question of deference to the llama typical of the Collao during royal initiation and other ceremonies. That the Inca, from the time of the first emperor Manco Capac, acknowledged their origin to have been in Titicaca, is made quite clear by Poma de Ayala (1936, pp. 84, 265); they favoured the white llama that the Collas themselves had gone so far as to revere as a tribal ancestor, and they would deck out such llamas, known as *napas*, with red shirts and necklaces and gold earrings. In the chapter of his *Nueva corónica* which he devotes to the festivals typical respectively of the Inca and of the four *suyus* of the empire, Guaman Poma also shows how pastoral songs of the Colla and Collasuyu were held in particular esteem at the royal court of Cuzco, to the extent of being sung in Aymara, the Colla language (1936, pp. 129, 319). The Inca connection with Titicaca is also supported by certain architectural parallels, while in his *Royal Commentaries* (I, xviii) Garcilaso el Inca notes that the political arrangement of a Tahuantinsuyu, or 'four districts', of empire had been previously elaborated at Tiahuanaco. A striking indication of how powerful the Colla with their massive llama herds had become as the Inca's predecessors lies in the fact that of all four *suyus* theirs was the one which retained most rights to local herd ownership under Inca rule; after the European conquest and the collapse of Cuzco, the Colla even recovered some of the llama wealth that had been alienated by the Inca.

Turning now to the pastoralism of the Inca and following Poma de Ayala's chapter on their rise (1936, pp. 79–193), we can trace a trajectory that runs from small beginnings in Cuzco to the latter-day Tahuantinsuyu, which spread thousands of miles between the present-day states of Colombia in the north and Chile in the south; and in so doing we can detect a turning-point in the 15th-century emperor Pachacuti, conqueror and acquirer of territory especially at the expense of the Colla, and Tupac Yupanqui, the great consolidator under whom herd units ran into many millions. Yet for present purposes, Inca practice, which in certain details no doubt echoed Colla and other precedents, can more conveniently be considered as one, especially insofar as it constitutes a system which went quite beyond any previous system in the Andean area in organization and scale.

In Tahuantinsuyu the domesticated Andean camelid, from now on referred to by the general term 'llama' (the alpaca not being separately noted except where the difference matters), can be seen to have combined an impressive array of functions. For the Inca and their subjects ate its flesh, fresh or as charqui; wore its skin as sandals, or cut it into thongs to secure the foot-plough (taclla) or into bottles to carry water across the desert; turned its fat into tallow; spread its dung as manure or gathered it for fuel; made its tendons into slings for the scarecrow, the herder, and the soldier; shaped its bones into weaving instruments; and span and wove its wool into cloth both coarse (auasca) and fine (cunbe), and into the threads and main cord of the quipu. As well as singly providing these organic and commodity resources, and in the absence of any analogue except the human being itself, the llama was also widely exploited by the Inca as transport, along the roads for which Tahuantinsuyu is famous, and in particular to support the crop-carrying farmer (Fig. 22.2) and the campaigning soldier.

For all these and yet other purposes, the Inca instituted a programme of breeding, distinguishing and counting types and ages of beast down to the minutest detail by means of the quipu, and conducting a thorough census in the month Aya Marcay (November; Poma de Ayala 1936, p. 256). Also through running and other athletic trials (which conjoin the two meanings of 'race') they prized the strongest and fittest beasts. The Huarochirí Narrative further tells us (Ch. 10) that the prowess revealed by such trials was associated with the enhanced penis displayed by golden and silver llama statuettes that have survived from Inca times. By means of this controlled reproduction, the llama came to acquire another order of value, one of permanency more like that of the precious metals with which it was equated and into which it was cast ceremonially. Beyond this again, through multiplication and increase, the llama became literally capital, perhaps its most distinctive role of all in the plan of Tahuantinsuyu, as we shall see, and one presided over by Yacana, the celestial llama at the centre of the sky (Huarochirí Narrative, Ch. 29, Zuidema 1983).

A prime Incaic use of the llama was the military one. By mobilizing troops and ranks of llamas, for both transport and food on the hoof, little affected by season and harvest, the Inca disposed of a state army unparalleled in America, whose campaigns related less to ritual than to policies of permanent territorial gain, and which in its day proved largely irresistible. In his chapter on the world-ages that preceded the Inca (1936, pp. 48–78), Poma de Ayala makes much of the military potential that went with the llama breeding characteristic of the latter two ages of the Purun and the Auca (whose name means 'war-like'). Then, after military conquest and as part of Inca policies of pacification and colonization, the llama had no less critical a role. In the case of people already in possession of herds, these and all their reproductive goods, in Murra's words: 'became the property of the Inca crown, which then reissued some of it back to the inhabitants and set public boundaries'; again, after conquest 'all

Figure 22.2 Llamas carrying potatoes (Poma de Ayala 1936, p. 1149).

llamas were defined at least in theory as state property' (1980, pp. 94, 96). In practice, no other interest group was permitted to keep herds which in any way might rival those of the Inca, the former being linguistically distinguished from the latter by the respective terms *huaccha* (poor) and *capac* (mighty).

This alienation of ethnic property facilitated, in turn, the distinctively Inca policy of granting llamas as capital to settlers in conquered territory where previously there had been few or no llamas; known as *mitima*, these grantees were further encouraged to migrate by being made exempt from labour tribute and other state obligations (Murra 1980, p. 178). Through the device of the llama grant, the Inca secured their hold on the coastal valleys, filling in and completing Condesuyu as an imperial quarter; and between the larger and repeatedly extended Colla and Chinchasuyu they removed populations over thousands of miles – no less than 4000 *mitima* families were seen journeying up to occupy former Canari lands in Ecuador and Chinchasuyu when Pizarro was already at Caxamarca. Herding was specified as the first of the skills required of a *mitima*, and the head-shepherds among them and the longest standing came from the Collao. In return for their official generosity, the Inca expected to gain from the multiplication of their grantees' animals, taking, as it were, interest from capital. They also expected the supply of cloth woven from llama and alpaca wool which was universally recognized as 'one of the main bonds and symbols of citizenship' (Murra 1980, pp. 52, 55, 156, 174). They might even be seen to have created a market-demand for the *mitima* and llama-producers generally in the strict laws that they instituted, which, according to Poma de Ayala (1936, pp. 272–3), obliged every community however small regularly to sacrifice these animals in order to consume their blood and meat.

Above all, through the *mitima* (those who 'leave' their first homes) and their interest, in both senses, in their own displacement, the Inca achieved stability for Tahuantinsuyu as a continuous territory, within its frontier, or outer fence of pasture. In this respect it is highly significant that of the four *suyu* the one which most resisted conquest, the Antisuyu of the montana and valleys of the upper Amazon, was also the one known to be least adaptable to llama herding.

In all these respects, at the time of the European invasion nothing like Tahuantinsuyu existed elsewhere in America, this difference being directly attributable to the resource inherited by the Inca as Andeans and heirs to Tiahuanaco, and exploited by them as architects of Tahuantinsuyu: the domesticated camelid. In the tribute lists of Mesoamerica we find an official catalogue of commodities supplied in Peru by the llama; clothing as cotton capes, food as bushels of maize and beans, durables as precious metals (Brotherston 1979, pp. 233–6): besides having all these assets in one, the llama, as we have seen, supplied transport and, above all, increased and multiplied its value through breeding. This key economic difference doubtless relates, in turn, to the different ways in which the four-quarter model of tribute was developed in Mesoamerica and Tahuantinsuyu. In the former, the four quarters mapped on the title-page of the Aztec Mendoza Codex yielded only commodity tribute, which was collected via chains of head towns and along routes that often ran through neutral or hostile territory within the quarter. In the latter, despite an even less tractable terrain, the four quarters, *tahuantin-suyu*, were consolidated entire, like pasture within its fence.

Models of control and authority

So fundamental was herding to the Inca enterprise that in Tahuantinsuyu the social relations it implies as an activity were transferred ideologically to become a model for the state itself, indeed it arguably provided the enabling concept of 'state' in the first place. This is so for the two dominant social models of the Inca state, that which relates ruler to ruled, and that which relates ruler to authority. For both, a wealth of evidence is provided by the native sources used so far; in addition we have the 11 hymns or prayers, recorded in the 16th century in Quechua by Molina, which make up the liturgy of the *Situa*, a ceremony of cleansing and purgation held in the month *Coya raimi*, which was September (Rowe 1953, Lara 1969, pp. 179–86).

First, the *Situa* hymns repeatedly equate flock with folk, both as subjects of the Inca 'who founded Cuzco'. Requests are made, in the same terms and the same phrase, that under the Inca both people and animals (*runa llama*) should enjoy peace and safety; should increase and multiply; and should not fall into enemy hands or stray into sin.

Oh dew of the world
Viracocha
inner dew
Viracocha
you who dispose by saying
'Let there be greater and lesser gods'
great Lord
dispose that here
men do multiply
fortunately.

Let me live in peace and in safety,
Father Viracocha,
with food and sustenance,
with maize and llamas,
with all manner
of skills.
Abandon me not,
Remove me
from my enemies
from danger
from all threat
of being cursed, ungrateful
or repudiated.

The parallels here with Semitic liturgy and the logic of, say, 'The Lord is my Shepherd' (Psalm 23) are so strong that an influence via the Spanish Christian mission might be suspected were it not for numerous independent testimonies to the nature of the *Situa*. What is more, while obviously a piece of spiritual rhetoric, the flock–folk equation proves to have a firm material and economic basis in Tahuantinsuyu. For the sheer tally of the two orders of unit in question, animal and human, was consigned to the recording device used initially for the former: the *quipu*.

This prime piece of the herder's equipment (Poma de Ayala 1936, p. 351) registered animal units with decimal-place value notation; displayed colours as semantic variables which corresponded in the first instance to those of the actual llama wool; and had a structure of cord and dependent thread that even replicated the main-cord custom of llama tethering.

The competence of the *quipu* as a human, as opposed to just an animal, tally in Tahuantinsuyu can be judged not least from the fact, recorded in the *Nueva corónica*, that the llama census of the month *Aya marcay* was also a human census; Guaman Poma even makes an implicit comparison between the selection of males and females from both species for particular purposes, like male troops for warfare, and chosen virgins for wool production (1936, p. 257). In another chapter (pp. 193–234), the *Nueva corónica* details the categories of age and usefulness, ten for male and ten for female, according to which the human census or 'visit' was conducted. Extended to the minutest element of value in the state, reliably and retrospectively over the years, the *quipu* accounting system had, as a major feature, the noting of absence and non-performance. The Quechua term for this failure in conduct, *hucha*, was noted on the *quipu* as greater (*hatun*) or lesser (*huchuy*) (Poma de Ayala 1936, p. 361). For its part, *à propos* ceremonies that accompanied irrigation work, the *Huarochirí Narrative* (Ch. 31) notes darkly how absences of goods and of personnel were recorded on official *quipus*. By these means, and through a matching apparatus of police whose initially pastoral function is patent from the official title (*llama michic; michic*), the state was able precisely to gauge quantities not just of commodity but of labour tribute rendered.

Outraged by Spanish lawlessness and lamenting the demise of Tahuantinsuyu, Guaman Poma in his day significantly appealed to the notion of the *michic* as a last means of preventing disaster and of restoring order to society. At the same time he underpinned the traditional flock–folk equation of Tahuantinsuyu most succinctly when complaining to Philip III that in vice-regal Peru the Indians bore the burden of tax payments like domesticated animals, while *mestizos* and other mixed-bloods exempt from tax were allowed to remain wild like the vicuña (1936, pp. 215, 890, 1153).

From the non-performance and the non-compliance monitored by the *quipu*, it was but a step to the rhetoric of disobedience, crime, and sin, and corresponding retribution in the name of the state. Another of Poma de Ayala's chapters is devoted precisely to orders and types of official punishment (*Nueva corónica*, pp. 301–14); one such was reserved for those who simply moved without permission from their allotted place in the realm (Murra 1980, p. 110). In other words, like their flocks, the subjects of Tahuantinsuyu could be considered contained and penned, pastured elements of the great *Pax incaica*, safe as such from the threat of enemies and the barbaric wild beyond its rim.

When it comes to relations not so much between ruler and ruled as between ruler and authority, the *Situa* hymns further highlight the pastoral model. For here, in what emerges as a truly monotheistic impulse, monarchy is endorsed by the supreme spiritual principle known as

Viracocha, and as the 'creator' (*camac*) of earth and men, and so on.

Invoked in most of the hymns, this figure is asked to guard the Inca just as the Inca guards his flocks. Throughout Tahuantinsuyu the rites of this supreme shepherd can be shown to have been imposed over local deities and '*huacas*', a process examined by Poma de Ayala in another chapter of his *Nueva corónica* (pp. 261–73). Irreverence beneath the Inca imposition is shown up in the *Huarochirí Narrative* which basically remains loyal to the lightning god, *Pariacaca*, and other shamanist *huacas* of the region; this source also tells how the priests imported and appointed by the Inca left their posts on hearing about Pizarro's advance (Ch. 18). This last detail is significant also because it indicates how far religion, or 'the Church' as it is often called, had become subject to and regulated by the state, exposing the universally imposed *Viracocha* or divine shepherd to have been, in practice, a back-projection of and from the secular power of the Inca themselves. Similarly, though formally distinct and guarded by special herders, who possibly included the *aclla*, or chosen virgins of the sun (another exclusively Inca institution), the Church herds relied on the state for allocations of pasture, just as ritual llama sacrifice imposed by the Inca served *mitima* and state-herding interests.

On the same subject of Inca-appointed priests, the *Huarochirí Narrative* specifies, in passing, their period of service as 15 days, or half the official Inca month. This indicates, in turn, how the appropriation of divine authority coincided in practice with the institution of a state year-calendar, one which could encompass in a single standardized whole the various rhythms of agriculture, curing, and the myriad other rituals of society, as well as the demands of material tribute. According to Poma de Ayala's chapter on the subject (1936, pp. 236–60), between the solstitial and equinoctial celebrations in honour of the divinely sanctioned Inca and his queen, this calendar deferred thematically to the tasks of the pastoral year with its llama census in *Aya marcay* (November) and intervening one in *Aimoray* (May), and with its regular sacrifices of llamas and alpacas throughout, like that actually depicted for *Pacha pucuy* (March; see Fig. 22.3).

From these further sets of evidence, Tahuantinsuyu appears to have been as distinctive ideologically within native America as it was economically. The sort of entreaty made in the *Situa* hymns, to a single god that can guarantee monarchic power and guide its course like a shepherd, goes beyond anything that can be found in comparable native American religious texts, which likewise are entirely devoid of equations in principle between human and animal herds that are faithful to their keeper. In Mesoamerican sources, such as the Twenty Sacred Hymns of the Aztecs and the Aztec address to the Franciscan missionaries of 1524 (Brotherston 1979, pp. 63–9), the divine, the ruler, and the ruled interrelate quite differently, a typical model being the tripartite one which places the archetypally opposed social groups of planters and hunter-warriors under the aegis of the aristocrat-priests; and that these last were no mere appointees of the emperor is clear from a whole range of evidence, like the generic difference between priestly and secular texts, the

Figure 22.3 Llama sacrifice in the month *Pacha pucuy* (Poma de Ayala 1936, p. 240).

plural calendrics of the divine *tonalpohualli* and the tribute-year, and the sheer layout of Tenochtitlan's pyramids.

In Tahuantinsuyu, how far flock-economy shaped ideology is well caught in these lines from another prayer to *Viracocha*, in which learning obedience to State and Church is directly equated with animal domestication (Lara 1969, p. 192):

> the vicuna of the wilds
> the viscacha of the rocks
> become domesticated
> in his presence
> so too my heart
> with each dawn
> renders you its praise
> my father and creator

(*purun wikuna / qaqa wiskacha / uywaman tukun / paypaj qayllapi / sunqoypas kikin / sapa paqarin / anayniskuni / yayay kamaqey*)

The larger native American context

Exploring Tahuantinsuyu texts in their New World context helps us to recognize further principles of difference which impinge directly on our assessment of the nature of the Inca state and, by extension, of pastoralism generally. For Andean and Inca herding evidently modified whole cultural paradigms that were otherwise common to America, by virtue of their common source in shamanism, and in agricultural practice whose antiquity is universally testified to, not least by the *Popul vuh* in Mesoamerica and Poma de Ayala in Tahuantinsuyu (1936, p. 48); and just as evidently it produced whole new modes and genres of expression otherwise absent from the continent.

In formulating human–animal relations as such, a wealth of evidence suggests that Andean society continued to share certain beliefs and codes of behaviour with other parts of the New World, despite its own advance into pastoralism. In the cosmogonical context of the world-ages or suns found in much of native America, for example, we may detect in Tahuantinsuyu texts traces of the animal-other that is typically invoked in the hunter's propitiation, and which is best known in the Mesoamerican philosophy of the '*nahual*'. In the case of the Flood, the catastrophe that ends the first world-age, the *Huarochirí Narrative* (Ch. 3) tells us that humans were warned of its onset by none other than a llama, one possessed of just the sixth sense, and that psychic intimacy, if not solidarity, that characterizes the *nahual*. In the subsequent catastrophe of the eclipse, Tahuantinsuyu's consistency with native American belief is yet more striking, since here we are involved specifically with domestication, and the domestic contract that binds humans to creatures who have left the wild to fall under their control. In fact, the eclipse in question is said to have been occasioned by heartless human exploitation of domestic creatures: regaining their wildness these turn on their masters and help to destroy them. In Mesoamerican texts, such as the *Popol vuh*, the upheaval is led by the mortars and kitchen utensils tired of unfeeling use, and not least by the dog and turkey, the two creatures whose long-standing alliance with American man is elsewhere celebrated in rituals that survive even today, for example in the southwest of the United States (Brotherston 1979, pp. 111–13). That in Tahuantinsuyu this same contract was perceived to have extended to llamas is clear from the *Huarochirí Narrative* (Ch. 4), where during the eclipse the revolt of the utensils is accompanied by one on the part of these creatures, who turn on their masters in savage herds; the same moral is then reinforced in the parable of the few wild vicuña who help the poor Huatyacuri to outdo the teams of llamas enlisted by his rich and insensitive antagonist (Chs 4 & 5).

Consistent with the terms of this domestic contract is the inclusion of the tamed creature as a ritual ally and even social companion. The actual process of inclusion is well illustrated in Poma de Ayala's report of the ceremonies proper to the month *Uma Raimi* (October) (p. 254), in which supplicants for rain included not just human children and dogs, whose tears and howls were elicited as sympathetic magic (as in Mexico), but a black llama as well.

Taken together, all this suggests that at one level of its culture, the Andes shared a native American recognition of the moral problem raised by the use and exploitation of animals, and of the need to regulate and ritualize it. And, in practice, the Inca would refrain from eating llama meat at certain periods, out of that contractual respect which in Mexico extended even to maize, doctrinally the source of human flesh. Yet, as we have seen, as objects of Tahuantinsuyu's pastoral economy, llamas were none the less bound to be treated in ways that could not be contained within these traditional contractual limits. For they were obliged to function as mere units of value, items of exchange devoid of particular status or rights, and were transacted on the grand decimal scale of the *quipu* as the indispensable capital-plus-interest of the state. This is certainly the role they are assigned in Guaman Poma's account of the world-ages, which lies closer than does the shamanist *Huarochirí Narrative* to official Inca thinking, and emphasizes the material and statistical value of llamas and their wool as the factor that distinguished the age of the Purun and the beginnings of state power.

One response to this discrepancy or moral dilemma appears to have been the selection of individuals on which to bestow, in a sacred-cow logic of compensation, as it were, the regard that could not be practicably accorded to the species generally. In this respect, the single black llama who sang at *Umu raimi*, while hundreds of his lesser brethren were slaughtered and consumed by state order, resembles the royally chosen and lavishly attired *napa* of the Colla ceremonies; or the red llama which the emperor took as a singing companion and musical guide in court performances of the *Quechua yaraví* (see Frontispiece). Indeed, only in terms of such privileging could individual llamas have come to serve humans as their representatives in their deeper spiritual needs, which is what they undoubtedly did as bearers of guilt, a paradoxical and terminal privilege. In the *Situa* ceremony human sins and failings were transferred on to eviscerated llamas which were washed out of the capital via its river. Once again, no other American religion offers anything like this parallel to the well-known scapegoat and sacrificial-lamb practices developed in the Old World by the Semitic pastoralists.

Finally, besides modifying cosmogonical and other paradigms found widely in native America, Inca herding led to a literary pastoralism that was otherwise quite absent from the continent. While the courts at Tenochtitlan and Cuzco alike cultivated poetry in modes that derived from social practice, planting songs from the farmer, battle songs from the warrior, funeral songs from the mourner, and so on, only in the Inca case were the pastoral songs of the herder also heard, for reasons that by now will be obvious (Brotherston 1979, pp. 260–5). Themes characteristic of this pastoral mode included the nameless yearning inspired by the landscape of the high puna, lonely and sublime (unruly sons of royal families were customarily exiled to pastoral life on the puna); love between male and female herders, in which the desired one is compared with the elusive untamed vicuna, or is revealed as the inaccessible princess of the *aclla* who guarded the Church's flocks; and the herders' flutes whose plaintive sound could presage the suicide of one or both victims of an impossible love.

So strong are these Andean pastoral conventions that they survive even today in *Quechua* poetry, despite the influence of their more sexist European analogues (Yaranga Valderrama 1986). They likewise inform a whole series of legends ('*Hirtenmärchen*' in Kelm 1968) and even dramas, which tend, however, to allude more directly to the social conditions and constraints under which herders actually worked. A pastoral love-story recounted by Murua in this way reveals the pastures beyond the edge of town and agriculture as the place of transformation and fantasy: the metamorphosis of shamanism is here made to serve the particular needs of a lonely or alienated individual, who changes into a staff, a snake, a bear, in the cause of sexual fulfilment (Basadre 1938, pp. 31–8, Arguedas 1949, Kelm 1968). The drama *Apu Ollantay*, where the rebellious Antisuyu hero strays from the right path like a disobedient llama lamb, foregrounds Inca policies of confining high-born daughters to the *aclla* (Brotherston 1986).

Like the encyclopaedic *Nueva corónica* and *Huarochirí Narrative*, and like the *Situa* hymns, these poems, legends, and dramas offer their own perspective on the complex workings of Andean pastoralism, and on those grounds deserve more attention than they have received. Indeed, as more self-consciously literary products, they offer special insights into Inca pastoral ideology as it involved such concepts as obedience, alienation, and love. Yet for now, the point must simply be to indicate their consistency in turn with the economics of Tahuantinsuyu, and their consequent uniqueness in the New World.

Conclusions

Privileging native texts as evidence and respecting their New World context, this enquiry started by suggesting Tahuantinsuyu as a valuable term of reference by which to assess the ideology, no less than the economics, of pastoralism. In the first place, there can be little doubt about the huge significance of pastoralism for and within Tahuantinsuyu: the practices associated with it, particularly the *mitima*, radically impinged on the lives and movements of beast and human alike, to an extent which it is, incidentally, hard to reconstruct, in the absence of the Inca system itself, from llama herding as it has survived in the Andes today. From this base, pastoralism was shown to have permeated principal aspects of Inca ideology. It patently shaped models of political control and authority, along with their extension into the divine; it modified traditional codes of interaction between humans and animals, such as the domestic contract of American world-age cosmogony; and it inspired a gamut of literary pastoralism. The internal consistency of this evidence, its distinctiveness within native America, and its wider socio-political resonance, surely endorse it as consequential for any comprehensive analysis of pastoralism.

Acknowledgements

In preparing this chapter I have been helped by Olivia Harris, who read through an early draft, and by the discussion at the World Archaeological Congress 1986.

Note

1 The page-numbers given for Poma de Ayala's *Nueva Corónica* are his original ones (Poma de Ayala = Guaman Poma).

References

Arguedas, J. M. 1949. *Canciones y cuentos quechuas*. Lima: Huascaran.
Arguedas, J. M. 1975. *Dioses y hombres de Huarochiri*. Mexico: Siglo XXI.
Ascher, M. & R. Ascher 1981. *Code of the Quipu. A study in media, mathematics, and culture*. Ann Arbor: University of Michigan.
Basadre, J. 1938. *Literatura inca*. Paris: Desclée de Brouwer.
Bökönyi, S. 1989. Definitions of animal domestication. In *The walking larder*, J. Clutton–Brock (ed.), ch. 2. London: Unwin Hyman.
Brotherston, G. 1979. *Image of the New World. The American Continent portrayed in native texts*. London & New York: Thames & Hudson.
Brotherston, G. 1986. The royal drama *Apu Ollantay*. *Comparative Criticism* **8**, 70–94.
Browman, D. L. 1989. Origins and development of an Andean pastoralism: an

overview of the past 6000 years. In *The walking larder*, J. Clutton–Brock (ed.), ch. 23. London: Unwin Hyman.

Ducos, P. 1989. Defining domestication: a clarification. In *The walking larder*, J. Clutton–Brock (ed.), ch. 3. London: Unwin Hyman.

Kelm, A. 1968. *Vom Kondor und Vom Fuchs*. Berlin: Gebr. Mann.

Lara, J. 1969. *La literatura de los quechuas*. La Paz: Editorial Juventud.

Murra, J. V. 1980. *The economic organization of the Inka state*. Greenwich: JAI Press.

Murra, J. V. & R. Adorno (eds) 1980. Guaman Poma, *El primer nueva corónica y buen gobierno*. 3 vols. Mexico: Siglo XXI.

Pease, F. (ed.) 1980. *Guaman Poma, El primer nueva corónica y buen gobierno*. Caracas: Ayacucho.

Poma de Ayala, F. G. 1936. *El primer nueva corónica y buen gobierno*. Facsimile edition. Paris: Musée de l'Homme.

Rowe, J. H. 1953. Eleven Inca prayers from the Zithuwa Ritual. *Kroeber Anthropological Society Papers* nos 8 & 9.

Shimadu, M. & I. Shimadu 1985. Prehistoric llama breeding and herding on the north coast of Peru. *American Antiquity* **50**, 3–26.

Spalding, K. 1984. *Huarochirí. An Andean society under Inca and Spanish rule*. Stanford: Stanford University Press.

Trimborn, H. 1939–1941. *Damonen und Zauber in Inkareich*. Berlin: Gebr. Mann.

Yaranga Valderrama, A. 1986. The Wayno in Andean civilization. In *Voices of the first America*, G. Brotherston (ed.), 178–95. Santa Barbara: New Scholar.

Zuidema, R. T. 1983. Towards a general star calendar in ancient Peru. In *Calendars in Mesoamerica and Peru*, A. F. Aveni & G. Brotherston (eds), 235–61. Oxford: BAR.

23 Origins and development of Andean pastoralism: an overview of the past 6000 years

DAVID L. BROWMAN

Introduction

Substantial advances have been made in the 1970s and 1980s in deciphering the complex history of the domestication of Andean camelids. Four closely related taxa are involved: the wild vicuna (*Vicugna vicugna*) and guanaco (*Lama guanicoe*), and the domestic llama (*Lama glama*) and alpaca (*Lama pacos*).

Franklin (1978, 1983), Raedeke (1979), and Jefferson (1980) have done much to refine the work by Koford (1957) and others detailing the ethology of the wild species. As one or both of the wild camelids have been suggested as ancestors of the domestic varieties, such knowledge is indispensable to our theories of domestication. While some contemporary researchers (such as Cardozo 1975, Craig 1985) argue for Pleistocene ancestors for these domestic species (such that the guanaco and vicuna would not be involved as progenitors), I follow the majority who believe that the llama and alpaca are directly descended from the wild species.

Vicuna are sedentary, obligate drinkers of water, and altitudinally restricted to between 3700 and 4900 m above sea-level, while guanaco can be either sedentary or migratory, are both grazers and browsers, are periodic drinkers of water, and range from sea-level to over 4000 m in altitude. In addition, guanaco have a longer period of parental care of their young, extending over two growing seasons, compared with one season for the vicuna (Franklin 1983, pp. 573–4). The greater flexibility of the guanaco allows it to occupy a wider array of zones, including the semi-arid *puna* and *altiplano* grasslands, where the archaeological sites, discussed here, have been excavated.

In the Old World, pastoralism is often seen as a secondary spin-off of agriculture, as a specialization resulting from farmers being forced to adapt to environments on the margins by population pressures (as described, for example, by Lees & Bates 1974). In contrast, the pattern that is being recovered from the Andes suggests that llama and alpaca pastoralism, and what Wing (1973, 1977) calls the 'pastoring' of guinea-pigs, developed much earlier than plant cultivation.

Studies of domestication in the Andes have four major foci:

(a) distributional studies (target animal occurs outside its previous normal range, implying human intervention);
(b) utilization studies (a sudden shift or increase in the importance of the animal, suggesting management of the stock by human curators);
(c) population studies (a shift in the mortality, suddenly more young or old animals being cropped; and
(d) morphometric studies (occurrence of distinctive morphological features that result from human selective pressures).

Distribution and intensification studies

Historical records as well as present-day distribution indicate the Andean highlands to be the primary habitat for domestic camelids and thus research focused there. Gade (1969, p. 341) is among several noting that the majority of herding centres for alpaca today are in an area within a *c.* 150 km radius of Lake Titicaca, and thus suggested the Titicaca basin as the centre of alpaca domestication. To test this, and other hypotheses, we need appropriate faunal analyses of prehistoric periods. It has been in this area that major contributions have been made in the past two decades.

The basic presumption is that a high degree of dependence on one animal is correlated with domestication and herding. A significant difference exists in the methodologies employed to assess species importance. For example, Wing uses MNI (minimum number of individuals) while Wheeler and I use NISP (number of individual specimens present). NISP skews as follows:

(a) it ignores the difference in number of bones in the skeletons of different species;
(b) it is affected by cultural practices such as fragmenting bones for marrow extraction;
(c) it results in an over-emphasis of species butchered on site; and
(d) it reflects collection techniques (e.g. more bone fragments recovered through finer mesh screens, etc.).

MNI skews as follows:

(a) it is very sensitive to aggregation procedures, i.e. how the collection unit is defined and what its size is;
(b) it does not provide information on the relative abundance of fauna at the site as a whole;
(c) there are intrinsic problems in how to match or pair elements; and
(d) it overestimates the contributions of rare species.

MNI is more appropriate if one wants to calculate estimates of biomass or meat contribution, and if the diversity in the number of species represented is high. However, it is my argument that in the case of domestication studies, where species diversity is low, and dependence on a

Table 23.1   Chronology for camelid dependence at Telarmachay (sources: Lavallee *et al.* 1984, Wheeler 1984).

Date	Camelid percentage
7000–5200 BC	65
5200–4800 BC	78
4800–4000 BC	82
4000–3500 BC	87
3500–2500 BC	86
2500–1750 BC	89

single species is high, that MNI is less desirable than NISP as MNI tends to place too much emphasis on rare species, and masks the relative abundance of the target species (the potential domesticate).

The percentages that follow are, where possible, calculated from the numbers of individual specimens. In some cases, however, original data are only available as MNI percentages.

Central highlands

Three sites near Lake Junin, Panaulauca, Pachamachay, and Uchcumachay, and one site slightly further north in Cerro de Pasco, Lauricocha, provide evidence that the pampas of Junin was one centre of domestication.

Faunal analyses at Panaulauca by Wheeler (Wheeler *et al.* 1976 [1977], Pires–Ferreira *et al.* 1977) indicated a shift in species utilization there at around 5500–5000 BC. Camelids contributed 26 per cent of the assemblages dating to 7000–5500 BC, but increased to 86–87 per cent of the assemblages dating between 5500 and 2500 BC. Subsequent work by Moore (1982, 1985a, 1985b) has confirmed this general pattern at Panaulauca. Initial analyses of Pachamachay by Wheeler indicated 96–98 per cent of the collections were camelid in the phases between 4200 and 1750 BC. Subsequent analysis by Kent (1982a) indicated camelid percentages in more recent levels of the site, ranging from 93 per cent at 2200 BC to 82 per cent by 400 BC. Recent work at Telarmachay by Wheeler (1984) and Lavallee *et al.* (1984) provided a finer-tuned chronology for increasing camelid dependence (Table 23.1).

For Lauricocha, Wing (1972) placed domestication in the 6000–3000 BC period. Cardich (1983) argued that the shift to domestication occurred between Lauricocha II and Lauricocha III. Wheeler and colleagues were able to re-examine some of the Lauricocha collections, and estimate a proportion of some 59 per cent for Lauricocha I, before 6000 BC, and an increase to about 85 per cent by Lauricocha III, around 3000 BC.

The pattern for the Junin area thus appears to be one of increased dependency upon camelids, beginning at least as early as 7000 BC. The recovered assemblages prior to 7000 BC are too small to provide good statistics, but suggest little dependence upon camelids. By 4500–4000 BC, up to 80–90 per cent of the faunal assemblage in some highland sites may be made up of camelid species, and there appears to be little change in the emphasis on quantity of camelids for the next 4 millenniums.

North highlands and north coast

As one moves further north, remains of domestic camelids occur later. At two sites near Pampa de Lampas, just north of Junin, camelid percentages are still high (listed at 64–80 per cent at *c.* 3000 BC, Wing 1978, p. 169). But further north at Guitarrero Cave (the locus of the earliest claimed occurrence of the domestic *Phaseolus* bean), the importance of camelids is dramatically less. Initial estimates ranged from 10 to 17 per cent for all levels, but later re-analyses indicated no more than 10 per cent for levels earlier than 5000 BC, and no more than 33–35 per cent for the next 6 millenniums (Wing 1978, 1980).

Studies of the Chavin contemporary temple sites in Ancash and Cajamarca have allowed some refining of our understanding of the later spread of camelids. At Chavin de Huantar, there is a sharp shift in dependence on camelids *c.* 500 BC. For the Urubarriu phase (850–450 BC), camelids made up about 70 per cent of the faunal remains, but there was an increase in the subsequent Chakinani and Janabarriu phases (450–200 BC) to 96–97 per cent camelid (Miller 1979, 1981). At Huaricoto, there is a shift in camelid utilization from 26–31 per cent for levels pre-700 BC up to 56–69 per cent in post-700 BC levels (Burger 1985). This shift is more dramatically reflected in the site of Huacaloma near Cajamarca (Shimada 1982, 1985) where the average contribution per phase can be assessed as in Table 23.2.

Northern highland sites indicate an increase in the number of camelids about 700–500 BC, indicating a shift in human and camelid interaction. The precise nature of the shift is debated: was it evidence of southern pastoralists moving into the area, of local herding adaptation, or of new trade simply bringing in meat animals?

The occurrence of domestic camelids becomes progressively later further north. While the Chavin contemporary site of Pacopampa exhibited a shift in camelid dependence after 800 BC (Rosas & Shady Solis 1974, p. 24), unpublished work by A. Meyers and U. Oberem suggests that it is not until *c.* AD 1000 that substantial numbers of camelids occur at Cochasqui, Pinchincha, near Quito.

Table 23.2 Camelid content of faunal remains at Huacaloma (source: Shimada 1982, 1985).

Date	Phase	Camelid percentage
1200–900 BC	Early Huacaloma	15
900–600 BC	Late Huacaloma	10
600–500 BC	EL	41
500–300 BC	Layzon	88
post –600 BC	various Cajamarca	95

North-coast sites reflect this late northward expansion of domestic camelids. At Cholupe, Shimada (1981) first suggested introduction of camelids by the Peruvian Initial Period (1750–1000 BC), although later (Shimada *et al*. 1983) she indicated that domestic camelids did not occur at Huaca Lucia–Cholupe until 700 BC. Pozorski (1976, p. 101) suggested an Initial Period/Early Horizon period (*c*. 1000 BC) for introduction of camelids at Cabello Muerto, with subsequent extensive dependence upon camelids during the Moche phases (beginning 200 BC). Recently, Pozorski (Lynch 1986, p. 174) suggested that an invasion of new peoples brought maize, new ceramic and architectural styles, guinea-pigs, and camelid husbandry to the north-coast Casma Valley about 900 BC. Both highland and coastal evidence suggests a northward expansion of domestic camelids sometime between 1000 and 500 BC.

Southern highlands

Ayacucho appears to represent an area transitional between possible domestication centres in the Junin and Titicaca grasslands. In his initial analysis, MacNeish (1969, pp. 26, 38) believed that he had isolated evidence for the llama at Jaywamachay Cave between 6300 and 5000 BC, with domestication clearly evident in the subsequent phase from 5000 to 3800 BC. This initial assessment was subsequently revised, with the first appearance of domestic camelids not seen until 3100–1750 BC (MacNeish *et al*. 1975, p. 46). Both initial calculations by Wing (1978, p. 169), and my subsequent computations based on the final report (MacNeish *et al*. 1983) suggest relatively little dependence on camelids, especially as contrasted with either Junin to the north or Titicaca to the south.

One of our major problems in terms of clearly identifying the Titicaca basin as a second centre of domestication is that we have almost no good data from sites prior to 1500 BC. Minaspata and Marcavalle are representative of data from the Cuzco area. The initial analysis of Minaspata assemblages, for a unit *c*. 1000 BC to AD 500 (Wing 1973, 1978), indicated an MNI estimate of *c*. 45 per cent camelids. Subsequent analysis with finer chronological units, provided the NISP estimates given in Table 23.3. The apparent shift in camelid percentage actually reflects a drop in guinea-pig dependence at the site; the camelid percentage might be argued to be rather constant if guinea-pig-'corrected'. The sequence from Marcavalle is much finer in time-scale, with four phases from 1000 BC to

Table 23.3 Camelid content of faunal remains at Minaspata (based on Dwyer & Wheeler 1985).

Date	Original report %	Guinea-pig-'equalized' %
1000–0 BC	55	73
AD 1–500	79	79
AD 500–1000	83	83

Table 23.4 Camelid content of faunal remains of Pikicallepata (source: Wing 1973, 1978).

Date	Camelid percentage
1350–1150 BC	56
1150–950 BC	48
950–750 BC	46
750–250 BC	40

Table 23.5 Camelid content of faunal remains at Chiripa.

Date	Camelid percentage
1350–1000 BC	98
1000–850 BC	91
850–600 BC	92
600–350 BC	91
350 BC–AD 100	88

600 BC (Mohr–Chavez 1982, p. 244); MNI-based camelid percentages in these phases range from 84 to 97 per cent.

In the Cuzco sites, there is good match between MNI and NISP estimates, further south, in the Pucara vicinity, the MNI and NISP estimates are not in as good concordance. For the site of Pikicallepata, the number of guinea-pigs, as well as the MNI computations, result in lower estimates of camelids (Table 23.4).

For the period 1450–1050 BC the site of Qaluyu had 41 per cent camelid (corrected for human and fish), and the site of Q'ellokaka had 28 per cent, using MNI estimates. But for Pucara, using NISP estimates, Wheeler & Mujica (1981) computed 96 per cent camelids in the pre-Pucara levels, 93 per cent in the 850–500 BC levels, and 97 per cent in the 500–200 BC levels. I suspect that the disparity in percentage is an artefact of the NISP versus MNI approaches, rather than a sudden shift in camelid importance at the site.

At the southern end of Lake Titicaca, the Chiripa NISP-based sequence also covers no earlier than 1350 BC (Table 23.5). The gradual decrease in camelid proportions is due to an increase in numbers of birds from the lake, which also correlated with an increase in fish bone, suggesting a probable increase in exploitation of lacustrine resources.

The evidence from Ayacucho, Cuzco, and Titicaca indicates a shift to herding prior to 1500 BC; the revised Ayacucho data indicate that herding can be identified no earlier than c. 3000 BC which is roughly in line with the Junin data.

The evidence from further south indicates a later spread of herding into Argentina and Chile. The data here are much more tentative. Jensen and Kautz (1974, p. 46) reported domestic camelids at Tarapaca 2A, which they estimated to be *c.* 2000–1500 BC. Subsequent analyses by Simons (1973, 1980) indicated that the Tarapaca 2A materials were clearly guanaco, but there might be some domestic camelids at Tarapaca 12 & 18, *c.* 2700–1800 BC. Hesse (1982) reports identification of domestic camelids from the Atacama area by 3000–2500 BC. Clear evidence of domestic camelids can be found in Rio Loa by 500 BC. In Argentina, Yacobaccio (1979) reports domestic camelids at Huachichona, Humahuaca, dating to 1450 BC and further south at Las Cuevas, Salta, by 500 BC. The Chilean and Argentine evidence is sketchy, but appears to support a southerly diffusion of domestic camelids by the 2nd millennium BC.

Population and management shifts

The idea that shifts in patterns of mortality are indicative of human management or domestication has been current in studies of herding for some time, and is one presently emphasized in the Andes. Collier & White (1976) have challenged the idea that a shift to many immature animals in an archaeozoological sample indicates domestication. They show wide variation in population structures in non-domestic herds, in vulnerability of age/sex classes to predators, and in selective hunting impact. While these are valid criticisms, herding practices in the Andes make it appropriate to consider the mortality argument seriously. Contemporary herders in the Andes experience high mortality amongst neonate animals of both camelids and introduced European stock, due to climatic variables and disease.

Wheeler (1982, 1984) and Lavallee *et al.* (1984) have argued that neonate mortality is principally due to enterotoxaemia diarrhoea caused by *Clostridium* sp. bacteria. Risk of this disease is very high in modern herding practices, as the disease is spread in corrals, and in specialized grazing areas called 'bofedales'. Hence Wheeler argued that a shift to a high number of neonate deaths can be interpreted as evidence of these herding practices, and thus evidence of domestication. Similarly, Kent (1982a) has pointed out that occupation of a site during the calving season can result in high numbers of neonate remains. Wheeler has suggested that the occurrence of a single burial of 13 neonate camelids from Telarmachay is clear evidence of diarrhoea-related mortality; but Franklin (1982) and Malo Anccasi (1982) point out that among both wild and domestic camelids, climate is a major cause of mortality. In my experience with Bolivian herders, enterotoxaemia did not kill in a single catastrophe, rather deaths were spread out over several weeks. On the other hand, an unexpected snowfall during February resulted in the immediate catastrophic death of nearly 80 per cent of the camelid offspring in one herd. Obviously, caution must be exercised in interpreting the cause for shifts in neonate numbers in mortality curves.

The generally accepted interpretation of a substantial number of juvenile camelids in an assemblage is that the primary purpose of herding was for meat animals. However, if there are substantial numbers of adult domestic animals it is assumed that a shift in management has occurred, with an emphasis on keeping older animals for both wool production and for use as caravan animals. Sites in the Andes provide evidence for both types of herding. The Pucara and the Chiripa patterns indicate a clear shift toward keeping older animals for caravan and wool purposes in the Titicaca basin. Contrasting the patterns from Kent's (1982a) analyses of a Junin pampa site (Pachamachay) and a Titicaca basin site (Chiripa), it appears that two different management patterns are being employed. The Junin focus on pastoralism seems primarily meat-oriented, while that of the Titicaca area seems primarily wool- and caravan animal-oriented.

The meat-orientation emphasis appears to have been maintained in the northern part of Peru for an extensive period. When camelid utilization expanded north c. 1000–500 BC, it was the meat aspect which appeared to be important; juvenile animals dominated at Chavin after 400 BC, with 40–50 per cent of the animals slaughtered at 2–3 years of age; and a similar pattern is observed for Huacaloma, where the shift from a meat orientation to a cargo/wool orientation does not occur until the Middle Cajamarca phase c. AD 500 (Shimada 1985).

The 'charki' factor

The well-known 'schlepp' effect of Perkins & Daly (1968) describes the practice of bringing only the edible parts of a carcass back to the home base. Dried and salted camelid meat is known as 'charki' in the Andes. Miller (1979, 1981) observed an absence of foot elements at the archaeological site of Chavin; he also noted that modern herders near Cuzco only traded charki made from body and limb parts, and never used foot or head elements. Thus, he hypothesized that an absence of foot elements could be correlated with meat traded as charki, and coined the term 'charki effect', which is synonymous with the 'schlepp effect' of the Near East.

However, there is a major problem with this charki effect : it simply does not exist uniformly. In markets of Bolivian Aymara, charki is often composed exclusively of flattened heads (broken, dried and stacked in 0.5 m diameter sheets) and foot elements. Stanish (1985) and Williams & Clark (1986) report a number of Late Intermediate Period (AD 1000–1450) tombs in the Moquegua coastal zone, immediately west of the Titicaca basin, which include large numbers of once-articulated llama feet (mainly phalanges 1, 2, and 3) but no other llama bones. Similar offerings of llama feet are reported for tombs of the Maitas culture (AD 750–1200) in the Azapa valley of Chile (Munoz Ovalle & Focacci Aste 1985, p. 26), just southwest of the Titicaca basin. Knowing this Bolivian pattern for trade in charki, where foot and head elements dominated, Miller's assertion that head and foot elements are 'never used' for charki is not true, and caution must be exercised in arguing that a shift in foot/limb elements implies either a management shift to meat production, or to cargo/wool orientation.

Some reassessment might be needed in the argument for the appearance of coastal llama herds by 700 BC. Shimada *et al.* (1983) and Shimada & Shimada (1985) assume that since the *charki* effect pattern is not observed on the north coast sites, therefore trade is not responsible for the presence of llama in these sites, and that the bone distribution is evidence of coastal herding. However, as the *charki* effect is not universal, this argument collapses. Moreover, today in the southern coastal areas, live animals brought in by highland herders are a very important trade item. Faunal assemblages from recent coastal sites would include not only bones from entire animals, but the sites would also include the corrals where the caravan animals were kept. Thus, the occurrence of corrals and entire animal skeletons on the coast is not evidence against trade and for local herding, but might be seen as *prima facie* evidence for trade.

The limb/foot ratio, and the neonate/juvenile/adult ratios do provide us with techniques to identify shifts in utilization and management. Together with other evidence they can be utilized to demonstrate certain aspects of domestication.

Kent (1982a,b) suggests that one ought to be able to distinguish highland from coastal camelids based on osteon (Haversian system) densities, but this idea has not been tested.

Morphological traits

The most appropriate measure of domestication might be argued to be the occurrence of discrete morphological traits only present in domestic animals. There are some size distinctions between the llama and alpaca, but unfortunately the same is true of the guanaco and the vicuna. Wing (1972) distinguished small camelids from large camelids on the basis of the measurements of femur, humerus, and scapula, and later refined this (1973) to measurements of the astragalus, calcaneum, and distal widths of the humerus and tibia. While this provided a mechanism for determining the number of small versus large camelids on a site, it did not permit differentiation of wild from domestic (alpaca and vicuna are both small; guanaco and llama are large). Independently, Miller (1979) and Kent (1982a) worked out a series of metrical criteria to differentiate all four species. Kent's set is the most complete, and apparently the most successful.

Rick (1980, 1983) proposed that Pachamachay had been inhabited by a settled population of vicuna-hunters, living year around at the site, with seasonal utilization first occurring in Period 7, when he suggested a dry-season period of hunting. Kent's (1982a) re-analysis of the fauna, using his morphological parameters, indicated no evidence for vicuna being the major camelid, and that, if anything, after Rick's Period 5 (*c.* 1500 BC), domestic camelids, particularly alpaca, dominated; and that if there was any seasonality in Period 7 (800–400 BC), it was likely wet-season occupation, not dry-season. Thus, a faunal analysis using better diagnostic criteria contradicted most of Rick's model for sedentary vicuna-hunting. For those of us believing the domestication of alpaca to have taken place in

Junin, possibly as early as 4000 BC, Rick's model was uncomfortable, as it flatly contradicted the domestication postulates; thus I prefer Kent's hypothesis.

Apart from the relative sizes of the bones, another morphological trait that can be used to identify camelids is the shape of the incisors. Alpaca incisors are thought to be intermediate in form between vicuna and guanaco/llama. Vicuna incisors are parallel-sided, with open roots; guanaco and llama incisors are spatulate-sided with closed roots. Alpaca incisors have an enamel distribution like that of the vicuna, but the cross-section is more rectangular than it is in the vicuna incisor, where it is square. Wheeler (1984) has used this trait to argue for alpaca domestication as early as 4000 BC at Telarmachay. While this seems to be clearly an alpaca trait, not all alpaca exhibit such incisors (Kent 1982a, p. 142, Shimada & Shimada 1985, p. 18), so it is only useful in identifying the presence of alpaca, not in quantifying their numbers.

Final remarks

Our current understanding suggests that there was at least one centre of domestication in the Lake Junin area, and possibly a second in the Lake Titicaca area. Consideration of all parameters indicates a probable period of intensification of utilization of camelids c. 7000 BC, with sufficient management taking place that domestic animals may be recognized by c. 4000 BC. The centre for alpaca domestication may be the Lake Junin area. In measures of small versus large camelids, the ratio of small camelids in the Lake Junin archaeological sites frequently goes as high as 9 or 10 to 1, while, in contrast, large camelids frequently dominate the Lake Titicaca environs, being 3 or 4 times as frequent in some sites as small camelids. A shift in the management of camelids from meat to wool and cargo-bearing appears to occur first in the Lake Titicaca region. The two zones are for a time distinguishable in terms of focus on large versus small domestic camelids, and on meat versus wool/cargo-bearing attributes.

References

Burger, R. L. 1985. Prehistoric stylistic change and cultural development at Huaricoto, Peru. *National Geographic Research* **1**(4), 505–34.

Cardich, A. 1983. A propósito del 25 aniversario de Lauricocha. *Revista Andina* **1**(1), 151–73.

Cardozo, A. 1975. *Origen y filogenia de los camélidos sudamericanos*. La Paz: Academia Nacional de Ciencias de Bolivia.

Collier, S. & J. P. White 1976. Get them young? Age and sex inferences on animal domestication in archaeology. *American Antiquity* **41**(1), 96–102.

Craig, A. K. 1985. Cis-andean environmental transects: late Quaternary ecology of northern and southern Peru. In *Andean ecology and civilization*, S. Masuda, I. Shimada & C. Morris (eds), 23–44. Tokyo: University of Tokyo Press.

Dwyer, E. B. & J. C. Wheeler 1985. *Animal utilization in the southern highlands:*

Early horizon to early intermediate at Minas Pata, Lucre Valley, Cuzco. Paper presented at the 50th annual meetings, Society for American Archaeology.

Franklin, W. L. 1978. *Socioecology of the vicuna.* Unpublished PhD dissertation, Wildlife Sciences, Utah State University, Provo, Utah.

Franklin, W. L. 1982. Biology, ecology, and relationship to man of the South American camelids. *Mammalian Biology in South America. Pymatuning Symposia in Ecology* 6, M. A. Mares & M. H. Genoways (eds), 457–90. Pittsburgh: University of Pittsburgh Press.

Franklin, W. L. 1983. Contrasting socioecologies of South America's wild camelids: the vicuna and the guanaco. In *Advances in the study of mammalian behavior,* J. F. Eisenberg & D. G. Kleiman (eds), 573–629. Special Publications of the American Society of Mammalogists 7.

Gade, D. W. 1969. The llama, alpaca and vicuna: fact vs. fiction. *Journal of Geography* **68**(6), 339–43.

Hesse, B. 1982. Archaeological evidence for camelid exploitation in the Chilean Andes. *Säugetierkundliche Mitteilungen* **30**(3), 201–11.

Jefferson, R. T. Jr. 1980. *Size and spacing of sedentary guanaco family groups.* Unpublished MS thesis, Animal Ecology, Iowa State University, Ames, Iowa.

Jensen, P. M. & R. R. Kautz 1974. Preceramic transhumance and Andean food production. *Economic Botany* **28**(1), 43–55.

Kent, J. D. 1982a. *The domestication and exploitation of the South American camelids: methods of analysis and their application to circum-lacustrine archaeological sites in Bolivia and Peru.* Unpublished PhD dissertation, Anthropology, Washington University, St. Louis, Missouri.

Kent, J. D. 1982b. *Osteon population density and age in South American camelids.* Unpublished paper presented at the 47th annual meeting, Society for American Archaeology.

Koford, C. B. 1957. The vicuna and the puna. *Ecological Monographs* **27**(2), 153–219.

Lavallee, D., M. Julien & J. Wheeler 1984. Telarmachay: niveles precerámicos de ocupación. *Revista del Museo Nacional* **46**, 55–133.

Lees, S. H. & D. G. Bates 1974. The origins of specialized nomadic pastoralism: a systemic model. *American Antiquity* **39**(2), 187–93.

Lynch, T. 1986. Current research: Andean South America. *American Antiquity* **51**(1), 171–6.

MacNeish, R. S. 1969. *First annual report of the Ayacucho archaeological–botanical project.* Andover: Robert S. Peabody Foundation for Archaeology, Phillips Academy.

MacNeish, R. S., T. C. Patterson & D. L. Browman 1975. The central Peruvian prehistoric interaction sphere. *Papers of the Robert S. Peabody Foundation for Archaeology* **7**. Andover: Phillips Academy.

MacNeish, R. S., R. K. Vierra, A. Nelkin–Terner, B. Lurie & A. Garcia Cook 1983. *Prehistory of the Ayacucho Basin, Peru.* Vol. 4: *The preceramic way of life.* Ann Arbor: University of Michigan Press.

Malo Anccasi, M. 1982. Causas de mortalidad en crías de alpaca. *Allpak'a* **1**(1), 31–4.

Miller, G. R. 1979. *An introduction to the ethnoarchaeology of Andean camelids.* Unpublished PhD dissertation, Anthropology, University of California, Berkeley, California.

Miller, G. R. 1981. *Subsistence and social differentiation at Chavin de Huantar: some insights from the preliminary analysis of the faunal remains.* Paper presented at the 46th annual meetings, Society for American Archaeology.

Mohr Chavez, K. L. 1982. The archaeology of Marcavalle, an Early Horizon site in the valley of Cuzco, Peru, Pt. 1. *Baessler–Archiv* **28**, 203–329.

Moore, K. M. 1982. *Prehistoric animal use in Junin: preliminary results of the 1981 season at Panaulauca Cave.* Paper presented at the 10th annual meeting, Midwest Conference on Andean and Amazonian Archaeology and Ethnohistory.

Moore, K. M. 1985a. *Current research in the animal economy of the central Andes: teeth and bones from Panaulauca Cave, Junin.* Paper presented at the 13th annual meeting, Midwest Conference on Andean and Amazonian Archaeology and Ethnohistory.

Moore, K. M. 1985b. *Hunting and herding economies on the Junin puna: recent paleoethnozoological results.* Paper presented at the 50th annual meeting, Society for American Archaeology.

Munoz Ovalle, I. & G. Focacci Aste 1985. San Lorenzo: testimonio de una comunidad de agricultores y pescadores Post–Tiwanaku en el Valle de Azapa (Arica, Chile). *Chungara* **15**, 7–30.

Perkins, D. & P. Daly 1968. A hunter's village in Neolithic Turkey. *Scientific American* **219**(5), 96–106.

Pires–Ferreira, E., J. Wheeler Pires–Ferreira & P. Kaulicke 1977. Utilización de animales durante el período precerámico en la cueva de Uchcumachay y otros sitios de los Andes centrales del Perú. *Journal de la Société des Américanistes, Paris* **64**, 149–54.

Pozorski, S. G. 1976. *Prehistoric subsistence patterns and site economies in the Moche Valley, Peru.* Unpublished PhD dissertation, Anthropology, University of Texas, Austin, Texas.

Raedeke, K. J. 1979. *Population dynamics and socioecology of the guanaco (*Lama guanicoe*) of Magallanes, Chile.* Unpublished PhD dissertation, Zoology, University of Washington, Seattle, Washington.

Rick, J. W. 1980. *Prehistoric hunters of the high Andes.* New York: Academic Press.

Rick, J. W. 1983. *Cronología, clima y subsistencia en el precerámico peruano.* Lima: Instituto Andino de Estudios Arqueológicos.

Rosas, H. & R. Shady Solis 1974. Sobre el período formativo en la sierra del extremo norte del Perú. *Arqueológicas* **15**, 6–35.

Shimada, I., C. G. Elera & M. J. Shimada 1983. Excavaciones efectuadas en el centro ceremonial de Huaca Lucia–Cholope, del Horizonte Temprano, Batán Grande, costa norte del Perú:1979–1981. *Arqueológicas* **19**, 109–210.

Shimada, M. J. 1981. *Ethnozooarchaeology of north Peru: highland-coast comparisons.* Paper presented at the 46th annual meetings, Society for American Archaeology.

Shimada, M. J. 1982. Zooarchaeology of Huacaloma: behavioral and cultural implications. In *Excavations at Huacaloma in the Cajamarca Valley, Peru, 1979*, K. Terada & Y. Onuki (eds), 303–36. Tokyo: University of Tokyo Press.

Shimada, M. J. 1985. Continuities and changes in patterns of faunal resource utilization: Formative through Cajamarca periods. In *The Formative Period in the Cajamarca Basin, Peru: excavations at Huacaloma and Layzon, 1982*, K. Terada & Y. Onuki (eds), 289–310. Tokyo: University of Tokyo Press.

Shimada, M. J. & I. Shimada 1985. Prehistoric llama breeding and herding on the north coast of Peru. *American Antiquity* **50**(1), 3–26.

Simons, D. D. 1973. *Faunal remains from northern Chile.* Paper presented at the 38th annual meetings, Society for American Archaeology.

Simons, D. D. 1980. Man and guanaco at an early site in northern Chile. In *Prehistoric trails of Atacama: archaeology of northern Chile*, C. W. Meighan & D. L. True (eds), 189–94. Monumenta Archaeologica **7**. Los Angeles: University of California at Los Angeles.

Stanish, C. 1985. *Mortuary architecture and interregional elite alliance in the*

Post–Tiwanaku south central Andes. Unpublished PhD dissertation, Anthropology, University of Chicago, Chicago, Illinois.

Wheeler, J. C. 1982. *Lamoid domestication and early development of pastoralism in the central Peruvian Andes*. Paper presented at the 47th annual meetings, Society for American Archaeology.

Wheeler, J. C. 1984. On the origin and early development of camelid pastoralism in the Andes. In *Animals and archaeology*. Vol.3: *Early herders and their flocks*, J. Clutton–Brock & C. Grigson (eds), 395–410. Oxford: BAR International Series S202. (Spanish version [1984]: La domesticacíon de la alpaca (*Lama pacos*) y la llama (*Lama glama*) y el desarrollo temprano de la ganadería autóctona en los Andes centrales. *Boletín de Lima* **36**, 74–84.

Wheeler, J. C., E. Pires–Ferreira & P. Kaulicke 1976. Preceramic animal utilization in the central Peruvian Andes. *Science* **194**(4264), 483–90. (Spanish version [1977]: Domesticación de los camélidos en los Andes centrales durante el período precerámico: un modelo. *Journal de la Société des Americanistes, Paris* **64**, 155–65.

Wheeler, J. C. & E. J. Mujica 1981. *Prehistoric pastoralism in the Lake Titicaca basin, Peru, 1979–80 field season*. Final report to the National Science Foundation.

Williams, S. R. & N. R. Clark 1986. *Investigations in the burial areas of the late prehistoric Estuquina site near Moquegua, far southern Peru*. Paper presented at the 14th annual meeting, Midwest Conference on Andean and Amazonian Archaeology and Ethnohistory.

Wing, E. S. 1972. *Preliminary report on the prehistoric use of animal resources in the Peruvian Andes*. Paper presented at the 37th annual meeting, Society for American Archaeology.

Wing, E. S. 1973. Utilization of animal resources in the Peruvian Andes. In *Andes 4: Excavations at Kotosh, Peru*, S. Izumi & K. Terada (eds), 327–52. Tokyo: University of Tokyo Press.

Wing, E. S. 1977. Caza y pastoreo tradicionales en los Andes Perúanos. In *Pastores de Puna: Uywamichiq punarunakuna*, J. Flores Ochoa (ed.), 121–30. Lima: Instituto de Estudios Perúanos.

Wing, E. S. 1978. Animal domestication in the Andes. In *Advances in Andean archaeology*, D. L. Browman (ed.), 167–89. The Hague: Mouton.

Wing, E. S. 1980. Faunal remains. In *Guitarrero Cave: early man in the Andes*, T. F. Lynch (ed.), 149–72. New York: Academic Press.

Yacobaccio, H. D. 1979. *Arte rupestre y tráfico de caravanas en la puna de Jujuy: modelo e hipótesis*. Paper presented at the Jornadas de Arqueológia del Noroeste Argentino, Universidad del Salvador, Buenos Aires.

24 *Are llama-herders in the south central Andes true pastoralists?*

MARIO A. RABEY

Llamas (*Lama glama*) are currently classified as 'domestic' animals, the human societies that interact with them as 'pastoralists', and the set of interactions between these camelids and such human societies is often named 'pastoralism'. In this chapter, I will present evidences that suggest that these classifications may be not fully appropriate. The information used here was obtained from my own fieldwork among the llamas' herders in the farthest part of northwest Argentina, between 21°S and 23°S. I have used data published by Flores Ochoa (1968, 1975) and Custred (1977) about herders in Peru, and by Aldunate *et al.* (1981), Castro *et al.* (1982), and Gunderman (1984) with reference to Chile.

It must be pointed out that alpaca (*Lama pacos*), the other Andean domestic camelid, does not exist in the Argentine Andes, where the herds are only llamas. But in most parts of the Andes, present-day herding practice is to breed llamas and alpacas together. Inhabitants of *punas* – the land situated above 3000–3500 m, in the zone where llamas are found – also breed sheep and goats, but these are managed separately. Unmixed llama breeding is of great theoretical interest, because the llama is a more generalized animal than the alpaca, and is probably more similar to the wild ancestor. Consequently, unmixed llama-herding provides a model which may be more appropriate in understanding the domestication process and the relationship between humans and animals in that process, than the mixed herding of llamas and alpacas.

Although there are several ways of breeding llamas, depending on the type of habitat involved, in the Argentine *punas* there are two main systems, called by the local herders: *del cerro* (hill breeding) and *del campo* (plains breeding).

Hill breeders (Figs. 24.1 & 24.2) live between 3900 and 4600 m above sea-level (ASL). These heights, and those which follow, are approximate and can vary in the order of 300 m, according to the specific ecological features of different areas. Their social unit is a family, or a coalition of kindred families, which usually has possession of the territory surrounding a small stream, and whose boundaries are the peaks separating it from neighbouring small basins.

The hill breeders' main homestead is situated on the lower part of their territory, at about 4100 m, since they can thus control the fringe between 3900 and 4300 m. They remain there during the rainy season, from late

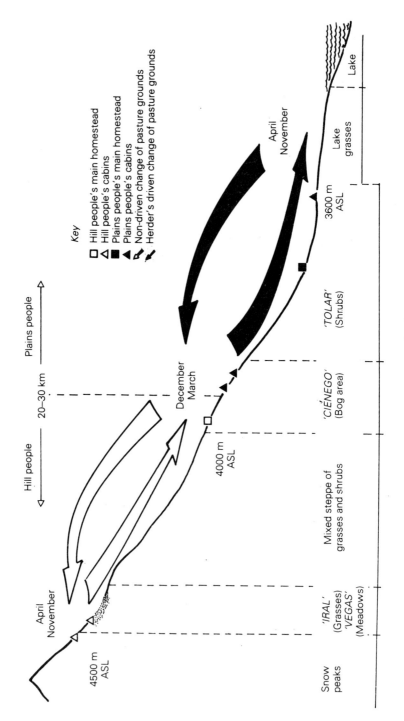

Figure 24.1 The two herding systems.

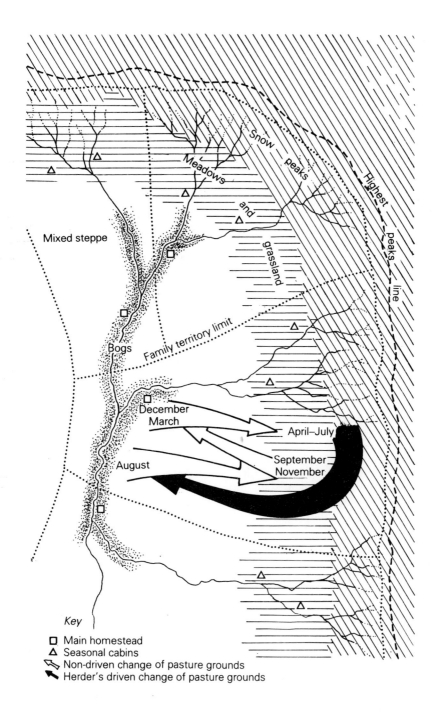

Figure 24.2 The hill people's herding system.

November to the beginning of April. Their llamas make use of two foraging zones in that area:

(a) the small plains and light slopes covered with a sparse steppe of shrubs, the *tolar*, which in the rainy season is covered with a carpet of small plants;
(b) the *ciénegos* (bogs) bordering the streams, where there is an unbroken vegetation cover.

When the rains come to an end, the seasonal pastures become exhausted, and the *ciénego* pastures cannot support the herds of llamas and other domestic animals.

At that time of year, the peasants and their flocks carry out one of the changes of ground in the annual cycle currently named 'transhumance'. From the grazing grounds near the main homestead, they move to the zone situated between 4300 and 4800 m, near the steep mountain peaks which are covered by eternal snow. The animals here again make use of two grazing areas:

(a) the highest slopes covered with a scattering of grass, and also carpeted by small plants during the rainy season and immediately afterwards, the *iral*;
(b) the *vegas*, on the permanently moist ground formed by daily thawing of ice, an area where llamas prefer to forage.

Sheep and goats often remain on the lower levels, because they cannot withstand the cold as the llamas do. The llamas go up alone to the high mountains, around 15 April, without any driving by the herders but, on the contrary, followed by them.

The herders return to their main homestead at the end of July, to make *chayacos*, a basic ritual in their indigenous religion and adaptive strategy (Merlino & Rabey 1983, Rabey & Merlino 1985). They take some llamas with them: those they can drive easily, usually young animals. In fact, it is the only moment of the annual cycle in which llamas do not move spontaneously, but must be driven by the herders. The herders do not worry about animals which remain in the uplands because all are adults, including a high percentage of males. Pumas (*Felis concolor*), the only adult llamas' predator, cannot easily make such animals their prey. The herders and their young flocks remain a month at the main homestead, in whose neighbouring bogs there is sufficient pasture for the occasion. Then control over the llamas is relaxed, and they afterwards return to the high area, followed by the herders. A new descent is made in December, when the rainy season generally starts. All the llamas now go down, and they do so without any human driving. Animals and herders then remain in the lowland area, until the beginning of the following dry season.

The hill herders' control over their llamas is very lax. With the exception of pregnant females and those with offspring, the llamas forage alone over well-defined natural territories, migrating periodically without any human driving. The herders' care is restricted to:

(a) selecting males as breeders, habitually those that protect females and young without a great amount of aggression;
(b) protecting young llamas from predators;
(c) feeding orphans and baby llamas whose mothers have no milk;
(d) periodically keeping the herd together.

The herding system is very different among the plains breeders (Fig. 24.1). The annual cycle of seasonal movements is reversed here (Rabey *et al.* 1985, p. 25). During the rainy season, the herders live on the mountainous fringe of shrubby steppe that surrounds the *puna*'s fluvial and lacustrine plains. By the beginning of April, when the dry season starts, they move to their main homesteads on the bottom slopes. Llamas then graze on the temporarily moist ground, as the waters dry out. Some family members occasionally settle on the lower fringe at the edge of rivers or lakes, where vegetation is similar to that of the *ciénegos*. Finally, at the end of the dry season, the animals graze on the bottom of the depressions. When the dry season finishes and the rains begin, the herders and their animals go up again to the highlands.

Contrary to what happens in the hills, in the plains breeding system the movements of llamas are guided by the herders. This difference is clearly apparent in their going down at the end of the rainy season. The llamas then try to go up to higher pasture grounds. The plains herders must then exercise a great deal of effort to make their llamas go to the lower pasture grounds.

With regard to the relationship between the llamas' behaviour and that of the human population linked to them, some conclusions can be reached from comparing the two herding systems. First, it seems that there is an innate behavioural pattern that guides the seasonal movements of llamas according to a basic rainy season–lower areas and dry season–higher areas scheme. This pattern was probably fixed early in the behaviour of llamas' wild ancestors, even before their first contact with human populations; although some research must be done in order to eliminate the alternative hypothesis of a behavioural pattern learned from humans. The control in hills breeding is only directed towards: (a) obtaining a product; (b) diminishing the mortality in young animals; and (c) strengthening the territorial features of llamas' behaviour, like the demarcation of feeding territories. The structural elements in the more generalized hills pattern are two of the llama's behavioural features:

(a) the selection of a feeding and sleeping territory by a single family group;
(b) the annual cycle of movements.

The first feature is, according to Franklin (1983, p. 573), common with the contemporary American wild camelids; the second one is characteristic of llamas, but it may be also characteristic of the (still not well-known) behaviour of wild guanacos (*Lama guanicoe*) which, according to Raedecke (1978), Franklin (1983, p. 573) and Cajal (1985, p. 90), are either migratory or sedentary. The seasonal ground-changing cycle of the *punas*' inhabitants must have been built around their camelids' behavioural pattern. The Andean ground-changing cycle could be a cultural pattern basically adapted to the innate behavioural pattern of llamas or of their wild ancestors, and based on a proper knowledge of this.

On the contrary, on the plains surrounding the rivers and lakes, herders must make a strong effort to use the natural resources available there for their llamas. The inversion of the annual cycle of movements which they must practise results in a much stricter human control over the llamas' activities than that practised by their neighbouring herders in the hills. Likewise, in Peru and northern Chile, where mixed herds of llamas and alpacas are the normal system, the herding scheme is strongly modified by the presence of the alpacas. In fact, the alpaca is an animal with much more strict ecological, nutritional, and management needs than the llama. For instance, as Gunderman (1984, p. 107) says, herders must keep the males and females separated from each other. Therefore, herds containing alpacas require intense human care, which includes management of mating and continuous herding over good pasture grounds.

It is, therefore, evident that in the more generalized herding system of the hills, llamas and human societies do not constitute a pastoralist system, at least in the conventional meaning of the word, because llamas forage almost freely and – except in the case of young individuals and their mothers – they require almost no care. The herders' family activity consists of accompanying, but not leading, them on their annual grazing cycle, and practising a set of minor controls designed to reduce natural pressures. Therefore, the relationship between llamas and human societies in the upper Argentine *punas* should not be termed 'pastoralism'.

Furthermore, the set of selective pressures on llama populations created by human control in the environment of the high *punas* cannot be named 'domestication' if we use the most current definitions. For instance, Bökönyi (1969, 1985, ch. 2, this volume) stresses that the essence of domestication includes basically the removal of the animal species from their natural habitat and natural breeding community, and their maintenance under controlled breeding conditions. Zeuner (1963) defined domesticated species as those whose breeding is completely controlled by humans; based on this, Eisenberg (1986) says that domestication 'reaches a point where the gene pool of a species is under the total control of humans'. In the *punas*, llamas are not removed from their natural habitat, neither are they maintained under controlled breeding conditions, except the above-mentioned selection of males; even less is their gene pool under the total control of humans. Moreover, the behavioural changes postulated by Meadow (1984) as a consequence of the domestication process are not apparent in the llama's case. The man–llama relationship can only be included in the 'domestication' category if we use a broader definition, such as that proposed by Ducos (1978, p. 54, ch. 3, this volume), for whom domestication is linked with human appropriation of living animals as economic objects of ownership, inheritance, and exchange.

The distinction between hunting and pastoralism, or between wild and domesticated animals, may therefore become non-operative. Ingold (1980, 1986) has presented data about the man–reindeer relationship which also do not enter into the above-mentioned definitions; reindeer are domestic because they belong to the Lapp households, this domestication being a consequence of taming, i.e of a man–animal social relation, and not of a biological process. From my own point of view, this set of problems in the classification of man–animal relationships will require much stronger theoretical discourse than that made to date, at least on the 'domestication' side of the problem.

Acknowledgements

The fieldwork on which this paper is based was possible thanks to CONICET (National Council for Scientific and Technological Research, Argentina) financial aid. An important part of this fieldwork and the subsequent theoretical analysis has been carried out in co-operation with the biologist R. Tecchi. I am also grateful to R. Merlino with whom I discussed various subjects developed here. J. García Fernández had the idea for the figures and C. Colarich converted them into their final form. Pedro Edmunds and Norberto Méndez contributed to the English version of this chapter.

References

Aldunate, C., J. Armesto, V. Castro & C. Villagrán 1981. Estudio etnobotánico en una comunidad precordillerana: Toconce. *Boletín del Museo de Historia Natural, Chile* **38**,183–223.

Bökönyi, S. 1969. Archaeological problems and methods in recognizing animal domestication. In *The domestication and exploitation of plants and animals*, P. J. Ucko & G. W. Dimbleby (eds), 219–29. London: Duckworth.

Bökönyi, S. 1985. Problèmes archéozoologiques. In *La protohistoire de l'Europe*, J. Lichardus & M. Lichardus–Itten (eds), 571–81. Paris: Presses Universitaires de France.

Bökönyi, S. 1989. Definitions of animal domestication. In *The walking larder*, J. Clutton–Brock (ed.), ch. 2. London: Unwin Hyman.

Cajal, J. L. 1985. Comportamiento. In *Estado actual de las investigaciones sobre camélidos en la República Argentina*, J. L. Cajal & J. M. Amaya (eds), 87–100. Buenos Aires: Secretaría de Ciencia y Técnica.

Castro L., M. C. Villagrán & M. K. Arroyo 1982. Estudio etnobotánico en la precordillera y altiplano de los Andes del Norte de Chile (18–19°S). In *El ambiente natural y las poblaciones humanas del Norte Grande de Chile* 2, A. Veloso & E. Bustos (eds), 133–203. Montevideo: UNESCO.

Custred, G. 1977. Las punas de los Andes Centrales. In *Pastores de Puna*, J. Flores Ochoa (ed.), 55–85. Lima: Instituto de Estudios Peruanos.

Ducos, P. 1978. 'Domestication': definition and methodological approaches to its recognition in faunal assemblages. In *Approaches to faunal analysis in the Middle*

East, R. H. Meadow & M. A. Zeder (eds), 53–6. Harvard University: Peabody Museum Bulletin 2.

Eisenberg, J. F. 1986. The selection of mammalian species for domestication. Precirculated paper. In *Cultural attitudes to animals including birds, fish and invertebrates*. Vol. 1, World Archaeological Congress (mimeo).

Flores Ochoa J. 1968. *Los pastores de Paratía: una introducción a su estudio*. Mexico: Instituto Indigenista Interamericano.

Flores Ochoa, J. 1975. Pastores de alpacas. *Allpanchis* **8**, 5–23.

Franklin, W. L. 1983. Contrasting socioecologies of South America's wild camelids: the vicuña and the guanaco. In *Advances in the study of mammalian behaviour*. Special Publication, American Society of Mammalogists, Vol. 7. J. Eisenberg & D. Kleiman (eds), 573–629.

Gunderman K. H. 1984. Ganadería aymara, ecología y forrajes: evaluación regional de una actividad productiva andina. *Chungará* **12**, 99–124.

Ingold, T. 1980. *Hunters, pastoralists and ranchers*. Cambridge: Cambridge University Press.

Ingold, T. 1986. Reindeer economies and the origin of pastoralism. *Anthropology Today* **2**(4), 5–10.

Meadow, R. H. 1984. Animal domestication in the Middle East: a view from the eastern margin. In *Animals and Archaeology*. Vol. 3: *Early herders and their flocks*, J. Clutton–Brock & C. Grigson (eds), 309–37. Oxford: BAR International series 202.

Merlino, R. & M. A. Rabey 1983. Pastores del altiplano andino meridional: religiosidad, territorio y equilibrio ecológico. *Allpanchis* **21**, 141–79.

Rabey, M. A. & R. Merlino 1985. El control ritual-rebaño entre los pastores del sur de los Andes Centrales. Presented at the 45 Congreso Internacional de Americanistas, Bogotá. *América Indígena*, in press.

Rabey, M., R. Rotondaro & R. Tecchi 1985. El ecosistema laguna de Pozuelos. *Ambiente* **45**, 18–25.

Raedecke, K. J. 1978. *El guanaco de Magallanes, Chile: distribución y biología*. Santiago, Chile: Corporación Nacional Forestal.

Zeuner, F. E. 1983. *A history of domesticated animals*. London: Hutchinson.

PREDATION

Introduction to predation

JULIET CLUTTON-BROCK

Predators feed on prey; that is they are normally carnivores which kill other living animals for food, but a parasite may also be termed a predator on its host. It has been argued that all human activities involving the exploitation of animals are predation and, in the social sense, parasitism. But it is only when humans are subsisting as hunters, in fact like carnivores, that there is no element of protection in the interactions between predators and prey. As soon as animals are herded or domesticated humans become protectors rather than predators. The relationship ceases to be one of simple predation and becomes a form of symbiosis in which individual animals benefit from the association until the moment of their death, and the species benefits genetically by becoming much more widespread than it was in the wild.

In the development of cultural systems it is generally agreed that humans were hunter-gatherers in the Palaeolithic period. In the ensuing Mesolithic they began to change towards the cultivation of plants, the storage of grain, and the keeping of dogs and maybe a few herd animals. The full domestication of livestock animals followed, on a world-wide scale, during the Neolithic period, around 7000 years ago. Hunting did not end with the Palaeolithic, however, and people in almost all societies past and present have been and are hunters, fishers, and gatherers of wild foods. The behavioural patterns of the hunter-gatherers, which evolved in hominids over, say, 3 million years, are still present in humans today, in the same way that the Pekingese still has the same predatory behaviour as the wolf.

In all interactions between predators and their prey there is an attack on the part of the predator and a defence on the part of the prey. The defence may be active, as when a moose is attacked by a wolf, and it will fight back, sometimes even killing the predator, or it may be passive and achieved by the concentration of numbers, as in a colony of tightly packed shellfish or the bunching together of herd animals. Wilson (1980, p. 23), in his, now classic, work on sociobiology, which is the study of the biological basis of social behaviour, quotes the Ethiopian proverb that says, 'When spiders' webs unite, they can halt a lion.' Wilson gives a succinct summary of the tenets that are now well known on the evolution of social behaviour, territoriality, and defence against predation. From these tenets it can be seen that humans, unlike most other animals, have evolved all the strategies of both predators and prey, and indeed in the Pleistocene they did play both roles. As hunters they developed the physical ability to range over large territories and the co-operative skills to kill prey much larger than themselves, but their gregarious behaviour and tolerance of crowding also provided a defence against predators such as wolf, bear, and the big cats. The need for this is explained by Geist in Chapter 25 who suggests a new scenario for predator–prey interaction in the Pleistocene of North America.

Hunting may appear to the anthropologist to be a straightforward activity, a matter of applying the necessary skills to searching for prey, killing it, and then sharing the meat and other products amongst the nuclear family or band. This so-called contented world of the hunter-gatherers has been described by Sahlins (1974) as 'the original affluent society'. From the biological point of view, however, the ecological relationships of predators and their prey are very involved and inter-dependent. With animal predators, a balance is normally maintained by the controls of environment and population dynamics so that prey species are not exterminated. With human hunting at the end of the Pleistocene, these controls appear to have broken down so that 'overkill' became a common occurrence in many parts of the world.

Although some will argue against the supposition that humans were responsible for the extermination of the mammoth, the mastodon, North American equids, and other large mammals, it is certain that hunter-gatherers do deplete their resources. Evidence for this is discussed by Cooke & Ranere in Chapter 26 for prehistoric Panama where they have carried out a comprehensive study of broad-spectrum hunting as well as predation on invertebrates.

Sloan, in Chapter 27, postulates how the massive exploitation of shellfish, and in particular the oyster, in the Mesolithic of northern Europe postponed the need for the walking larder and made possible the development of sedentary hunter-gatherer systems. The terminology of these shellfish economies is also discussed by Uerpmann in Chapter 9. Even the resource of shellfish could not last indefinitely, however, as is demonstrated, on the other side of the world, by Spennemann in Chapter 28 on the over-exploitation of molluscs in Tonga, around 1500 BC.

Chapter 29 by Colley & Jones demonstrates how renewed study of the fish remains from Rocky Cape has revealed a much greater diversity of species than was identified in the preliminary work on this prehistoric site in Tasmania. The question still remains, however, of why did the Tasmanians stop eating fish?

The obvious partner for humans in the hunt is the dog, the first animal to be domesticated, but Hooper in Chapter 30 puts forward the intriguing suggestion that the honeyguide has had a much more ancient symbiosis in leading early hominids to nests of bees. Another ancient interaction with birds is described by Eastham in Chapter 31 who gives an account of the large number of species killed by the Mousterians 40 000 years ago in Spain. The killing of large numbers of small birds, ostensibly for food, still continues today in Mediterranean countries, although it has become a subject of great concern to conservationists.

As with the chapters on domestication and pastoralism, those on predation cover a wide variety of species in many parts of the world in ancient times and at the present day. The intention of bringing them together in this book is to demonstrate the great diversity of interaction between humans and animals, and the continuity of this diversity through time. Even today, a wolf may be looked on as a dangerous predator, or as a much-loved companion, or as a prey to be killed to make a fur hat.

The anciently formed ecological and behavioural links between humans and animals have not changed in the modern world and it is essential for our survival that we conserve them. It is not so much that, 'if we do not study the past we shall be doomed to live it', but that if we do not study the past we may be doomed to live like battery chickens, enclosed in small, clean buildings, protected from predators, and with an enormously reduced perception of the living world.

References

Sahlins, M. 1974. *Stone age economics*. London: Tavistock Publications.
Wilson, E. O. 1980. *Sociobiology*. Cambridge, Mass. & London: Belknap Press of Harvard University Press.

25 Did large predators keep humans out of North America?

VALERIUS GEIST

Introduction

The notion of human supremacy is so deeply rooted in our culture that it colours our views of prehistory, including how humans colonized America. The most popular conception of this is Martin's (1967) overkill hypothesis, upgraded (1973, 1984) and discussed in the affirmative by a number of authors in Martin & Klein (1984), the latest attempt to deal with the problems of megafaunal extinctions. Martin's conception is one of appealing simplicity and great attractiveness in that one can make a most plausible case for it. Human colonization and extinctions of numerous large, huntable, and edible mammals did go hand in hand in many localities, so why not in North America? As Martin depicts it, humans entered North America probably 12 000–13 000 years BP, then expanded rapidly southward, colonizing all of North America by 11 000 BP and South America by 10 000 BP. In the process they brought, to use Martin's term, a 'blitzkrieg' upon the megafauna and in a very short time, too short to be adequately recorded in the fossil record, exterminated the Rancholabrean megafauna.

Using much the same evidence, plus some data based on more realistic assumptions than Martin made, one can turn those conclusions on their head: humans, and other members of the Siberian Pleistocene fauna *failed* to gain entry into North America *because* of the Rancholabrean megafauna, and colonized North America only after the megafauna had begun to collapse. The Siberian fauna is made up of cold-climate generalists. They faced in the Rancholabrean fauna a densely packed fauna of specialists, characterized by large-bodied herbivores and carnivores. The largest carnivore was a bear (*Arctodus simus*), much larger in size than the largest brown bear (Stock 1953, Kurtén & Anderson 1980), a long-legged cursorial species with a foreshortened, 'bull-dog' face. The predator fauna was diverse, containing in the latter part of the Wisconsinian glaciation a huge lion (*Panthera leo atrox*), various sabre-toothed cats (*Smilodon fatalis, S. gracilis, Homotherium serum*), a cheetah-like large cat (*Miracinonyx studeri*), the mountain lion (*Felis concolor*), the jaguar in the warmest latitudes (*Panthera onca augusta*), several short-faced bears (*Arctodus pristinus, A. simus, Tremarctos floridanus*), black bear (*Ursus americanus*), and dire wolf (*Canis dirus*). Man, as a super-predator, would have faced stiff competition, and would have faced at his kills direct confrontations with at least the largest predators. Since humans have had historically such great difficulties dealing with the much smaller, omnivorous brown or grizzly bear (as shown below), how could they have handled the much larger carnivorous *Arctodus*, particularly in an open landscape without trees to climb? How is it that various Siberian species appear in Alaska early in the Pleistocene, but – with few exceptions – do not colonize lower North America until the Rancholabrean fauna had begun to collapse?

When did humans enter North America?

There is little dispute that after 12 000 BP humans colonized lower North America (Haynes 1967), nor do I contend that human hunters after successful entry hastened the extinction of the remnants of the Rancholabrean herbivores (Martin 1984). I accept Guthrie's (1984) ecological hypothesis of megafaunal extinctions as most likely; Guthrie (1984) and King & Saunders (1984), point out that the proboscidians killed by Palaeo-Indians were small, indicating a decline in net-energy and nutrients for ontogenetic growth. That is, the mammoth and mastodons were declining in size at the time they were hunted, an indication that ecological conditions were probably not favourable. Other mammals also became dwarfed (Edwards 1967, Wilson 1980, Guthrie 1984). That extinctions coincided with a pause in deglaciation has been proposed by Ruddman & Duplessy (1985). Also some extinctions preceded the entry of humans (i.e. *Glyptotherium*, 13 970 ± 310; *Eremotherium*, 15 900; *Miracinonyx*, 12 770 ± 900; *Bootherium* and *Equus complicatus*, 17 200 ± 600; *Mammuthus primigenius* in Alaska, 15 380 ± 300, see Kurten & Anderson 1980, Table 19.6 & p. 364, Anderson 1984).

However, the controversial dates indicating an earlier entry of humans into North America are of special interest. It is as if humans pecked around the edges of North America, but failed to colonize the continent. There are early, disputed dates in Alaska, the Yukon Territory, and Venezuela (Bryan 1973; Rouse 1976; Stanford et al. 1981, Bischoff & Rosenbauer 1981, see also Kurtén & Anderson 1980). Linguistic evidence also suggests an early intrusion of humans in the south (Rogers 1985). Since there was a major expansion of upper Palaeolithic people 40 000–30 000 years BP, and the Alaska dates approach that range, it may well be that humans came to America's door in an early wave, but failed to gain entry.

If humans did indeed come earlier than 12 500 BP, they could have travelled in boats along the west coast (boats may have been used to colonize Australia some 40 000–35 000 years ago, see Geist 1978). If so, then we expect the faunas of the Channel Islands to show extinctions before 12 500 BP. There were dwarf mammoth on the Channel Islands (Cushing et al. 1984); such island dwarfs should readily fall victim to human hunters, as so well documented by Martin (1984). From Haynes (1967) we know that the frequency of human finds was 11 sites before

Table 25.1 Human sites and megafaunal extinctions 14 000–8000 years BP.

	Time-span (years BP)						Authors
	14 000 –13 000	12 000	11 000	10 000	9000	8000	
Number of human sites	0	11	10	15	26	31	Haynes (1967)
Number of extinct species	3	2	10	10	14	11	Kurtén & Anderson (1980)
Percentage of human sites	0	11.9	10.8	16.3	28.2	33.7	
Percentage of species alive	94	90	70	50	22	0	

Human sites, time to doubling approx. 3000 years, annual growth of human population 0.69/3000 × 100 = 0.02% per annum.
Percentage human sites vs. percentage species alive, $r = 0.956$, $r^2 = 0.914$, $t = 6.5$, $P < 0.01$.

12 000 BP, ten in the following millennium, 15 in the next, 26 in the next, and 31 in the millennium 9000–8000 BP. Megafaunal extinction spans from about 14 000–8000 BP (Kurtén & Anderson 1980). Thus human colonization of North America is inversely related to the presence of the megafauna, with human sites increasing after megafaunal collapse. Human sites keep pace with megafaunal extinction. This is not expected if humans came and conquered. Occupation should not pace extinction, but should shoot up following colonization. It does not. The pace of human colonization and megafaunal extinction is nearly perfectly correlated over 6000 years, and the rate of increase in human sites is extremely slow (assuming occupation sites from 13 000–8000 BP are as likely to be found) (Table 25.1).

One site only, that from Lubbock Lake in Texas, indicates that humans met the large *Arctodus* (Fig. 25.1). The site is dated about 11 100 BP (Johnson & Shipman in press; the 12 650 ± 350 BP date given by Anderson [1984] is in error). The remains indicate human cut-marks. Whatever the fate of that bear, and the remains indicate a large specimen (Johnson pers. comm.), humans and *Arctodus* did co-exist for a few centuries in lower North America. At that time the domestic dog was already in America (Anderson 1984), which has some bearing on how humans and *Arctodus* could have co-existed.

Human expansion along with other Siberian species

Humans colonized North America as a late part of a Siberian megafauna, members of which had existed in Alaska for hundreds of thousands of years. With megafaunal collapse, expanding populations of Siberian

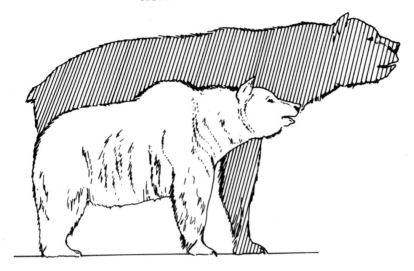

Figure 25.1 A comparison of approximate body sizes of *Arctodus simus* (cross-hatched) and *Ursus arctos horribilis*. Large, male grizzly bears measure about 1.2 m in shoulder height. Mounted skeletons of California grizzly and *Arctodus* from Rancho La Brea measure about 1.0 m and 1.3 m respectively (Stock 1953). In Alaska and the Yukon *Arctodus* reached gigantic proportions (Kurtén & Anderson 1980).

colonizers mixed with expanding populations of a few, eurytopic American species to form the new, Holocene American fauna.

Humans entered lower North America along with grizzly bears (*Ursus arctos*), moose (*Alces alces*), caribou (*Rangifer*), and glutton (*Gulo luscus*). The wapiti (*Cervus elaphus canadensis*) and extant *Bison bison* are probably both recent Siberians. I cannot find a single unequivocal and soundly dated find of wapiti prior to megafaunal extinction in lower North America. A Sangamon find turned out to be a jaw from a young bison; so called *Cervus* teeth are not acceptable evidence in the absence of more diagnostic parts (see Churcher 1984). The North American bison is today a Holocene dwarf (Wilson 1980), and most likely of recent Alaskan origin; one would expect *Bison* to colonize along with wapiti and thinhorn sheep (*Ovis dalli*). Prior to about 11 000 BP *Bison b. antiquus* is found; it is quite different from *B. b. occidentalis /bison* found after that date (Wilson & Churcher 1984). Iso-enzyme and DNA analyses show that all wapiti in North America are not only closely related, but also closely related to wapiti from the Altai in Siberia. If the hypotheses pertaining to wapiti and bison are valid, then bison, be they the plains or woodland forms, should be virtually identical.

Another Siberian species that expanded with megafaunal extinction is the timber wolf (*Canis lupus*), which may well have colonized along glacial margins in late glacial times (see Martin & Gilbert 1978). This species appears sporadically in the Rancholabrean tar pits (Kurtén & Anderson 1980), but expanded roughly with the demise of *Canis dirus*, the large 'dire wolf'. An old Siberian with limited success during the Rancholabrean, the bighorn sheep (*Ovis canadensis*) expanded just before megafaunal extinction (17 000–14 000 BP, Geist 1985).

As the Rancholabrean specialists vanished, so primitive members of the fauna expanded in range and numbers (i.e. *Odocileus* sp., *Tayassu tajacu*, *Antilocapra americana*, and *Ursus americanus*). Human colonization was part of a larger faunal event that shaped today's impoverished, almost certainly poorly adapted fauna of North America (see Geist 1985).

However, if humans only hastened the extinction of America's Rancholabrean fauna, and if the findings of earlier entries by humans into America are valid, then we must ask what kept the Beringian or Siberian fauna out of lower North America? Even if the earlier dates for humans are disputed, there is no dispute that wapiti arrived in Alaska as early as the Irvingtonian (Guthrie & Matthews 1971) and existed at least sporadically in Alaska until late glacial times (Guthrie 1966, Kurtén & Anderson 1980). Brown bears existed in Alaska in the Wisconsinian period, but not in lower North America (Kurtén & Anderson 1980); they did exist in Alaska along with *Arctodus*, maybe via ecological segregation since *Arctodus* finds are in the unglaciated areas. Post-glacially, the brown bear occupies the range once held by *Arctodus simus*. The Wisconsinian fauna of Beringia contains other Siberian species which appear in lower North America only in late Wisconsinian/Holocene times (i.e. *Gulo gulo*, *Alces alces*, *Rangifer tarandus*). However, the Alaskan fauna does contain a good number of typical Rancholabrean species (see Kurtén & Anderson 1980).

Sporadically, Siberian species did gain entry during mid-Pleistocene times into lower North America (i.e. *Panthera leo atrox*, *Bison latifrons*, *Oreamnos* sp.), while elephants (*Mammutus* sp.) and black bear (*Ursus americanus*) came in early Pleistocene times. These became successful members of the Rancholabrean fauna; others did not. The dhole (*Cuon alpinus*) came and went extinct (Kurtén & Anderson 1980). The bighorn sheep (*Ovis canadensis*) lingered a long time before becoming successful (see Geist 1985).

On the Siberian side of Beringia Upper Palaeolithic people appear about 14 000 BP (Vereshagin 1967, Rouse 1976) and shortly thereafter in America (Haynes 1967). In Europe, the first expansion of Upper Palaeolithic people, 40 000–35 000 BP, coincides with a glacial recession and a brief interstadial (Geist 1978). Members of this early dispersion could have reached Beringia before the last glacial advance, but if so, they failed to establish themselves.

Human anti-predator strategies

At this point it is useful to turn to a subject addressed by very few investigators. How do humans prevent themselves from becoming prey? What did we do in our long evolutionary history to avoid predation? In

Australopithecus, arboreal adaptations indicate that it probably depended on climbing abilities, particularly when escaping from a predator. In dire straits, humans climb, a throw-back to ancient adaptations.

Since little remains of arboreal adaptations in *Homo*'s morphology, excepting maybe in enigmatic Neanderthals who hardly used that capability for climbing (Geist 1981), we suspect that anti-predator strategies other than those of *Australopithecus* were employed. How does one safely spend the night on the ground without fire, guns, or axe? Kortland (1980) provides an answer: in Africa, use a packaging of thorns! Lions will not penetrate through thorns to get at prey. In short, *Homo erectus*, using a thorn-covered ground-nest, a corral of thorns, could spend the night safely away from trees. And that method is still employed to this day (Post 1974). It can be improved by fire, as well as the use of a stabbing weapon, and probably the use of a loud voice that mimics a predator's threat call (Geist 1978). From the perspective of a predator, being confronted by a visual wall, behind which sounds a peculiar, angry conspecific, mixed with incongruous smells and sounds may be awe-inspiring. Maybe such suffices even for a thin wall *without* thorns, but we do not know. Our safety currently relies, in the wild, on destruction of carnivores that kill humans, so that the *habit* of eating humans cannot be passed on. That is an important point: if food habits are indeed *initiated* as a tradition, then stopping that tradition is important to human safety. We do kill man-eating carnivores (Corbett 1946) and we may well have acted like that in the past if the behaviour of native Americans is an indication (Storer & Tevis 1955), provided, of course, we could kill carnivores at an acceptable cost in injuries and lives. The upper limit in this ability may well have lain with the brown bears (and polar bears – provided sufficient dogs were at hand).

Bears and native North Americans

The relationship of the great brown and polar bears to humans that co-existed with them has been one of unease. A study of interactions between bears and men such as undertaken by Herrero (1985), or as one can pursue oneself in such accounts as are available (i.e. Wright 1909, Storer & Tevis 1955, Nelson 1969, 1973, Kurtén 1976, Craighead 1979, Young & Beyers 1980), or in reflecting on one's own encounters with grizzlies, allows one to unbraid certain patterns in the behaviour of grizzlies. One also notes that in pristine North America, grizzly bears were locally very common west of the Mississippi. And it was indeed possible, in season, to travel along rivers and see hundreds of bears, as did Pattie (1831) along the Arkansas River. He counted 220 in one day, eight of which his party killed as these bears attacked, while two more were killed in camp that night. In spring these rivers were lined with the carcasses of drowned buffalo, but also elk (see Roe 1951, Prince Maximilian von Wiede, in Thomas & Ronnefeldt 1982) which was one reason why grizzly bears would be so numerous along rivers. Travellers in early California reported seeing

30–50 bears per day (Storer & Tevis 1955). In 1871 alone, 750 grizzly-bear hides were bought by the Hudson's Bay Company in the Cypress Hills of what came to be southern Alberta: an area of only 2500 km^2 (Ondrack 1985).

That native people co-existed with the great bears, often tenuously, is amply recorded from the American west (Storer & Tevis 1955). The evidence was in the form of some natives killed (David Thompson 1787, in Hopwood 1971) and many natives and some voyagers with lacerated, deformed faces, missing ears, nose and eyes as reported by Ross (1831) in his journals from British Columbia, and others (Storer & Tevis 1955). There is further evidence to be seen in the awe in which grizzlies were held by natives, their reluctance to hunt the bears then as now, the honours that accrued to one who had killed grizzlies, the fact that a grizzly was not usually attacked except by going ceremoniously on a war-path against it, with all the preparations such a decision entailed, and never except in company of 4–10 warriors, or – depending on tribe and locality – after great ambush preparation, council, and shamanistic rites (Wright 1909, Storer & Tevis 1955, Prince Maximilian von Wiede 1833, in Thomas and Ronnefeldt 1982). Californian Indians avoided good bear habitat as much as possible. Some would not allow young men to go hunting alone, for fear of bears; they delighted in the fact that the Spaniards had the capability to readily kill grizzly and expressed their pleasure at this with gifts; they rapidly colonized land cleared of grizzly bears by Spanish soldiers; some tribes would not hunt grizzlies or trap them but readily sought the aid of white hunters to do the task for them; tribes might join to avenge the killing of one of their members by a grizzly, hold dance festivals when a grizzly was killed or even erect a cairn on the spot where a grizzly expired at their hands. Yet at least to one tribe, the Monos, an offshoot of the Shoshoneas, grizzlies were beasts to be hunted and killed, a task at which they excelled (Storer & Tevis 1955). However, in aboriginal times California was very sparsely settled by Indians (Kroeber, in Storer & Tevis 1955), a reflection possibly of the fact that grizzlies were not only avoided, but were also direct competitors of man for food.

The central valleys of California had the right characteristics to sustain an advanced agricultural culture, similar to the Aztecs. These valleys contained fertile, flooding rivers traversing open landscapes, essential ecological determinants of all great civilizations, occidental, oriental, or American (Carneiro 1970). Today's California is a densely settled land with thriving agriculture. Yet the Spanish found it a land with a low native population, but many large grizzly bears (Storer & Tevis 1955).

The accounts of many witnesses indicate that the great brown bears were, in all but exceptional cases, avoiding interactions with humans, excepting the female with cubs if surprised, or the occasional starving bear, be it in Spring after hibernation, in Fall after berry crop failures, from debilitating old age, or the occasional wounded bear (Nelson 1973, Young & Beyers 1980, Herrero 1985). The manner in which the bears acted indicated that they did not regard humans as prey, although they might investigate humans and their activity probably from sheer curiosity, a curiosity which some observers misinterpreted as attacks. When wounded or provoked and attacking humans, the brown bears treated humans as if they were some kind of bear. That is, they tended to rise on their hind legs – as against a conspecific – and tried to 'disarm him'. To disarm a conspecific, a grizzly bear bites and holds the opponent's snout, disabling him from biting. They do the same to humans, crushing face and jaw. Many bears after punishing with bites left the offending humans. A few covered him with branches, as if to 'cache him'. Since bears do eat conspecifics, it does not surprise me that an occasional bear prepares a human, whom he would treat like a conspecific, for food. Standing still and upright, facing the grizzly, is advocated by the Kutchin people of sub-Arctic Alaska (Nelson 1969) and by Zenas Leonard (1804–57) a mountain man and trapper (in Froncek 1974, p. 280), which implies that the lack of flight may intimidate the grizzly; flight provokes attack. Once aggressively pursued by humans, the great brown bear becomes very secretive and, except for his signs, virtually invisible. The accounts of the stories of the stock-killing bears that inhabited the west between about 1880–1920 present tragic evidence of this (Young & Beyers 1980). The brown bear gives every indication of willingness to avoid and learn, as if it had been second to some larger bear in the past, and indeed it had, to the cave bear, *Ursus spelaeus* in Europe, and to the great *Arctodus simus* in Alaska.

Killing a bear safely with weapons tipped with stone or bone points is a very difficult task. The first technical problem is that flint and obsidian points on arrows or spears shatter when they hit a bone, while bone points are likely to chip and also fail to penetrate (see David Thompson in Hopwood 1971, pp. 193–293, Guthrie 1983). Stone points do cut very well through soft tissues, as good as iron points or better, but if the projectile is aimed at the heart, it is not at all certain to reach it. From the front, the heart and lungs of a large mammal are so well protected that a projectile would strike bone 90 per cent of the time; from the side, it is still some 50 per cent or so protected (Fig. 25.2). Half the spears thrown will not penetrate to their mark – if such is the heart.

If arrows or throwing-spears are the prime weapons, then even if one does penetrate the animal's chest, the narrow wound channel of such a weapon is not likely to disable a bear, nor kill it quickly. In fact, grizzly bears wounded with narrow cavalry lances remain long capable of sustained attack and die very slowly. Coronado's soldiers lanced a grizzly, pushing the shaft to half its length into a bear. This bear still caught the rider's horse, and while mauling it was run through with a second lance, after which it was apparently lassoed and finally dispatched (Thomas 1935). A similar incident, in which three Mexicans lanced a grizzly, was reported by Lieutenant Z. M. Pike in 1808 (Wright 1909, p. 33); two of the Mexicans were killed by the grizzly and the third was wounded.

Captain Clark of the famous Lewis and Clark expedition used a flint-lock rifle on grizzly bears. His largest bear received five balls through the lungs and five balls in other parts of the body, and took 25 minutes to die after this wounding (Lewis 1804–1806). Tales such as this are legion (Wright 1909, Young & Beyers 1980). A projectile through the brain or

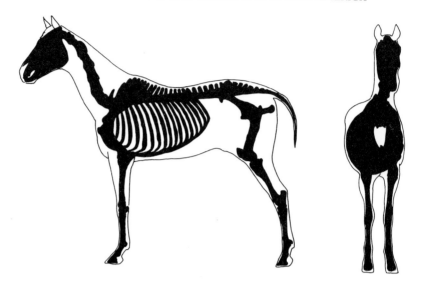

Figure 25.2   Distribution of bones in a large mammal, in lateral and frontal view, as illustrated by a domestic horse (after Ellenberg *et al*. 1956). The lungs and heart are well protected by bone. Few light throwing-spears are expected to reach vital organs from the side, let alone from the front.

the spinal chord will anchor a grizzly, but not a body wound inflicted by lance, arrow, knife, or small-bore, low velocity rifles. The broad-bladed, thick, hand-held boar-spear of medieval European hunters, with a blade about 10–12 cm wide and 2–4 cm thick, mounted on an eight-foot pole, cross-wound with leather, is a weapon used on bears. Such bears were distracted by a pack of hounds and then speared by a team of hunters. That even so formidable a weapon as a boar-spear, which opens a wound 12 cm wide by about 40 cm in depth does not kill instantly, is attested to by the damage the speared animal inflicted on the shafts of hunting spears exhibited in European hunting museums today. A group of determined men, accompanied by brave and equally determined large dogs, and armed with stout, broad lances could probably dispatch brown bears if not in safety, then with reduced risk. Inuits in pre-firearm days apparently did just that, but still considered such hunting very dangerous (Nelson 1969). Indians in the west expected injury and death on grizzly hunts (Lewis 1804–1806, Wright 1909). They did have lances as well as dogs, but I cannot find whether they used the latter; Indian people in Alaska do use dogs to warn of bears (Nelson 1973).

   The experimental work on the performance of reconstructed throwing-spears tipped with bone points fashioned after late Palaeolithic material by Guthrie (1983) shows both an appalling lack of penetration on moose carcasses, and the great fragility of the bone tips. Upper Palaeolithic hunters used light throwing-spears on bears, as revealed in cave art (Kurtén 1976), but the bear they hunted was small compared to the grizzly, and possibly more timid, judging from what is known of today's European brown bears.

Kurtén (1976) studied Pleistocene bear remains in great detail, searching among other things for evidence of interactions between men and bears. He concluded that whereas the European brown bear had been hunted by Neanderthal and Upper Palaeolithic hunters, no evidence exists that the much larger cave bear had shared a similar fate. Kurtén (1976) was critical of earlier interpretations of cave-bear remains, showing how such misinterpretations arose. It appears that Palaeolithic hunters stayed away from bears larger than brown bears, that is, from cave bears.

The *Arctodus* problem

Had humans reached North America when it had an intact Rancholabrean fauna, they would have met, among a number of large carnivores, the large *Arctodus simus*. Large specimens of this bear, such as were typical of Alaska or the Yukon (Kurtén & Anderson 1980) dwarfed even the giant among extinct brown bears, the large Alaskan coastal brown bear (Kurtén 1976). Skeletons of *Arctodus simus* from California are a foot taller at the shoulders than the skeletons of large California grizzly bears (Stock 1953). Moreover, *Arctodus simus* differed in several important respects from the grizzly bear: it was carnivorous. *Arctodus simus* possessed large, functional carnassials, a snout which shortened in evolution into a tool for grabbing and holding onto prey, and a cursorial body form. It had long legs and was yet massive in build as behoved a 'bulldog bear' (Kurtén & Anderson 1980).

Also, *Arctodus* apparently differed behaviourally from brown bears: like sabre-toothed cats, dire wolf, and American cheetah, it is found in relatively high numbers in natural traps, in which its prey also came to grief. Such we find in Natural Trap cave in Wyoming (Martin & Gilbert 1978), and in the tarpits of Rancho La Brea (Stock 1953). One does not find similar depositions of brown bears or black bears, a possible suggestion that *Arctodus simus* was more ready to go after prey or carrion in chance circumstances that would tempt neither the brown nor the black bear. This also means that, as the largest Rancholabrean carnivore, it was probably much more ready to approach potential prey, the kills of other carnivores or of human hunters. Naïve polar bears do as much, and even hunted ones may become belligerent if hungry (see Nelson 1969).

The distribution of grizzly in America overlaps in good part that of the extinct *Arctodus*; grizzly bears expanded into lower North America with the extinction of *Arctodus*, although in Alaska they apparently were contemporaneous (Kurtén & Anderson 1980). Lower North America proved to be superb grizzly bear habitat. Since grizzly exclude black bears in open plains, except where the latter have recourse to trees, one has reason to suspect that *Arctodus* excluded grizzly bears, except in the coldest periglacial regions. The grizzly could not outclimb *Arctodus*, and in areas with few trees, neither could Upper Palaeolithic hunters. Such hunters, facing an approaching *Arctodus*, an eager, carnivorous bear twice the mass of a grizzly, would be faced with a difficult problem. Their weaponry, good for caribou, would not have been of much use, but would only enrage the giant bear. There is no evidence for broad lances in the Beringian late glacial record, and one questions the use of narrow-bladed spears in view of Cornados' experience of lancing bears. Whether dogs would have been an aid in dispatching this giant bear is unknown, but dogs probably were available to Upper Palaeolithic hunters arriving in Siberia 14 000 BP (see Kurtén & Anderson 1980).

That predation on America's megafauna was historically quite heavy is indicated by several factors: the diversity of predators was high, and they were individually much larger than their Eurasian or African counterparts. The prey grew large and cursorial, and early horned and antlered artiodactyls grew huge horn-like organs. The best examples of the latter are *Bison latifrons, Cervalces,* and *Ovis canadensis*. Large horns and antlers are a function of the security strategy for the neonate: if the young must grow rapidly to survivable size then the female must produce a milk rich in solids, that is she must be capable of saving a large fraction of nutrients from growth towards production. In males this capability is reflected in large antler-size, which is proportional to the female's perinatal investment (Geist 1986, 1987). *Ovis canadensis* in its expansion, following the peak of the Wisconsinian maximum, was closely associated with both Rancholabrean predators and Siberian ones (Martin & Gilbert 1978). In *Bison latifrons*, the relatively small molars and premolars indicate a deviation from grazing towards softer foods. Grasses have a lower protein and mineral concentration than foliages (Vogt 1948), and small teeth suggest foliage and forbes feeding. That is, high milk solid production requires 'concentrate feeding' (Hoffmann 1973). Such a diet is also indicated in the dentition of *Cervalces*, which carried antlers rivalling the largest antlers of the Alaskan bull moose. In short, predation appears to have been heavy in the Rancholabrean megafauna, and I suggest that Siberian hunters moving into Alaska would have found the large carnivores an insurmountable foe.

References

Anderson, E. 1984. Who's who in the Pleistocene: a mammalian bestiary. In *Pleistocene extinctions*, P. S. Martin & R. G. Klein (eds), 40–89. Tucson: University of Arizona Press.

Bishoff, G. L. & R. G. Rosenbauer 1981. Uranium series dating of human skeletal remains from the Del Mar and Sunnyvale Site, California. *Science* **213**, 1003–5.

Bryan, A. L. 1973. Paleo-environments and cultural diversity in late Pleistocene South America. *Quaternary Research* **3**, 237–56.

Carneiro, R. L. 1970. A theory of the origin of state. *Science* **169**, 733–8.

Churcher, C. S. 1984. *Sangamona*: the fugitive deer. In *Contributions in Quaternary Vertebrate Paleontology*, H. H. Grenoway & M. R. Dawson (eds), 316–31. Pittsburgh: Carnegie Museum of Natural History, Special Publications No. 8.

Corbett, G. 1946. *Man-eaters of Kumaon*. Oxford: Oxford University Press.

Craighead, F. C. Jr. 1979. *Track of the Grizzly*. San Francisco: Sierra Club Books.

Cushing, J., M. Daily, E. Noble, V. L. Royh & A. Wenner 1984. Fossil mammoth from Santa Cruz Island California. *Quaternary Research* **21**, 376–84.

Edwards, W. E. 1967. The late Pleistocene extinction and diminution in size of many mammalian species. In *Pleistocene extinctions*, P. S. Martin & H. E. Wright Jr. (eds), 141–54. New Haven: Yale University Press.

Ellenberg, W., H. Bau & H. Dittrich 1956. In *An atlas of animal anatomy for artists*, L. S. Brown (ed.). New York: Dover.

Froncek, T. (ed.) 1974. *Voices from the wilderness.* New York: McGraw–Hill.

Geist, V. 1978. *Life strategies, human evolution, environmental design.* New York: Springer.

Geist, V. 1981. Neanderthal the hunter. *Natural History* **90**, 26–36.

Geist, V. 1985. Pleistocene bighorn sheep; some problems of adaptation, and its relevance to today's American megafauna. *Wildlife Society Bulletin* **33**, 351–9.

Geist, V. 1986. The paradox of the great Irish stags. *Natural History* **95** (3), 54–64.

Geist, V. 1987. On speciation in Ice Age mammals with special reference to deer and sheep. *Canadian Journal of Zoology* **65**, 1067–84.

Gould, S. G. 1974. The origin and function of 'bizarre' structures: antler size and skull size in 'Irish elk' *Megaloceros giganteus. Evolution* **28**, 221–31.

Guthrie, R. D. 1966. The extinct wapiti of Alaska and Yukon Territory. *Canadian Journal of Zoology* **44**, 45–7.

Guthrie, R. D. 1983. Osseous projectile points: biological observations affecting raw material selection and design among Paleolithic and Paleoindian peoples. In *Animals and archaeology*. Vol. 1: *Hunters and their prey*, J. Clutton–Brock & C. Grigson (eds), 273–94. Oxford: BAR International Series 163.

Guthrie, R. D. 1984. Mosaics, allochemicals and nutrients. In *Quaternary extinctions*, P. S. Martin & R. G. Klein (eds), 259–98. Tucson: University of Arizona Press.

Guthrie, R. D. & J. V. Matthews 1971. The Cape Deceit fauna – early Pleistocene mammalian assemblage from the Alaskan Arctic. *Quaternary Research* **1**, 474–510.

Haynes, C. V. 1967. Carbon-14 dates and early man in the New World. In *Pleistocene extinctions*, P. S. Martin & H. E. Wright Jr. (eds), 267–86. New Haven: Yale University Press.

Herrero, S. 1985. *Bear attacks.* New York: Lyons Books, Winchester Press.

Hoffman, R. R. 1973. The ruminant stomach. *East African Monographs in Biology.* **2**, 1–354.

Hopwood, V. G. (ed.) 1971. *David Thompson. Travels in western North America, 1784–1812.* Toronto: Macmillan of Canada.

Johnson, E. & P. Shipman in press. Scanning electron microscope studies of bone modification. *Current Research in the Pleistocene* **3**.

King, Y. E. & J. J. Saunders 1984. Environmental insularity and the extinction of the American mastodon. In *Quaternary extinctions*, P. S. Martin & R. G. Klein (eds), 315–39. Tucson: University of Arizona Press.

Kortland, A. 1980. How might early hominids have defended themselves against large predators and food competitors? *Journal of Human Evolution* **9**, 79–112.

Kurtén, B. 1976. *The cave bear story.* New York: Columbia University Press.

Kurtén, B. & E. Anderson 1980. *Pleistocene mammals of North America.* New York: Columbia University Press.

Lewis, M. 1804–1806. Reprinted 1980 in *Man meets grizzly*, F. H. Young & C. Beyers (eds). Boston: Houghton Mifflin.

Martin, L. D. & B. M. Gilbert 1978. Excavations at Natural Trap Cave. *Transactions of the Nebraska Academy of Science* **6**, 107–16.

Martin, P. S. 1967. Prehistoric overkill. In *Pleistocene extinctions*, P. S. Martin & H.

E. Wright Jr. (eds), 75–120. New Haven: Yale University Press.

Martin, P. S. 1973. The discovery of America. *Science* **179**, 969–74.

Martin, P. S. 1984. Prehistoric overkill: the global model. In *Quaternary extinctions*, P. S. Martin & R. G. Klein (eds), 354–403. Tucson: University of Arizona Press.

Martin, P. S. & R. G. Klein (eds) 1984. *Quaternary extinctions*. Tucson: The University of Arizona Press.

Nelson, R. K. 1969. *Hunters of the northern ice*. Chicago: University of Chicago Press.

Nelson, R. K. 1973. *Hunters of the northern forest*. Chicago: University of Chicago Press.

Ondrack, J. 1985. *Big game hunting in Alberta*. Edmonton: Wildlife Publishing.

Pattie, O. G. 1831. Reprinted 1980 in *Man meets grizzly*, F. H. Young & C. Beyers (eds). Boston: Houghton Mifflin.

Post, L. van der 1974. *A story like the wind*. London: Penguin.

Roe, F. G. 1951 (2nd edn 1970). *The North American buffalo*. Toronto: University of Toronto Press.

Rogers, R. A. 1985. Glacial geography and native North American languages. *Quaternary Research* **23**, 130–7.

Ross, C. 1831. Reprinted 1980 in *Man meets grizzly*, F. H. Young & C. Beyers (eds). Boston: Houghton Mifflin.

Rouse, J. 1976. Peopling of Americas. *Quaternary Research* **6**, 597–612.

Ruddman, W. F. & J. C. Duplessy 1985. Conference on the last deglaciation: timing and mechanism. *Quaternary Research* **23**, 1–17.

Stanford, D., R. Bonnichsen & R. E. Morlun 1981. The Ginsberg experiment: modern and prehistoric evidence of a bone-flaking technology. *Science* **212**, 438–40.

Stock, C. 1953. *Rancho La Brea*. Los Angeles County Museum, Series no. 20. Paleontology no. 11 (5th edn).

Storer, T. J. & L. P. Tevis Jr. 1955. *California grizzly*. New York: Promontory Press.

Thomas, A. B. 1935. Reprinted 1980 in *Man meets grizzly*, F. H. Young & C. Beyers (eds). Boston: Houghton Mifflin.

Thomas, D. & K. Ronnefeldt (eds) 1982. *People of the first man*. New York: Promontory Press.

Vereshagin, N. K. 1967. Primitive hunters and Pleistocene extinctions in the Soviet Union. In *Pleistocene extinctions*, 369–98. New Haven: Yale University Press.

Vogt, F. 1948. *Das Rotwild*. Vienna: Osterreichischer Jagd und Fischerei Verlag.

Wilson, M. 1980. Morphological dating of late Quaternary bison on the northern plains. *Canadian Journal of Anthropology* **1**, 81–5.

Wilson, M. C. & C. S. Churcher 1984. The late Pleistocene Bighill Creek formation and its equivalents in Alberta: correlative potential and vertebrate palaeofauna. In *Correlation of Quaternary chronologies*, 150–75. Norwich: Geo Books.

Wright, W. H. 1909. *The grizzly bear*. Reprinted 1977, Lincoln: University of Nebraska Press.

Young, F. H. & C. Beyers (eds) 1980. *Man meets grizzly*. Boston: Houghton Mifflin.

26 *Hunting in Pre-Columbian Panama: a diachronic perspective*

RICHARD G. COOK
and ANTHONY J. RANERE

Introduction

Linares (1976) has demonstrated that even small-scale human modification of the tropical rain forest enabled Pre-Columbian hunters to concentrate on medium-sized caviomorph rodents and other mammal species, whose biomasses were enhanced by horticulture and its attendant disturbances. Berlin & Berlin (1983) report the same pattern of 'garden-hunting' among the contemporary Aguaruna and Huambisa in the Peruvian '*montaña*'. In this particular situation, the procurement of animal protein does not impinge upon carbohydrate production; rather, the two are complementary.

Linares' study site was Cerro Brujo, a small hamlet located on a steep–sided peninsula on the northwestern Caribbean coast of Panama (Fig. 26.1). Here year-round heavy rains permit continuous plant-cropping, but prevent effective burning. The hamlet's shifting cultivations could fallow their forest plots for long periods of time, ensuring the replenishment of soils and of the mammal populations (Linares 1976, 1980a, Linares & White 1980).

On the opposite (Pacific) side of the isthmus, the climate is very different. Dry seasons are long. Around Parita Bay (Fig. 26.1), 4–6 months can pass without effective precipitation. Evaporation is intensified by strong katabatic winds. Dry periods during the rainy season can ruin crops. Uncontrolled fires quickly destroy forest cover (Linares 1977, Cooke 1979).

In this region, human cultural development and demography have followed patterns dissimilar to those of the Caribbean coast. The recently surveyed Santa Maria drainage (Fig. 26.1) was settled by horticultural peoples as early as Period 2B (Preceramic B: 5000–2500 BC) (see Cooke & Ranere 1984 for a summary of the regional chronology). During this period, human populations, though small and scattered, were distributed across all the major ecozones from the cordillera to the coast (Weiland 1984, 1985). The cultivation of maize (*Zea mays*) and the intensive collection of tree crops, particularly palm-nuts, are documented for some sites (Ranere & Hansell 1978, Piperno 1984, Piperno & Clary 1984, Piperno *et al.* 1985).

By the beginning of the Christian era, permanent nucleated settlements,

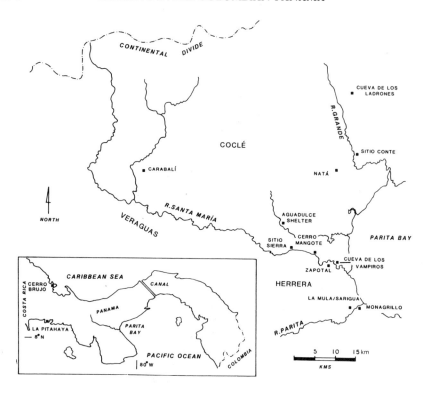

Figure 26.1 Map of the Parita Bay region of central Panama, showing the location of 13 archaeological sites which have provided faunal materials. Inset: map of the Republic of Panama, showing the location of Cerro Brujo and La Pitahaya, and the Tonosi valley.

for which maize was now a staple crop, are in evidence along the narrow but fertile coastal plan (Hansell 1987). At the time the Cerro Brujo hamlet was first occupied (about AD 600), these had become organized into antagonistic village federations, which vied for prestige, scarce resources, and, probably, arable land (Linares 1977, Cooke 1979, 1984a). Towns of over 1500 people were visited by the invading Spanish between 1515 and 1529. Their chronicles describe in detail the open and, in many parts, tree-less habitats adjacent to the population centres, such as Natá. They refer, too, to hunting techniques appropriate for capturing the prevalent prey species, such as netting doves (Columbidae) and driving deer (Cervidae) by firing the grass (Andagoya 1913, p. 197, Espinosa 1913, p. 178; see also Cooke 1979, 1984b).

How did faunal exploitation on the Pacific slopes differ from that of the under-populated Caribbean coast? Is it possible to distinguish between environmental and cultural influences, i.e. between the effects of a seasonally arid environment and human social and hunting behaviour? Did strategies develop to offset the depletion of terrestrial game resources? In this chapter, we will address these questions in a preliminary form by reference to samples of vertebrate bone (other than fish) from ten sites located within 55 km of the present-day coastline of Parita Bay.

Chronological and contextual notes on the studied sites

The central Panamanian sites whose faunas we consider are listed in Table 26.1 together with information on dating, precipitation, and excavation and recovery techniques. For comparative purposes, we also refer graphically and textually to the published bone samples from Cerro Brujo and from La Pitahaya, a large village located on the coast of Chiriqui (see Fig. 26.1), which was occupied from about AD 500–1200 (Linares 1980b, Linares & White 1980).

Four basic site types are represented in the Parita Bay area:

(a) three small, 'one-family' rockshelters: the Cueva de los Vampiros – hereafter 'Vampiros' (Cooke & Ranere 1984), the Aguadulce Shelter – hereafter 'Aguadulce' (Ranere & Hansell 1978), and Carabalí (Valerio 1985, 1987);
(b) one large, 'multi-family' rockshelter: the Cueva de los Ladrones – hereafter 'Ladrones' (Bird & Cooke 1978, Cooke 1984a),
(c) three coastal shellmounds, occupied by several families, but not necessarily permanent settlements: Cerro Mangote (McGimsey 1956, Ranere 1980), Monagrillo (Willey & McGimsey 1954, Ranere & Hansell 1978), and Zapotal (Willey & McGimsey 1954, Giausserand in prep.); and
(d) three large and permanent villages: La Mula–Sarigua (Willey & McGimsey 1954, Hansell 1987), Sitio Sierra (Cooke 1972, 1979, 1984a, and Natá (NA–8) (Cooke 1972).

Two of the small rockshelters – Vampiros and Carabalí – have light occupations that date to Period 2A (Preceramic A: 8000–5000 BC). The only non-fish vertebrate taxon that can be identified in the deposits that date to this period is the nine-banded armadillo (*Dasypus novemcinctus*), at Carabalí.

The shelters were most intensively utilized during Period 2B (Preceramic B: 5000–2500 BC), Period 3A (Early Ceramic A: 2500–1000 BC) and Period 3B (Early Ceramic B: 1000–300 BC). Carabalí and Aguadulce (during Period 2B) were probably intermittently occupied hunting and gathering stations. Ladrones and Aguadulce (during periods 3A and 3B) seem to have been more permanent settlements for families growing maize and rootcrops in neighbouring plots.

Cerro Mangote is a shellmound overlooking Parita Bay. At the time of its occupation (5000–3000 BC), it was between 3 and 1 km from the sea (Clary *et al.* 1984, and references therein). The depth and composition of the refuse and the presence of a cemetery (McGimsey *et al.* 1966) suggest that it was a regular place of residence, being used by several families. The

Table 26.1 Parita Bay faunal samples

Site	Type	Date excavated	Archaeologist	Faunal analyst(s)	Date of sample	Precip. (mm)	Mesh size (in.)
Vampiros	Shelter	1982	PSM	Cooke	1st mil. BC	1200	1/8
Carabalí	Shelter	1983–4	Valerio	Cooke	5000–300 BC	2700	1/8
Ladrones	Shelter	1974	Bird/Cooke	Cooke	5000–300 BC	1330	1/4[1]
C. Mangote	Shellmound	1956–79	McGimsey/Ranere	Cooke	5000–3000 BC	1500	1/8[2]
Aguadulce	Shelter	1973–5	Ranere	White/Cooke	5000–300 BC	1700	1/8
Monagrillo	Shellmound	1975	Ranere	Wing	2400–1200 BC	1230	1/8
Zapotal	Shellmound (? village)	1984	PSM	Cooke	1500–1000 BC	1500	1/8
La Mula	Village	1983–4	Hansell/Cooke	Cooke	c. 200 BC	1230	1/8
S. Sierra	Village	1975	Cooke	Cooke/Olson	AD 1–500	1500	1/8
Natá	Village or town	1970	Cooke	Cooke	AD 1300–1520	1500	1/4

1 A sloping sieve was used.
2 McGimsey's 1956 trenches were excavated without sieves.
PSM members of the Proyecto Santa Maria (Cooke & Ranere 1984).

samples from Ranere's test pits comprise the basal red zone ('R.Z.') [4900–4000 BC] and the overlying brown zone ('B.Z.') [4000–3000] (Ranere 1980).

Monagrillo is a smaller shellmound than Cerro Mangote. It was used by humans as early as 2400 BC, but witnessed its most intensive occupation between about 1700 and 1000 BC (Ranere, quoted in Cooke 1984a, p. 273). At this time, it was close to the active shore and at the mouth of a small river (the Parita). The faunal sample was collected by Ranere in 1975 in two small test pits and was analysed in preliminary fashion by E. Wing (Ranere & Hansell 1978).

Zapotal seems to have been a temporarily or seasonally occupied settlement located along an ancient beach line. A series of small and apparently ephemeral dwellings are being excavated by Giausserand. The site was occupied between about 2000 and 1500 BC. The faunal sample reported here comes from a 2 × 1 m test pit dug in 1984. It has only been partially analysed (only bones identifiable to the generic level are included in the figures).

La Mula–Sarigua was a sizeable settlement, probably a permanently occupied village, by the beginning of the 1st millennium BC. The sample here presented was recovered in a 2 × 1 m test pit excavated at the site by R. Cooke and A. Blanco in 1986. It has a single ^{14}C date 220 ± 90 BC (Beta–18863; shell, corrected for ^{13}C/^{12}C fractionation). At this time, the site was on a low ridge overlooking the sea, at the north of the Parita river.

Sitio Sierra, located on a flood-free knoll 12 km upstream from the mouth of the Santa Maria river, attained a maximum extent of 45 ha, probably between AD 500 and AD 700. Excavations conducted here in 1975 identified houses made of palm thatch and cane walls. The faunal samples we consider come from house features and rubbish-dumps whose ^{14}C dates fell between 65 ± 80 BC and AD 475 ± 110 (Cooke 1979).

The sample of nucleated settlements is completed by Natá (NA–8), arguably the largest town on the Pacific watershed at the time of the Spanish conquest (Cooke 1979). It is situated about 10 km from the coast, on the banks of a small river (the Chico). The bone sample was recovered from a 2 × 1 m test trench excavated in 1970 (Cooke 1972). Associated ceramics suggest a date of between AD 1300 and AD 1520.

The samples listed above give us information on the non-fish vertebrate taxa that were either used for food and other domestic purposes or were commensals discarded on middens. To complete the regional picture of faunal utilization, we also offer brief comments on materials from grave contexts which reflect the hunting of animals for ritual rather than diet.

Dietary contribution versus frequency of capture

In archaeofaunal analysis it is important to distinguish between dietary contribution and frequency of capture. Recent compendia of archaeofaunal techniques (e.g. Klein & Cruz–Uribe 1984) put undue emphasis on the use of large mammals as food, and ignore smaller fare. It is true that many

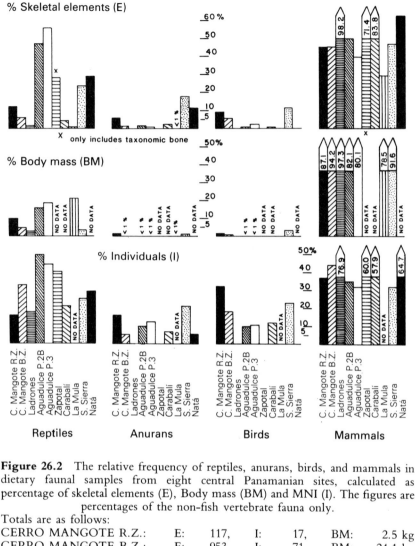

Figure 26.2 The relative frequency of reptiles, anurans, birds, and mammals in dietary faunal samples from eight central Panamanian sites, calculated as percentage of skeletal elements (E), Body mass (BM) and MNI (I). The figures are percentages of the non-fish vertebrate fauna only.

Totals are as follows:

CERRO MANGOTE R.Z.:	E:	117,	I:	17,	BM:	2.5 kg
CERRO MANGOTE B.Z.:	E:	953,	I:	71,	BM:	24.1 kg
LADRONES (P.2B & 3):	E:	279,	I:	13,	BM:	4.8 kg
AGUADULCE (P. 2B):	E:	818,	I:	40,	BM:	8.4 kg
AGUADULCE (P. 3):	E:	1134,	I:	47,	BM:	9.0 kg
ZAPOTAL:	E:	21,	I:	10,	BM:	No data
CARABALÍ (P. 2B & 3):	E:	272,	I:	19,	BM:	No data
LA MULA:	E:	778,	I:	ND,	BM:	1.4 kg
SITIO SIERRA (one feature only):	E:	1231,	I:	105,	BM:	53.2 kg
NATÁ:	E:	68,	I:	17,	BM:	No data

Neotropical hunting groups that live close to productive forests acquire most of their meat through the hunting, by males, of large- and medium-sized mammals and birds (e.g. Hill & Hawkes 1983); smaller animals are simply ignored. Where mammal biomasses are low, however, or where an unpredictable environment leads to periodic shortages of terrestrial organisms, small and apparently unpalatable fare (such as frogs and toads, snakes, mud turtles, rats, and passerine birds) can be important dietary elements, especially when they are seasonally abundant and/or gregarious. Weiss' recent work on the Campa of Peru has shown that this group regularly consumes several species of toxic amphibians (Cooke in press). Berlin & Berlin (1983, p. 306) report that 30 of the 85 species of reptiles and amphibians that are recognized by the Aguaruna and Huambusa of Peru, are actually eaten by them. In terms of the long-term dietary well-being of a community, the regularity of animal meat and fat consumption is more important than feasting on the products of occasional hunting trips (see Hugh–Jones 1979, pp. 170–80 for a succinct discussion of the unpredictability of male food acquisition versus the stability of female production in tropical Colombia).

In order to give a broad, if somewhat crude, impression of the discrepancy between dietary contribution and frequency of capture in the Parita Bay samples, we have presented, in Figure 26.2, the relative abundance of reptiles, anurans, birds, and mammals in the non-fish vertebrate bone samples at eight sites, expressed in terms of (a) number of skeletal elements (E); (b) body mass (BM) (calculated by skeletal mass/body mass allometry, Wing & Brown 1979); and (c) minimum number of individuals (I). All fragments attributable to each class have been included, except in the case of Zapotal where only bone identifiable to the generic level has been included.

The body mass conversions give the impression that mammal meat represents a far more important contribution to the regional diet, at all periods of time, than do reptiles, anurans, or birds. The minimum numbers of individuals, on the other hand, indicate that these three groups together make up a higher percentage of individuals taken than mammals, at the three sites with the largest samples: Cerro Mangote, Aguadulce, and Sitio Sierra. In a single large refuse-dump at Sitio Sierra, for example, reptiles, anurans, and birds each make up over 20 per cent of the non-fish vertebrates. At Aguadulce, the most frequently taken non-fish taxa are small turtles of the genera *Kinosternon* and *Chrysemys*.

The dietary contributions of fish and other aquatic resources

Any evaluation of the harvesting of terrestrial organisms should take into account the availability and productivity of fish, and of coastal and aquatic resources in general. Fishing and the collection of molluscs and crustaceans were undoubtedly profitable enough at several Parita Bay sites to have affected the scheduling, duration, and geographical range of hunting trips. Unfortunately, the data on the dietary contribution of shellfish and crustaceans have not yet been fully quantified (though see Hansell 1979 for a detailed analysis of the mollusc fauna at Monagrillo). In Figure 26.3, we present the proportions of fish skeletal elements (E) and body mass (BM) relative to those of the non-fish vertebrates. Quite understandably, the general trend is for fishing to decline in importance the further one goes from the coast. Thus, at Vampiros, the site nearest the coast, 99.6 per cent of the elements are fish, while at Carabali, the site furthest from Parita Bay, the value is a mere 5 per cent. This last site is located near several streams and a small river (the Gatú); nevertheless, the cool, fast-flowing water supports a low fish biomass and sustains few species (Cooke 1986).

The one site where the trend seems to be reversed is Sitio Sierra, located near freshwater 12 km from Parita Bay. Here fish comprise 97 per cent of the elements and 59 per cent of the body mass. By way of comparison, the Preceramic B Cerro Mangote site, which was situated much nearer the active shore, was less reliant upon fish: in the basal red zone, fish contribute 46 per cent of the body mass.

Natural environments, habitat modification, and human behaviour

We will now consider some aspects of the taxonomic composition of the central Panamanian non-fish vertebrate fauna, in an attempt to identify the distributions that might have been affected by natural environmental conditions, those that imply anthropogenic disturbances of some kind, and those that could have been influenced by less tangible cultural factors, such as socially induced restrictions upon hunting, and weapons technology.

In Figure 26.4, we present the relative abundance of 12 mammalian taxa, calculated according to their percentage minimum numbers of individuals from seven Parita Bay sites, La Pitahaya, and Cerro Brujo. Figure 26.5 gives the percentages of ten taxa in the herpetofaunal samples from three Parita Bay sites and Cerro Brujo. Table 26.2 summarizes the avian fauna from Cerro Mangote. (See Cooke 1984b for a discussion of the Sitio Sierra avifauna and Cooke 1984a and Cooke & Ranere in press for complete mammalian, avian, and herpetofaunal species lists.)

The white-tailed deer (Odocoileus virginianus)

The white-tailed deer is the only mammal species which occurs in all the Panamanian samples, both Atlantic and Pacific. The archaeological distributions show that it maintained its population levels along the Pacific watershed, from Cerro Mangote's occupation until the Spanish conquest, thus confirming ethnohistoric sources, which indicate that it was still locally abundant in the 16th century. At Cerro Brujo, it was taken less regularly (5.7 per cent of the MNI). The brocket deer (Mazama americana), which shuns open conditions, is present at this site, but is absent from all the Pacific-side samples.

Linares' scenario of 'garden-hunting' could be applied to the Parita Bay

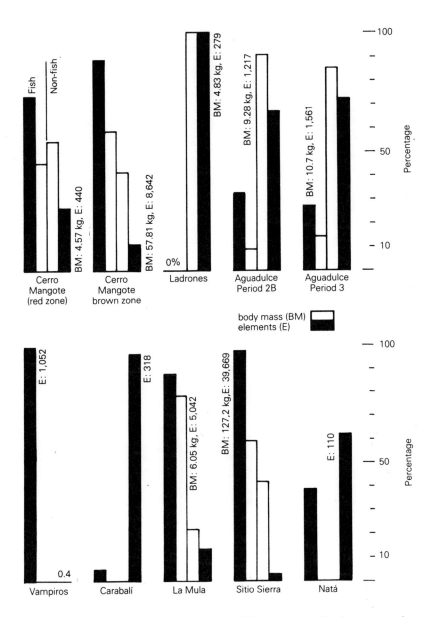

Figure 26.3 The proportions of fish to non-fish remains in the dietary vertebrate faunal samples of eight central Panamanian sites, calculated as the percentage of the body mass and skeletal elements. (Fish: the two bars to the left; non-fish: the two bars to the right.) The Sitio Sierra figures refer to the same feature as that in Figure 26.2 (date: ± AD 300). The Ladrones and Carabalí samples combine Period 2B and 3 levels.

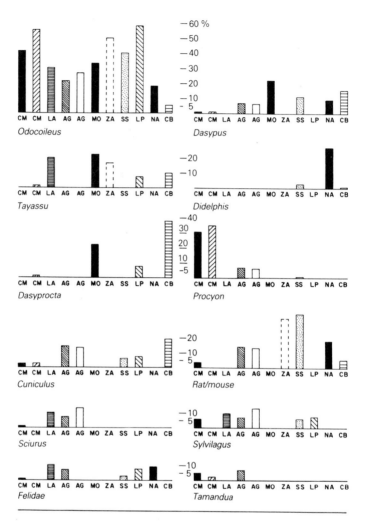

Figure 26.4 The relative frequency of mammalian taxa at seven Parita Bay sites, La Pitahaya, and Cerro Brujo, expressed as the percentage of the mammalian MNI (I) calculated for each context. CM 1 (left): Cerro Mangote, 1979 excavations, three features; CM 2 (right): Cerro Mangote, 1956 excavations, all features; LA: Ladrones, Period 2B & 3 levels; AG 1 (left): Aguadulce, Period 2B levels; AG 2 (right): Aguadulce, Period 3 levels; MO: Monagrillo; ZA: Zapotal; SS: Sitio Sierra, four features; LP: La Pitahaya; NA: Natá; CB: Cerro Brujo. Total I values: CM 1 = 80, CM 2 = 74, LA = 10, AG 1 = 14, AG 2 = 15, MO = 9, ZA = 6, SS = 97, LP = 12, NA = 11, CB = 108. 'Rat/mouse' represents the following taxa: CM: *Liomys adspersus*, AG: *L. adspersus, Zygodontomys brevicauda, Oryzomys* sp., *Sigmodon hispidus*; ZA: *Z. brevicauda*; SS: *L. adspersus, Z. brevicauda, O. cf. concolor;* NA: undetermined; CB: *Hoplomys & Oryzomys.*

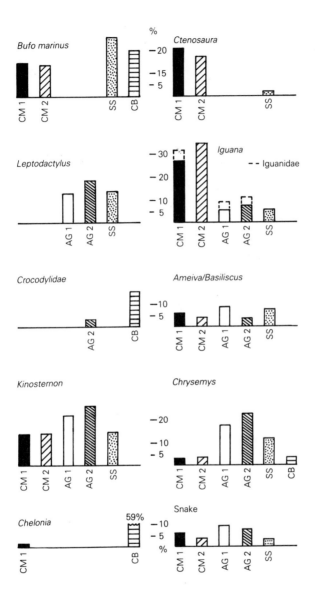

Figure 26.5 The relative frequency of ten taxa of reptiles and anurans from Cerro Mangote (CM), Aguadulce (AG), Sitio Sierra (SS), and Cerro Brujo (CB), expressed as the percentage of MNI (I) herpetofaunal samples only. CM 1: 1979 exc., all features; CM 2: Brown Zone, three features; AG 1: Period 2B levels; AG 2: Period 3 levels; SS: single feature (same as Figs 26.2 & 26.3); CB: all features. Total I values: CM 1: 67, CM 2: 29, AG 1 = 23, AG 2 = 27, SS = 50, CB = 50.

deer populations: in spite of rising human population densities (Weiland 1984) and increasing hunting pressure and deforestation, this species would have been favoured by the exponential disturbance of the primary forest cover (which it is unsuited to) and by the expansion of agricultural fields in different stages of regrowth (Bennett 1968). No more complicated an explanation may be necessary to explain the persistence of *Odocoileus*. In an unpublished paper, however, Cooke (1978) suggested that some kind of cultural management might have influenced the deer populations in the Parita Bay region. There is documentary evidence for deer *not* being consumed in one chiefdom (Parita), while in an adjacent territory (Natá) large supplies of dried and salted deer meat in special storehouses were encountered by the Spanish (Cooke 1979 and references therein).

In this competitive and politically unpredictable society, potlatching meat supplies is understandable behaviour: social controls over hunting the one available large mammal would protect stocks to ensure sufficient meat for periodic feasts. Another explanation could be that, in Parita's chiefdom, deer were tabooed in deference to cognitive prohibitions that are common among South American Amerindians today (Redford & Robinson 1987) and which were, until recently, followed by the Kuna of eastern Panama.

In contrast to *Odocoileus*, the distribution of the other mammals in the Parita Bay samples is more erratic. In some cases, a particular species' abundance can be explained simply as a function of the availability of habitats; in others, human predation would seem to be responsible for low representations.

One species, for example, whose presence and absence would have been conditioned by the proximity of a particular habitat, is the raccoon, *Procyon lotor*, which is the second commonest mammal at Cerro Mangote, where it represents over 25 per cent of the MNI. Raccoons are especially abundant in Panama along the coast, where they roost in mangroves and feed on mudflats and along beaches.

Other examples of 'serendipitous' hunting associated with local habitat availability can be seen in the herpetofaunal and avian samples. *Ctenosaura*, the black iguana, which in Panama prefers coastal habitats, is predictably more frequent at Cerro Mangote than elsewhere. The bird taxa recorded at this site (Table 26.2) can all be captured along the shore, in mangroves or in xerophytic scrub woodland. At Sitio Sierra, the archaeological avifauna includes several species, such as the white-tailed nightjar (*Caprimulgus cayennensis*), the Aplomado falcon (*Falco femoralis*) and the crested bobwhite (*Colinus cristatus*), whose present-day distribution throughout tropical America coincides with seasonal aridity and open habitats (Cooke 1984b).

In the above cases, the tenets of optimal foraging can be applied to explain the archaeological distributions: the Pre-Columbian hunters simply focused on those taxa which were easiest to obtain near their settlements.

Table 26.2   The avifauna from Cerro Mangote (5000–3000 BC). The figures have been aggregated from 18 different contexts identified by Ranere in the 1979 excavations (identifications by S. Olson and R. Cooke★).

Taxon	English name	I	% I
Egretta c.f. *caerulea* or *tricolor*	little blue or tricoloured heron	1	4
Eudocimus albus	white ibis	8	32
c.f. *Tringa melanoleuca*★	greater yellowlegs	1	4
Catoptrophorus semipalmatus	willet	5	20
Calidris canutus	knot	1	4
Calidris mauri or *pusilla*	western or semipalmated sandpiper	1	4
Calidris sp. indet.		+	
c.f. *Numenius phaeopus*	whimbrel	1	4
Geotrygon montana	ruddy quail-dove	1	4
Columbina talpacoti	ruddy ground-dove	1	4
c.f. *Zenaida asiatica*★	white-winged dove	1	4
Amazona ochrocephala	yellow-lored parrot	1	4
Passeriformes	passerines	3	12
Total		25	100

Caviomorph rodents

The diurnal agouti (*Dasyprocta punctata*) is either very rare or absent in the Parita Bay samples, except at Monagrillo. It is notoriously absent from the large samples at Sitio Sierra. At Cerro Mangote it is represented by one equivocal bone. The nocturnal paca (*Cuniculus (Agouti) paca*) is present at more sites, though it exceeds 10 per cent of the mammalian minimum number of individuals only at the Aguadulce Shelter.

Agoutis are most abundant in forests with plentiful supplies of fruiting trees, such as *Astrocaryum* palms and *Gustavia* (Smythe 1978). On Barro Colorado, they represent 6.1 per cent of the present-day mammalian population (Glanz 1982, Table 2). Around Parita Bay, where dry seasons are longer and more intense than on Barro Colorado, the agoutis' preferred food trees are scarce; hence this species may have had primevally low natural populations in this region. The paca, on the other hand, withstands drier conditions and can be very abundant in gallery forests. Linares (1976) showed that at Cerro Brujo the abundance of these two species was enhanced by the presence of man-made habitats associated with agricultural plots. Their distribution in the Parita Bay samples does not coincide with this pattern. It is likely, then, that human hunting pressure adversely affected these species, with the agouti coming off worse due to its lower population densities.

Peccaries

All the tayassuid bones that preserve distinguishing characters in the central Panama samples are from the collared peccary (*Tayassu tajaçu*).

This species has one of the largest ranges of any living wild ungulate and will thrive in habitats as different as primary rain forest and cactus scrub. Being a herd animal, however, its home range is large (60–800 ha) and its food requirements are prodigious. Its abundance, then, tends to be conditioned by the availability of suitable fruiting plants and by the unbrokenness of cover (Sowls 1984, Donkin 1985). At Monagrillo and Zapotal, both located to the south of the Santa Maria river, *T. tajaçu* represents 22.2 per cent and 16.7 per cent, respectively, of the mammalian minimum numbers of individuals. Just north of the river, however, peccaries have a different distribution: they are absent at Sitio Sierra and are very rare at Cerro Mangote. At Ladrones, *T. tajaçu* represents a quarter of the mammalian fauna. This site is situated at the edge of the foothills, where forest cover probably survived longer than on the coastal plain.

At Cerro Brujo, *Tayassu* represents 10.5 per cent of the minimum number of individuals and can be considered the third species in Linares' garden-hunting scenario. Peccary bones are also well represented at Mayan sites (Pohl & Feldman 1982 and references therein). The Miskito of Nicaragua take more white-lipped peccaries (*T. pecari*) than any other mammal on their long treks into the forest (Nietschmann 1973, p. 166). Various South American hunter-gatherers and horticulturalists prey heavily upon peccaries (Hames & Vickers 1983, Redford & Robinson 1985). Around Parita Bay, the collared peccary probably suffered heavy hunting pressure nearest the largest population centres and in areas where forest cover was least continuous. Even so, the incongruencies between the distributions of this species north and south of the Santa Maria river are difficult to explain.

Primates and other arboreal species

At all Panamanian sites, including Cerro Brujo, arboreal taxa are extremely scarce in midden bone samples. Primates and sloths are absent. Some facultatively arboreal species, like the coati (*Nasua nasua*), which can be extremely abundant in forest-edge habitats, are also absent. Squirrels are represented only at Cerro Mangote, Aguadulce, and Ladrones. The green iguana (*Iguana iguana*), which spends a large part of the day browsing in trees, is absent at Cerro Brujo where suitable habitats must have prevailed near the hamlet. At Cerro Mangote, on the other hand, the green iguana represents more than 20 per cent of the herpetofauna. Later in time, it becomes proportionately scarcer around Parita Bay, probably in response to human hunting pressure.

Dogs and technological limitations

Two aspects of hunting behaviour should be considered when the above distributions are analysed: the use of dogs for hunting and the kinds of hunting technologies that might have been employed by the inhabitants of Parita Bay.

Pre-Columbian communities of the Neotropics possessed dogs, however mangy and feeble they may have seemed to the hound-keeping Spanish (Gilmore 1950). In Panama, canid post-cranial elements are very rarely found in middens and can be usually attributed to the grey fox (*Urocyon cinereoargenteus*), a native species that is common in semi-open areas (Cooke 1984a). Their scarcity implies that they were not used as food, as they were in Mexico (Wing 1978, Hamblin 1984). Nevertheless, large quantities of dog teeth were used to make the necklaces that adorned some Panamanian cadavers (see following section).

At Cerro Mangote, a large canid humerus recovered from the 4th millennium BC 'brown zone' is probably from a hound-like dog (Cooke & Ranere in press). It is difficult to imagine that dogs of this size were *not* used for hunting. Certainly, hunting with dogs was practised by the Maya (Pohl & Feldman 1982). The specialization of certain Central and South American groups, such as the Boruca of Costa Rica and the Waiwai of Guyana, in the breeding of hunting dogs for exchange suggests the continuation of a pre-European custom (Fock 1963, pp. 5, 239, Fernández Guardia 1969, pp. 12–13).

The use of such efficient predators from the Preceramic B onwards could have had a drastic effect on mammalian and large reptile populations (Diamond 1984). The agouti and iguana for example, are very susceptible to dogs because of their diurnal habits and self-destructive escape strategies (Cooke 1979, Smythe pers. comm.). Collared peccaries can be effectively corralled by packs of dogs. In contrast, the fleet-footed white-tailed deer is efficient at escaping from dogs in broken terrain where dense undergrowth is available.

With regard to the scarcity of arboreal organisms, and especially primates, we find it unlikely that the archaeological distributions are due solely to habitat destruction or overhunting. Howler monkeys (*Alouatta alouatta*) survived in forest tracts near the Santa Maria river until recently (Bennett 1968) and there is still a remnant population at the southern edge of the watershed (Cooke, personal observation). Technological deficiencies may have been partly responsible: blowguns, which are the forest hunter's weapon *par excellence* (Yost & Kelley 1983), might never have been used for hunting in central and western Panama, making arboreal species less worthwhile to hunt (Linares 1976). The warriors of the Parita Bay chiefdoms fought the Spanish with shark-tooth studded wooden clubs, stones (with slings?), and spears with spear-throwers. They were not renowned as bowmen. Blowguns were used in Colonial times by the Talamancan groups in the forests of eastern Costa Rica for hunting birds; but their projectiles were clay pellets rather than wooden darts. The three species of highly toxic *Phyllobates* frogs, whose poison is used by the Emberá Chocó in Colombia, are not indigenous to Panama (Myers *et al.* 1978). Hence it is possible that a lack of rapidly acting dart poisons curtailed the effectiveness of the blowgun in prehistoric central Panama.

Exotic materials and exchange

The faunal samples from Parita Bay that we have considered so far come from kitchen middens and house floors. Artifacts made of bone are rare and are limited to deer metapodial needles – probably for net manufacture – Ariid catfish-spine awls, and, at Natá, a single jaguar (*Felis onca*) polished phalanx, which was probably used as an amulet (Cooke 1972).

The dietary and industrial use of animals, of course, only reflects a part of the total human impact upon native faunas. To complete the picture, we should consider, albeit briefly, possible ritual associations between animals and humans, especially as the recorded distributions in burial and other ceremonial sites are somewhat different from those found in domestic contexts (for a more detailed discussion, see Cooke & Ranere in press). The oldest burials, those of Cerro Mangote and Sitio Sierra's early cemetery (250–25 BC; Cooke 1979), contain only shell beads, sting-ray spines, and a human skull (a trophy?). In the later, Sitio Sierra cemetery (*c.* AD 1100; Cooke 1972, 1984b), the burial goods are no more extravagant and represent locally obtained animals: shark teeth, a flute made from a brown pelican (*Pelecanus occidentalis*) humerus, a scarlet macaw (*Ara macao*), a quail skeleton, and the calcined tibia of a paca.

On the other hand, the animal materials associated with the burials at Sittio Conte (Lothrop 1937) and other Pacific-side cemeteries (Ladd 1964, Ichon 1980) emphasize the importance of exotic or scarce items as symbols of social status (Linares 1977). Sitio Conte was used for high-rank individuals between AD 500 and AD 900, for whom the tusks, the teeth, bones, feathers, and skins of certain species were part of a metaphorical attire that could not be easily obtained within the territorial confines of the village.

Some males were adorned with necklaces made of hundreds of dog teeth and peccary tusks. The low frequencies of *Tayassu* bones in the local dietary samples and the sheer quantities of tusks used for the necklaces and aprons at Sitio Conte suggest that peccary teeth were traded into the Pacific plains from the forested regions of the interior. Manatee bone, which is ideal for carving, must also have been imported: sea-cows have not been recorded on the Pacific coast of tropical America during the Holocene. At Cerro Brujo, they were commonly hunted (Linares & White 1980). Sea-turtle remains are also rare in Parita Bay middens; few beaches appropriate for nesting are available along this section of the coast. Hence, it is likely that the whole shells placed in some graves were acquired outside the village territory.

Conclusions

In this chapter we have been careful not to stress a single pattern of causality to explain the proportions of non-fish vertebrates in archaeo-logical bone samples from central Pacific Panama. Some of the samples have low counts and should be interpreted with extreme caution (Grayson

1978). More information is also needed from inland sites.

We have noted that the medium and large mammal species that are archetypically hunted by Neotropical forest peoples – such as tapirs (*Tapirus*), capybaras (*Hydrochoerus*), white-lipped peccary (*T. pecari*), spider- and howler monkeys (*Ateles, Alouatta*), and coatis (*Nasua*) – are conspicuously absent. So too, are large forest birds like curassows, guans, and toucans (Rhamphastidae).

We have followed Linares (1976) in suggesting that the lack of an appropriate hunting technology – such as the blowgun – could account for the absence of some arboreal taxa. In spite of this possible limitation, however, the species composition of the samples suggests that the local hunters did not operate in unbroken tracts of forest, nor did they make periodic treks into distant sylvan habitats (perhaps because the constant threat of abduction or ambush by hostile groups made it too dangerous for them to do so). When exotic materials were needed to satisfy the whims of the local élites, they were probably imported through the tribal kinship networks from outlying villages.

The pattern of 'garden-hunting' described by Linares for the Caribbean slopes is not as easy to identify around Parita Bay as at Cerro Brujo. It is true that the white-tailed deer, naturally abundant in disturbed and semi-open areas, must have benefitted from the clearing of the forest and the expansion of agricultural plots. The erratic distributions of the caviomorph rodents and the collared peccary, however, indicate that the beneficial symbiosis between Pre-Columbian farmers and these prey species that was possible in the wetter, less heavily populated areas of Panama, was not achieved around Parita Bay, where the climate is seasonally arid and where human hunting pressure was more prolonged and intense than on the Caribbean slopes. We have suggested that these taxa may have been naturally scarce in this semi-arid environment.

The consumption of species which have restricted environmental requirements, such as raccoons and shore-birds, in addition to the frequent capture of food of small and less palatable items, such as passerine birds, amphibians, small lizards, and commensal mammals, vouch for an opportunistic element in regional hunting and collecting patterns, and suggest that women and children were responsible for acquiring certain types of meat. We have also proposed (Cooke 1986, in press) that the incorporation into the diet of toxic species like the toad *Bufo marinus*, and improved fishing techniques for small shoaling species with high seasonal biomasses, might have been internal responses to an increasingly depauperate terrestrial mammal fauna. Nevertheless, when one bears in mind the obvious productivity of the coastal biomes of Parita Bay, it is still dangerous to assume that the procurement of animal protein was in crisis. Caution is the better part of valour. We must await larger and geographically more varied bone samples before we can be confident that we have identified correctly the different cultural and environmental phenomena that act upon archaeological faunal assemblages.

Acknowledgements

We would like to acknowledge the contributions of the following archaeozoologists and palaeontologists to the analyses of the Parita Bay faunal samples: Storrs Olson (birds), Elizabeth Wing (the initial study of the Sitio Sierra fish and the Monagrillo samples), James Berry (the Sitio Sierra turtles), William Duellman and Arnold Kluge (anurans), Charles Handley (mammals, especially teeth), and R. Medlock (the 1956 Cerro Mangote deer bone). Laboratory assistance has been provided by: Marcela Camargo, Milton Collazos, Linda Cunningham, Julio Jaen, Gina Maduro, Annette Rolin, Carlota Rios, and Aureliano Valencia. We are indebted to all for their constancy and patience. Finally, we wish to thank Olga Linares, Stanley Rand, Nicholas Smythe, Neal Smith, and many other colleagues at the Smithsonian Tropical Research Institute for their helpful comments about the Neotropical fauna.

References

Andagoya, P. de 1913. Relación de los sucesos de Pedrarias Dávila, en las provincias de Tierra–Firme y de lo ocurrido en el descubrimiento de la Mar del Sur-. . .-(fragment). In *El Descubrimiento del Océano Pacífico: Vasco Núñez de Balboa, Hernando de Magallanes y sus Compañeros* Vol. II, J. T. Medina (ed.), 191–207. Santiago de Chile: Imprenta Universitaria.

Bennett, C. F. 1968. *Human influences on the zoogeography of Panama*. Berkeley: University of California Press.

Berlin, B. & E. A. Berlin 1983. Adaptation and ethnozoological classification theoretical implications of animal resources and diet of the Aguaruna and Huambisa. In *Adaptive responses of native Amazonians*, R. B. Hames & W. T. Vickers (eds), 301–28. New York: Academic Press.

Bird, J. B. & R. G. Cooke 1978. *La Cueva de los Ladrones: datos preliminares sobre la ocupación formativa*. Actas del V Simposium Nacional de Antropologia, Arqueologia y Etnohistoria de Panama, 283–305.

Clary, J. H., A. J. Ranere & P. Hansell 1984. The Holocene geology of Parita Bay, Panama. In *Recent developments in isthmian archaeology*, F. Lange (ed.), 55–83. Oxford: BAR International Series, 212.

Cooke, R. G. 1972. *The archaeology of the western Cocle province of Panama*. Unpublished PhD dissertation, London University, London.

Cooke, R. G. 1978. *Maximizing a valuable resource: the white-tailed deer in prehistoric central Panama*. Unpublished paper presented at the 44th Annual Meeting of the Society for American Archaeology, Tucson.

Cooke, R. G. 1979. Los impactos de las comunidades agrícolas sobre los ambientes del Trópico estacional: datos del Panama prehistórico. *Actas del IV Simposium Internacional de Ecologia Tropical*, Vol. III, 919–73.

Cooke, R. G. 1984a. Archaeological research in central and eastern Panama: a review of some problems. In *The archaeology of lower Central America*, F. Lange & D.Z. Stone (eds), 263–302. Albuquerque: University of New Mexico Press.

Cooke, R. G. 1984b. Birds and men in prehistoric central Panama. In *Recent developments in isthmian archaeology*, F. Lange (ed.), 243–81. Oxford: BAR International Series, 212.

Cooke, R. G. 1986. *Some social and technological correlates of in-shore fishing in Formative Central Panama*. Unpublished paper presented at the 9th Annual

Chac–Mool Conference, University of Calgary, 7–9 November.

Cooke, R. G. in press. Anurans as food in tropical America. *Archaeozoologica*.

Cooke, R. G. & A. J. Ranere 1984. The 'Proyecto Santa María': a multidisciplinary analysis of prehistoric human adaptations to a tropical watershed. In *Recent developments in isthmian archaeology*, F. Lange (ed.), 3–30. Oxford: BAR International Series, 212.

Cooke, R. G. & A. J. Ranere in press. Precolumbian influences on the zoogeography of Panama: an update. *Proceedings of the International Symposium on the Zoogeography of Mesoamerica, Merida, Yucatán*. New Orleans: Tulane University.

Diamond, J. M. 1984. Historic extinctions: a Rosetta Stone for understanding prehistoric extinctions. In *Quaternary Extinctions*, P. S. Martin & R. G. Klein (eds), 824–62. Tucson: University of Arizona Press.

Donkin, R. A. 1985. The peccary – with observations on the introduction of pigs to the New World. *Transactions of the American Philosophical Society*, **75**.

Espinosa, G. de 1913. Relación hecha por Gaspar de Espinosa, alcalde mayor de Castilla de Oro, dada à Pedrarias Dávilla. In *El Descubrimiento del Océano Pacífico: Vasco Núñez de Balboa, Hernando Magallanes y sus Compañeros*, J. T. Medina (ed.), Vol. II, 154–83. Santiago de Chile: Imprenta Universitaria.

Fernández Guardia, R. 1969. *Reseña histórica de Talamanca*, 2nd edn. San José de Costa Rica: Imprenta Nacional.

Fock, Niels 1963. Waiwai: religion and society of an Amazonian tribe. *National-museets Skrifter, Etnogafisk Raekke*, no. VIII. Copenhagen: National Museum.

Giausserand, M. In preparation. *Excavations at the Zapotal site in central Panama*. (Fieldwork for PhD dissertation, Yale University).

Gilmore, R.M. 1950. Fauna and ethnozoology of South America. In *Handbook of South American Indians*, **6**, J. Steward (ed.) (Bulletin of the Bureau of American Ethnology, **143**), 345–464.

Glanz, W. E. 1982. The terrestrial mammal fauna of Barro Colorado island: censuses and long-term changes. In *The ecology of a tropical forest: seasonal rhythms and long-term changes*, E. Leigh, A. S. Rand & D. Windsor (eds), 455–68. Washington DC: Smithsonian Institution Press.

Grayson, D. K. 1978. Minimum numbers and sample size in vertebrate faunal analysis. *American Antiquity* **43**, 53–65.

Hamblin, N. L. 1983. *Animal use by the Cozumel Maya*. Tucson: University of Arizona Press.

Hames, R. B. & W. T. Vickers (eds) 1983. *Adaptive responses of native Amazonians*. New York: Academic Press.

Hansell, P. 1979. *Shell analysis: a case study from Panama*. Unpublished MSc dissertation, Department of Anthropology, Temple University, Philadelphia.

Hansell, P. 1987. The Formative in central Pacific Panama: La Mula–Sarigua. In *Chiefdoms in the Americas*, R.D. Drennan & C.A. Uribe (eds), 119–38. University Press of America.

Hill, K. & K. Hawkes 1983. Neotropical hunting among the Aché of eastern Paraguay. In *Adaptive responses of native Amazonians*, R. B. Hames & W. T. Vickers (eds), 139–88. New York: Academic Press.

Hugh–Jones, C. 1979. *From the Milk River*. Cambridge: Cambridge University Press.

Ichon, A. 1980. *L'archéologie du Sud de la Péninsule d'Azuero, Panama*. Mission Archéologique Française au Méxique, Série III, Mexico.

Klein, R. & K. Cruz–Uribe 1984. *The analysis of animal bones from archaeological sites*. Chicago: University of Chicago Press.

Ladd, J. 1964. Archaeological investigations in the Parita and Santa Maria zones of Panama. *Bulletin of the Bureau of American Ethnology*, **193**.

Linares, O. F. 1976. Garden-hunting in the American tropics. *Human Ecology* **4**, 331–49.

Linares, O. F. 1977. *Ecology and the arts in ancient Panama*. Studies in Pre-Columbian art and archaeology, no. 17, Dumbarton Oaks.

Linares, O. F. 1980a. Ecology and prehistory of the Aguacate Peninsula in Bocas del Toro. In *Adaptive radiations in prehistoric Panama*, O. F. Linares & A. J. Ranere (eds), 57–66. Peabody Museum Monographs, no. 5. Cambridge, Mass.: Harvard University Press.

Linares, O. F. 1980b. Ecology and prehistory of the Chiriqui Gulf sites. In *Adaptive radiations in prehistoric Panama*, O. F. Linares & A. J. Ranere (eds), 67–77. Peabody Museum Monographs, no. 5. Cambridge, Mass.: Harvard University Press.

Linares, O. F. & R. S. White 1980. Terrestrial fauna from Cerro Brujo (CA–3) in Bocas del Toro and La Pitahaya (15–3) in Chiriqui. In *Adaptive radiations in prehistoric Panama*, O. F. Linares & A. J. Ranere (eds), 181–93. Peabody Museum Monographs, no. 5. Cambridge, Mass.: Harvard University Press.

Lothrop, S. K. 1937. Coclé: an archaeological study of central Panama, Part 1. *Memoirs of the Peabody Museum of Archaeology and Ethnology* **7**. Harvard University.

Lothrop, S. K. 1942. Coclé: an archaeological study of central Panama, Part 2. *Memoirs of the Peabody Museum of Archaeology and Ethnology* **8**. Harvard University.

McGimsey, C. R. III 1956. Cerro Mangote: a preceramic site in Panama. *American Antiquity* **22**, 151–61.

McGimsey, C. R. III, M. B. Collins & T. W. McKern 1966. *Cerro Mangote and its population*. Unpublished paper presented at the 37th International Congress of Americanists, Mar del Plata.

Myers, C., J. Daly & B. Witkop 1978. A dangerously toxic new frog (*Phyllobates*) used by the Emberá Indians of western Colombia with discussion on blowpipe fabrication and dart poisons. *Bulletin of the American Museum of Natural History* **161**, Article 2.

Nietschmann, B. 1973. *Between land and water: the subsistence ecology of the Miskito Indians, eastern Nicaragua*. New York: Seminar Press.

Piperno, D. 1984. A comparison and differentiation of phytoliths from maize and wild grasses: use of morphological criteria. *American Antiquity* **49**, 361–83.

Piperno, D. & K. H. Clary 1984. Early plant use and cultivation in the Santa Maria Basin, Panama: data from phytoliths and pollen. In *Recent developments in isthmian archaeology*, F. Lange (ed.), 85–121. Oxford: BAR International Series, 212.

Piperno, D., K. H. Clary, R. G. Cooke, A. J. Ranere & D. Weiland 1985. Preceramic maize in Panama. *American Anthropologist* **87**, 871–8.

Pohl, M. & L. H. Feldman 1982. The traditional role of women and animals in lowland Maya economy. In *Maya subsistence: studies in memory of Dennis E. Puleston*, K. Flannery (ed.), 295–311. New York: Academic Press.

Ranere, A. J. 1980. *Nueva excavación y re-interpretación de Cerro Mangote, un conchero precerámico en la Región Central de Panamá*. Unpublished paper presented at the 'III Simposium Nacional de Antropologia de Panamá', December.

Ranere, A. J. & P. Hansell 1978. Early subsistence patterns along the Pacific coast of central Panama. In *Prehistoric coastal adaptations*, B. L. Stark & B. Voorhies (eds), 43–59. New York: Academic Press.

Redford, K. H. & J. G. Robinson 1987. The game of choice: patterns of Indian and colonist hunting in the Neotropics. *American Anthropologist* **89**, 650–67.

Sowls, L. K. 1984. *The peccaries*. Tucson: University of Arizona Press.

Smythe, N. 1978. The natural history of the central American agouti. *Smithsonian Contributions to Zoology* **257**.

Valerio, W. 1985. Excavaciones preliminares en dos abrigos rocosos en la Región Central de Panamá. *Vínculos* (Costa Rica) **11**, 17–29.

Valerio, W. 1987. *Analisis funcional v estratigrafico de 5F–9 (Carabalí), un abrigo rocoso en la Región Central de Panamá*. BA thesis, University of Costa Rica.

Weiland, D. 1984. Prehistoric settlement patterns in the Santa Maria drainage of central Panama. In *Recent developments in isthmian archaeology*, F. Lange (ed.), 31–53. Oxford: BAR International Series, 212.

Weiland, D. 1985. *Preceramic settlement patterns in the Santa Maria drainage in central Panama*. Unpublished paper presented at the 45th International Congress of Americanists, Bogota, July.

Willey, G. R. & C. R. McGimsey, III. 1954. The Monagrillo culture of Panama. *Papers of the Peabody Museum of Archaeology and Ethnology*, **49**(2), Cambridge, Mass.

Wing, E. 1978. The use of dogs for food. In *Prehistoric coastal adaptations*, B. Stark & B. Voorhies (eds), 43–59. New York: Academic Press.

Wing, E. & A. Brown 1979. *Paleonutrition*. New York: Academic Press.

Yost, J. A. & P. M. Kelley 1983. Shotguns, blowguns and spears: the analysis of technological efficiency. In *Adaptive responses of native Amazonians*, R. B. Hames & W. Vickers (eds), 189–224. New York: Academic Press.

27 Shells and settlement: European implications of oyster exploitation

DEREK SLOAN

A feature of the food-quest which first becomes obtrusive in the north European archaeological record during the 4th millennium BC was the consumption of shellfish and more particularly of *Ostrea edulis, Mytilus edulis, Cardium, Nassa reticulata* and *Littorina littorea* on a scale sufficient to result in the accumulation of substantial shell-mounds. Why this should have happened at this particular period and not for instance during the earlier phases of marine transgression is one of many problems left unsolved (Clark 1975, pp. 192–3).

Introduction

Shell-middens are a world-wide feature of the archaeological record. They are the logical product of any society which exploits the margins of the sea as a major economic resource. Marine molluscs are abundant and easily replenished, and their products have many potential uses (Ceci 1984), although it is probably reasonable to assume that in prehistory their value as a human food outweighed all other uses. Sadly, the early evolution of archaeological thought meant that shell-middens were associated with the concept of 'strand-looping', in which miserable and degenerate societies scavenged a poor living from the seashore in degraded circumstances, and this view of shell-middens exists to this day (Sloan 1985).

Much research on shell-middens has been undertaken in the past twenty years, and ideas have changed radically. The interpretational barometer has swung wildly as shell-middens and their enveloping 'cultures' have been analysed: at the time of writing, it is fair to say that European shell-middens may or may not represent:

(a) seasonal camps of inland hunter/gatherers;
(b) processing stations of essentially coastal societies;
(c) the miserable strand-loopers of tradition;
(d) any combination of the above.

However, these studies have tended to concentrate on individual sites, or long-known groups of sites, and the results of such studies are inevitably insular. This chapter takes a broader view of one group of shell-midden sites, and demonstrates that there may be more important considerations to be evaluated than some researchers have allowed.

The sites under consideration have one common feature – they are massive mounds of the common oyster, *Ostrea edulis*. Three assemblages of such sites have so far been identified: in Denmark, in Ireland, and in Scotland. This chapter concentrates on the Scottish sites, as the other two areas have been fully discussed in other publications.

Early shell-middens

As briefly mentioned above, a traditional view of shell-middens has gone down into archaeological lore. One of the aspects of this has been the assumption that shell-middens belong to the Mesolithic period – that they are the product of essentially nomadic hunter/fisher/gatherer societies. The problem of interpretation is increased by the knowledge that early sites will have been lost in sea-level rises; this is a confusing factor as in some areas isostatic uplift has raised early shorelines well above modern sea-level, while in the majority of maritime Europe almost all of the prehistoric shorelines have been inundated. Generalizations on this topic are dangerous, but it would seem (certainly from Scotland and Ireland) that earlier ('Mesolithic') shell-middens were fairly small, and based on the exploitation of a fairly wide range of resources (Mellars 1978, Coles 1971, Woodman 1978). These sites may or may not have been seasonal coastal camps or, conversely, part of a truly coastal-based economy (Mellars & Wilkinson 1980). Of the molluscan resources exploited, oysters were rare in the site assemblages – although it must always be remembered that the species exploited is largely dependent on the nature of the shoreline under consideration – and the individual sites seem only to have been occupied for periods of, at most, a few hundred years. However, at around 4000 BC, a new phenomenon emerges – massive middens of oyster shell.

Denmark

The classic Ertebølle sites have been well documented, and there is much discussion as to their exact status. Only a brief summary of the physical natures of these sites will be given here.

'Ertebølle' sites come in a range of sizes, and not all 'Ertebølle' sites are shell-middens; they are classically assumed to be the product of a 'Mesolithic' culture. (This is because, when first identified, they lacked the polished stone axes which were taken to denote 'Neolithic'; if recognized at a later period, they would almost certainly have been regarded as 'Neolithic' because they contain pottery.) Only the larger shell-midden sites are of interest to this chapter – those of 2000 m^3 or more of deposits, the main bulk of these being the valves of *Ostrea edulis*. These sites also contain sizeable assemblages of fish, bird, and mammal bone, which in two fairly recent studies (Clark 1975, Bailey 1978) have been calculated as the dominant resources, with oysters representing a fairly minimal level of nutritional value. Other interesting aspects of these middens are that they contain large numbers of fireplaces, many artefacts (although microliths are absent) and pottery; it has also been recorded (Lubbock 1865) that the oyster valves were noticeably larger than modern specimens from the area.

The Ertebølle midden sites can be demonstrated to be part of a large and complex economy, and much work has been done on their analysis and interpretation. Many smaller sites can be linked to the major middens, as shown in the analysis recently undertaken by Rowley–Conwy (1983). The interesting aspect of these sites has little to do with these minutiae, however; it is that the advent of the Ertebølle middens seems to arrive at 4100 BC (and to last at least until 3200 BC), to coincide with a major marine transgression and that the period of the sites' use is also the period of the climatic optimum. Most interestingly, although this period is well within the chronological 'Neolithic', there is no convincing evidence from any of these sites to show that they had a 'Neolithic' economy – in particular, no evidence of arable farming – although pottery and 'lamps' of fired clay are among the type artefacts of this culture.

Scotland

It has long been known that there were shell-midden sites in Scotland; as a result of isostatic uplift, Scotland's early shorelines have been well preserved. Two major middens of oyster shell have been investigated in the past, both in the Forth Valley – Inveravon (Grieve 1874, Mackie 1972) and Polmonthill (Stevenson 1946), and both were presumed to be typical 'Mesolithic' hunter–gatherer sites. The radiocarbon dates recovered by Mackie (1972) of between 4060 ± 180 BC and 2250 ± 120 BC seemed to support this interpretation, although, as Mackie noted at the time, the later dates, for a site without visible evidence of any hiatus in occupation, suggested serious falsities in the standard Three Age system (ibid).

Recent work has identified a total of 18 middens of oyster shell in the Forth Valley. Similar sites exist around Inverness (Gourlay 1980), around Elgin (Lubbock 1865, Sloan 1986), and possibly under Glasgow (Sloan 1982a (Fig. 27.1) Only one of these sites (Nether Kinneil) has been extensively excavated, and even this excavation represents no more than a 4 per cent sample of the site (Sloan 1982b): the neighbouring sites of Polmonthill and Inveravon have been investigated on a superficial level (above), and one further site of this group – Cadger's Brae – has been radiocarbon dated. A small trial excavation of one of the Inverness sites (Muirtown) was undertaken in 1979 (Gourlay & Myers unpublished).

Despite the limited amount of excavation, the sites have an impressive number of common qualities. Almost all the the material in the middens consists of oyster valves, although there are small representations of other common littoreal species of mollusc (*Littorina littorea, Cerastoderma edule, Patella vulgata, Mytilus edulis*), and crabs. All the sites contain copious

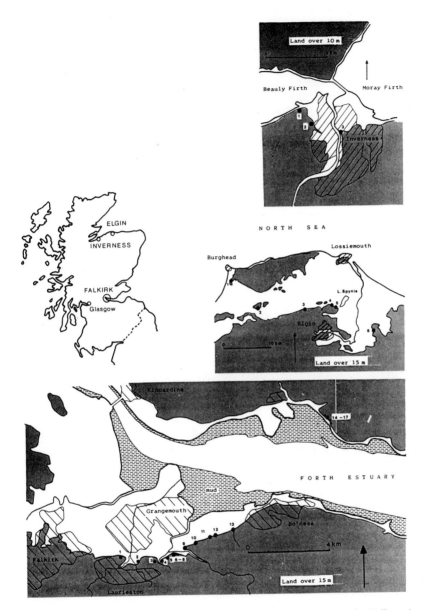

Figure 27.1 Location of 'Forth Valley' type shell-middens. Forth Valley sites: 1 Mumrills; 2 Cadger's Brae; 3 Millhall; 4 Little Kerse; 5 Piggery; 6–8 Polmonthill; 9 Inveravon; 10 Nether Kinneil; 11 East Kerse 2&3; 12 East Kerse 1; 13 Deil's Burn; 14–17 Torryburn. Loch Spynie sites: 1 Bennet Hill; 2 Easterton; 3 Findrassie; 4–5 Spynie; 6 Meft. Inverness sites: 1 Clachnaharry; 2 Muirtown; 3 Bank Street.

burning-levels and many stone-built hearths; in addition the Nether Kinneil site produced vast and enigmatic stone structures with suggestions of a complex and planned layout (Sloan 1986). Animal bone, including apparently domesticated cattle and possibly sheep, was present at Nether Kinneil, although only in very small quantities (about one piece for every cubic metre excavated), and there were small quantities of other materials – a very few stone artefacts, a few small sherds of coarse pottery, one elaborate bone pin, and one shell bead. Most unexpectedly, there is *no* evidence from any of the sampled sites for the exploitation of either wildfowl or fish; although the remains of hazel-nut and edible weeds have been recovered from Nether Kinneil, no trace of cereals has been found.

As a result of geomorphological factors and the limited nature of the investigations so far carried out on these sites, it is very hard to calculate the exact sizes of these middens. The three main Forth Valley sites (i.e. Nether Kinneil, Inveravon, and Polmonthill) probably contain in excess of 3000 m³ of deposits, and the site of Meft on Loch Spynie (originally attributed to the historic period on the basis of very flimsy evidence, see Lubbock 1865, p. 234) also seems to have been very large. (This site may not, strictly speaking, belong to the same grouping, as it does not appear to have been so predominantly comprised of oyster shells; only a re-investigation can make its status certain.) Nineteen ¹⁴C dates from four sites are available; all fall within the period 4060–2200 BC, although it is possible that this date range may be extended in both directions with further excavation. (Note: of the dates, only two, Muirtown and one from Inveravon, are on charcoal; the rest are on marine shell.) No matter what the exact significance of these dates in absolute terms, the important consideration is that, on the evidence available, these sites seem to have been continuously occupied. In the case of Inveravon, at least 1900 years of uninterrupted occupation is suggested, despite a transgressive episode which inundated part of the site.

The available data on vegetation and climate are, unfortunately, sparse; but it is possible to say that the inception of this economy – as with that of the Danish sites – seems to coincide with the climatic optimum and a major marine transgression (they date to the period immediately after the post-glacial maximum transgression), and to have been based on broad and shallow estuarine environments. Although analysis has been hampered by lack of funds, and it is still a long way from completion, it is possible to suggest that at Nether Kinneil the oyster valves show a decrease in size during the period of occupation of the site, and that they were originally of a very large size. This would argue for a selectivity in the process of collection of the shells, and possibly hint at a gradual over-exploitation of the resource. Another interesting observation is that there is evidence to suggest that the end of the use of Nether Kinneil may coincide with large-scale land clearance and the inception of arable farming; there are plough marks cut into the top of the midden at one point, and some parts of the site are covered with thick, burnt deposits, which also coat the attendant buried beach levels. There is a possibility that a similar event may have followed the abandonment of the Muirtown site (Gourlay & Myers unpublished).

Ireland

Recent investigations (Burenhult 1984) of a series of massive shell-middens in Ballysadare Bay, County Sligo (Fig. 27.2) provide interesting parallels with, in particular, the Scottish sites. The analysis of this work is still in progress, and, unfortunately, the earliest evidence would seem to have been lost to marine erosion. It is possible (ibid, p. 41) that the beginnings of a massive exploitation of Ostrea edulis may date from 3700 BC, although the earliest date from the excavated site (Culleenamore) is 2700 BC. The Culleenamore midden is some 100 m long, 30 m wide and up to 3 m deep, and forms part of an almost 3000 m long, more or less unbroken, chain of kitchen middens (ibid, p. 68).

The striking aspect of the Culleenamore site is its similarity to the Scottish middens. The vast majority of the shells in the midden are Ostrea edulis, with some representation of other common littoral species. The oyster valves have been convincingly demonstrated to decrease in size during the occupation of the midden, and there are copious burning levels and hearths within the bulk of the site. Despite the abundance of charred material, no cereals have been demonstrated to belong to the earlier parts of the site (Monk 1984, p. 212). Although one partial cereal grain was found in the later levels, this is most probably reflective of a much later, Iron-age, event. As with the Scottish sites, the bones of fish and wildfowl are absent, although a small sample of the bones of both domestic and wild animals was recovered (cattle, pig, sheep/goat, red deer), together with some artefacts, including pottery.

Other interesting parallels between this site and the Scottish middens lie in the possibility of structures within the mound, an apparently continuous occupation (Burenhult 1984, p. 133), and an association with a period of falling sea-level combined with a warm, wet climate (ibid, pp. 42, 338, Osterholme & Osterholme 1984). The occupation seems to have bridged the conventional 'Neolithic' and 'Bronze Age' periods, and it has been suggested that these middens formed 'home base' sites (ibid, p. 133).

Discussion

All the sites discussed here seem to be associated with periods of falling sea-level following major transgressions, all occur within the 'climatic optimum', and all are situated in shallow estuarine environments. Obviously they have the major attribute of being largely composed of oyster shell, a resource little seen in the archaeological record of earlier periods; they also all have plentiful evidence for activities taking place on the middens, in the form of burning layers, structures, etc. There is also a

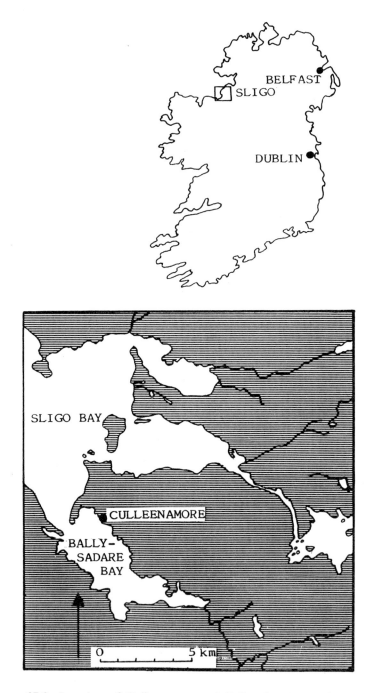

Figure 27.2 Location of Culleenamore and Ballysadare Bay (after Burenhult 1984, p. 23).

constant suggestion of over-exploitation of the basic resource, and uninterrupted occupation. All have evidence of 'Neolithic' activities, in the form of pottery and domesticated animals (although it can be argued that the 'domestic' faunas from the Ertebølle sites are intrusive, this may be a case of wish-fulfilment; concerted efforts were made to deny the existence of such faunas from the Forth Valley sites until the evidence became overwhelming).

Yet these groups of sites are widely separated geographically, and there *are* differences between them: the Ertebølle middens seem to represent more complex economies than the other two groups; however, it must be remembered that the Ertebølle complex has been studied in much greater detail than the Scottish or Irish sites, and that further fieldwork and excavation could fill in the gaps in the record in the other two areas, the fishing and wildfowling camps, etc., all of which are very hard to locate.

It has been variously suggested (e.g. Clark 1975, Woodman 1984, Sloan 1985) that each of these midden complexes represents a mere seasonal aspect of a mobile or semi-mobile economy. However, this may not be the real answer to the question of the status of the sites.

Rowley–Conwy (1983) has produced a powerful argument for regarding the large middens of the Danish Ertebølle as permanent sites, the home-bases of complex (i.e. sedentary) hunter-gatherers. These bases would, logically, be situated near the source of the most reliable food supply, and oysters could be most readily exploited during the lean periods of the year. Dating evidence for other Danish 'Mesolithic' cultures, suggesting that the coastal and inland sites may not be so closely related as previously thought, lends support to this theory (Jensen 1982, p. 64). There is a most attractive proposition here, if only we are willing to accept it; it is simply that we are used to thinking of 'hunters' as nomadic, and only farmers as 'sedentary', yet hunters *can* be sedentary, given sufficient resources. Another of Rowley–Conwy's points (1983, p. 112) that sedentary hunters have a more developed technology, with pottery, etc., also seems apposite, as does the contention that sedentism allows for a greater population. A greater population carries its own inertial force, possibly accounting for both the long occupations of the midden sites *and* the eventual over-exploitation of resources noted in all these midden complexes; yet a seasonally concentrated and non-mobile marine resource such as oysters is less vulnerable than, say, red deer, which might be driven away by the presence of a large and permanent human population.

This is an interesting theoretical approach, but it has only been applied in detail to the Danish sites. However, there *is* evidence from both Ireland and Scotland to fit this thesis. Recent research has shown a total lack of evidence for any sites contemporary with the Forth Valley shell-middens in the hinterland of the area (Sloan 1987); a similar observation has been made in respect of the sites in Ballysadare Bay (Burenhult 1984). Although little weight can be placed on such negative evidence, it seems a little odd that there should be no late 'Mesolithic' or early 'Neolithic' sites known, even from chance finds, in the intensely exploited Forth Valley, especially as there are many sites of Early Bronze Age and later dates known. So it does seem reasonable to suggest that the shell-middens themselves were the main foci of settlement during the 4th and 3rd millenniums BC.

It can be tentatively suggested that there is a common link here, and that a massive exploitation of *Ostrea edulis* made possible the development of sedentary hunter-gatherer systems in these three areas. This is apparently linked to more favourable marine and climatic conditions which began at around 4000 BC, making oysters a richer, more easily available, and more reliable resource. These complexes are restricted in geographical spread because it is only on temperate coasts that the optimal conditions could exist (Rowley–Conwy 1983, p. 118).

Acknowledgements

I am indebted to Paul Mellars and Peter Woodman for discussion of the ideas contained within this chapter. Also to Mrs D. W. Sloan for unfailing support in fieldwork, research, typing, for drawing Figure 27.2, and too many other areas to mention. To all volunteers, staff, and specialists who worked on the excavations and in post-excavation; to Frances Murray, David Devereux, and Patrick Ashmore; and to the SDD (Ancient Monuments), DES, the Russell Trust, the Royal Commission on Ancient and Historic Monuments of Scotland, and the British Archaeological Research Trust for their financial support. Not least to Dr Ted Luxon and Mr John Smith for their unfailing efforts to publicize and raise funds for this research.

References

Bailey, G. N. 1978. Shell middens as indicators of postglacial economies: a territorial perspective. In *The early postglacial settlement of northern Europe*, P. A. Mellars (ed.), 37–63. London: Duckworth.

Burenhult, G. 1984. The archaeology of Carrowmore. *Theses and Papers in North-European Archeology* **14**, 23.

Ceci, L. 1984. Shell midden deposits as coastal resources. *World Archaeology* **16** (1), 62–74.

Clark, J. G. D. 1975. *The earlier Stone Age settlement of Scandinavia*. Cambridge: Cambridge University Press.

Coles, J. M. 1971. The early settlement of Scotland: excavations at Morton, Fife. *Proceedings of the Prehistoric Society* **37**, 284–366.

Gourlay, R. 1980. Muirtown, Inverness. *Proceedings of the Prehistoric Society* **46**, 363–4.

Gourlay, R. & A. Myers unpublished. *A report on the preliminary investigation of a shell-midden*, Muirtown, Inverness.

Grieve, D. 1874. Notes on the shell-heaps near Inveravon, Linlithgow. *Proceedings of the Society of Antiquaries of Scotland* **9**, 45–52.

Jensen, J. 1982. *The prehistory of Denmark*. London: Methuen.

Lubbock, Sir J. 1865. *Prehistoric times*. London: Frederic Northgate.

Mackie, E. W. 1972. Radiocarbon dates for two Mesolithic shell heaps and a Neolithic axe factory in Scotland. *Proceedings of the Prehistoric Society* **46**, 412–16.

Mellars, P. A. 1978. Excavation and economic analysis of Mesolithic shell middens on the Island of Oronsay (Inner Hebrides). In *The early postglacial settlement of northern Europe*, P. A. Mellars (ed.), 371–96. London: Duckworth.

Mellars, P. A. & M. Wilkinson 1980. Fish otoliths as evidence of seasonality in prehistoric shell middens: the evidence from Oronsay (Inner Hebrides). *Proceedings of the Prehistoric Society* **46**, 19–44.

Monk, M. 1984. Charred plant remains, other than wood charcoal, from the Carrowmore excavations. In The archaeology of Carrowmore, G. Burenhult (ed.), *Theses and Papers in North-European Archaeology* **14**, 210–13.

Osterholm, I. & S. Osterholm 1984. The kitchen middens along the coast of Ballysadare Bay. In the archaeology of Carrowmore, G Burenhult (ed.), *Theses and Papers in North-European Archaeology* **14**, 326–45.

Rowley–Conwy, P. 1983. Sedentary hunters: the Erterbølle example. In *Hunter-gatherer economy: a European perspective*, G. Bailey (ed.), 111–26. Cambridge: Cambridge University Press.

Sloan, D. 1982a. A prehistoric site under Duke Street? *Glasdig* **2**, 7.

Sloan, D. 1982b, Nether Kinneil. *Current Archaeology* **84**, 13–15.

Sloan, D. 1985. Shell-middens and chronology in Scotland. *Scottish Archaeological Review* **3**, 73–9.

Sloan, D. 1986. Shell-middens of Scotland. *Popular Archaeology* **7** (1), 10–15.

Sloan, D. 1987. The puzzle of the shell mounds. *Scots Magazine,* **127** (4), 383–9, new series.

Stevenson, R. B. K. 1946. A shell-heap at Polmonthill, Falkirk. *Proceedings of the Society of Antiquaries of Scotland* **80**, 135–9.

Woodman, P. 1978. *The Mesolithic in Ireland*. Oxford: BAR British Series 58.

Woodman, P. 1984. Discussion. In the archaeology of Carrowmore, G. Burenhult (ed.), *Theses and Papers in North-European Archaeology* **14**, 389.

28 *Effects of human predation and changing environment on some mollusc species on Tongatapu, Tonga*

DIRK H. R. SPENNEMANN

Introduction

The Tongan Islands, being part of the western Polynesian triangle Fiji–Samoa–Tonga, were initially settled by people belonging to the Lapita cultural complex. The earliest radiocarbon dates show that Tongatapu was settled around 1500–1300 BC. The earliest sites, pottery-bearing shell-middens, are situated in a narrow band along the accessible shore (Groube 1971, Kirch 1978). As known from the midden deposits, the subsistence of these settlers relied predominantly on the exploitation of molluscs and fish offered by the lagoon and fringing coral reef. Although present, horticulture seems to have played a minor role only. At later times, by the first centuries AD, a change had taken place: pottery production was abandoned and horticulture and domestic animal-breeding (pigs, chicken) increased. In addition the settlements were shifted inland, apparently to be nearer to the gardens.

Most of the archaeological research carried out in Tonga has concentrated on the Lapita period (Spennemann 1986). Detailed studies on the exploitation of molluscs, however, have not been undertaken so far in Tonga, although some approaches have been made.

The problem

Onlookers at one of Jens Poulsen's excavations commented on the size of the excavated *Gafrarium* sp. shells and mentioned that such shells no longer occur in modern Tongatapu (Poulsen 1967, p. 299). This observation resulted in a series of measurements of the main mollusc species *Gafrarium tumidum*, *G. gibbiosum*, and *Anadara antiquata* (Table 28.1). If the Tongans were right in their statements, then two possible reasons for this change in overall size could be presumed: over-exploitation of the resources or change of the local environment. An initial graphical analysis of the length

Table 28.1 Species of shellfish from Tongatapu.

Scientific name	Common name	Tongan name	Habitat
Gafrarium tumidum			
Gafrarium gibbiosum	Venus shell	to'o	brackish water
Gafrarium pectinatum			lagoon
Anadara antiquata	arc shell	kaloa'a	reefs
Anadara cornes			
Ostrea sandvichiense	coral oyster	'sio	reefs
Chama iostoma	jewelbox shell	?	reefs
Turbo chrysostomus	turban shell	topulangi	reefs
Vasticardium sp.	coconutgrater shell	'to'iha	reefs
Periglypta puerpera	hardshell clam	tavatava	muddy sand/brackish

measurements failed to prove any significant reduction in shell size among the *Gafrarium* sample (Poulsen 1967, pp. 299–300), but a statistical re-analysis showed a decrease in size beyond expectation (Poulsen 1984, Spennemann 1985a 1985c).

This chapter discusses the effects of both human predation and changing environment on the predominantly exploited mollusc species *Gafrarium* and *Anadara*, as well as the consequences following this decrease in size for the subsistence of the early Tongans.

Shellfishing in ethnohistoric and modern Tonga is done by the women, who walk at low tide to the reef and mudflats or in the lagoon, search the ground with their toes, and dig out or pick up the shells with their hands. Due to overall size, *Anadara* shells offer more meat than *Gafrarium* and are therefore preferred. *Gafrarium* shells, however, are considered to be more tasty.

Under discussion are the sites excavated by Poulsen in 1963–64 and recently analysed shell samples from other sites. All are situated along the shore of the lagoon. The relative chronology of the pottery, derived from rim forms and decoration, allows clear sequencing of Poulsen's sites and major categorization into early, middle, and later Tongan Lapita (Poulsen 1983). Since the pottery series from the other sites are too small or not sufficiently worked on, the general division into early to late has to be adopted.

The changing environment

Covering 245 km², Tongatapu is the largest island of the Tongan archipelago (Fig. 28.1). It is a flat, tilted coral limestone block with only minor elevations, rising to a maximum height of some 65 m. While the

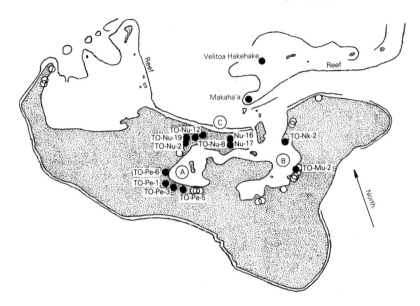

Figure 28.1 Map of Tongatapu, Tonga. Solid black dots: Lapita sites mentioned in the text; circles: other Lapita sites; A: Fanga 'Uta Lagoon; B: Fanga Kakau Lagoon; C: Nuku'alofa–Ma'ofanga peninsula.

northern shore slopes gently into the sea, the southern coast consists of inaccessible cliffs. Modern Tongatapu is dominated by the large Inner Lagoon, which consists of two pockets, Fanga 'Uta and Fanga Kakau.

As shown by Poulsen, based on an analysis of soil samples from site TO–Pe–1 (Crook 1967), the Fanga 'Uta Lagoon extended further inland during Lapita times than it does today. Geological research has shown that Tongatapu was either uplifted as a whole or tilted in its northern part (Taylor & Bloom 1977, Bloom 1980). Assuming that the ancient shorelines are represented by the Lapita shell-middens, that is the water level was 1.5–2.0 m higher than today, then the shoreline at the time of the arrival of the humans must have looked roughly like the sketch drawn in Figure 28.2a. Prehistoric Tongatapu was an island with a wide open bight and scattered islets within. This reconstruction is confirmed by some Lapita shell-middens which lie today within the Nuku'alofa-Ma'ofanga peninsula, and in Lapita times were on small islets (TO–Nu–2, –8, –12, –16, –18, –19). Taylor & Bloom (1977) argued for a former lagoonal entrance in the Nuku'alofa area, as coral heads found *in situ* in the streets of Nuku'alofa were dated to 4000 BP. This described environment closely resembles that typically preferred for Lapita settlement (Jennings 1980, p. 3).

During the tectonic changes, the present Nuku'alofa–Ma'ofanga penin-sula was formed and half of the bight cut off from sea water. This part of the lagoon (Fanga 'Uta) turned into a inner lagoon with brackish water

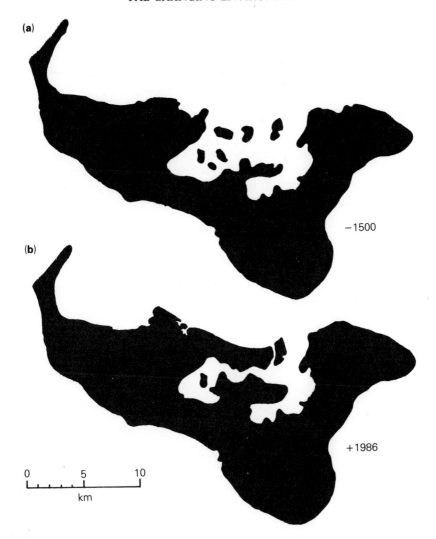

Figure 28.2 Tongatapu, Tonga: (a) coastline at about 1500 BC, (b) coastline at
AD 1986.

and almost no intake of pure sea water from the reef area. This can be
documented by the tidal range, which is currently some 1.2–1.4 m at the
sea side of Nuku'alofa, but at the inner lagoon is a mere 0.2 m (Braley
1979, Belz 1984).

This brackish water favoured species preferring this environment, like
Gafrarium, while putting other species, which are dependent on pure sea
water (i.e. *Anadara*) at a disadvantage. Theoretically this closure of one
half of the lagoon had the following effects on the two major mollusc
species *Anadara* and *Gafrarium*:

(a) while *Anadara* shells became scarce, *Gafrarium* shells increased in number;

(b) the *Gafrarium* shells increased in size, while the *Anadara* shells became smaller.

Proportion of species

The ratio between the lagoon species (*Gafrarium*) and the reef species (*Anadara*) within the shell-middens adjacent to the inner lagoon are set out in Tables 28.2 and 28.3. In general, the trend of changing environments seems to have started prior to the very beginning of Lapita settlements on Tongatapu, since the percentage of *Gafrarium* in the subsoil was already at 45–65 per cent.

As can be judged from the ratio between these two genera in the lagoon entrance site (TO–Nk–2) and in those sites at the inner lagoon where data on the underlying subsoil are presently known (i.e. TO–Pe–1, TO–Pe–5), the living conditions for *Gafrarium* shells were already rather favourable at the beginning of the Lapita occupation and were becoming more favourable during time.

Comparing ratios within the midden horizons of the individual sites (Table 28.2) it is clearly observable that the number of *Gafrarium* shells is steadily increasing. This is not only due to the environmental change, since contemporary adjacent sites show different ratios. It seems that the

Table 28.2 Percentages of *Gafrarium* sp. and *Anadara* sp. shells in various shell-middens from Tongatapu, Tonga with the chronology (after Poulsen 1983).

	Gafrarium	:	*Anadara*	*n*	Period
TO–Pe–6/III	66.9	:	33.1	1227	
TO–Pe–6/I	89.3	:	10.7	1695	
TO–Mu–2/L2 (S 9999)	92.3	:	7.7	26	
TO–Mu–2/L3	66.7	:	33.3	3	
TO–Mu–2/L4	33.3	:	66.7	6	
TO–Mu–2/L5	0.0	:	100.0	2	late Lapita
TO–Pe–5/III	85.0	:	15.0	60	
TO–Pe–1/II	96.0	:	4.0	430	
TO–Pe–3/I	66.0	:	34.0	3383	
TO–Pe–5/II	79.6	:	20.4	235	middle Lapita
TO–Nu–8/L3	86.2	:	13.8	145	
TO–Pe–1/I	92.0	:	8.0	1025	
TO–Pe–5/O–I	70.8	:	29.2	975	
TO–Nk–2	39.6	:	60.4	409	early Lapita
TO–Pe–1/subsoil	67.2	:	32.8	1489	natural death assemblage
TO–Pe–5/subsoil	45.4	:	54.6	379	subsoil

Table 28.3 Percentages of *Gafrarium* sp. and *Anadara* sp. shells in various shell-middens from Tongatapu, with the modern location of the sites (see Fig. 28.1).

	Gafrarium	:	Anadara	n
Reef islet sites				
TO–Vi–1	5.9	:	94.1	34
Lagoon entrance sites				
TO–Nk–2	39.6	:	60.4	409
Open lagoon sites				
TO–Mu–2	75.7	:	24.3	37
Inner lagoon sites				
TO–Pe–3	66.0	:	34.0	3383
TO–Pe–5	73.1	:	26.9	1270
Innermost sites				
TO–Nu–8	82.5	:	17.5	240
TO–Pe–1	93.2	:	6.8	1455
TO–Pe–6	79.9	:	20.1	2922

exploitable resources of *Anadara* shells were quickly depleted in the area. If this is true, then we can argue that neighbouring populations (like TO–Pe–3 and TO–Pe–5, roughly 1 km apart) exploited different shell beds. A detailed analysis and further sampling of the site might provide additional data and enable reconstruction of exploitation areas.

Shell size

The greatest length of all *Gafrarium* and *Anadara* shells derived from shell columns sampled by J. Poulsen was measured (see Fig 28.3). His data had been recorded to the nearest 0.25 cm and therefore vary slightly from the recently gathered data taken to the nearest millimetre. In addition to the data from archaeological contexts, some spits at TO–Pe–1 were sunk into the subsoil, providing information on the natural death assemblage before the arrival of humans. The other end of the sequence is represented by a market catch of *Gafrarium* shells bought in summer 1985 by the author (Spennemann 1985*b*). The recent background sample for *Anadara* was provided by a sample from a modern shell-midden (1970s) on Makaha'a Islet, offshore of Tongatapu (Spennemann unpublished fieldnotes 1985).

The interpretation of the data for the *Gafrarium* is straightforward as there is a steady decrease in size from early Lapita layers onwards. While the decrease from early to middle, and from middle to late Lapita is highly significant, the decrease from late Lapita to the recent sample is not. This shows that the major change in size had already taken place during Lapita times. In addition to the overall reduction of length, the modern *Gafrarium* shells are not as thick-walled as those found on the Lapita sites. Human activity – predation – is clearly observable. Instead of becoming larger, as might be expected from the favourably changed living conditions, the shells become smaller. A comparison of the natural death assemblage with

 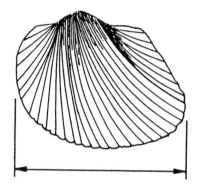

Figure 28.3   Greatest length taken. Left: *Gafrarium* sp.; right: *Anadara* sp.

the early Lapita layers shows that the shells from the archaeological layers are significantly larger. Human selection for the bigger specimens is clearly at work. The recent sample is very significantly smaller than the natural death assemblage, showing that human predation had a long-lasting impact on the *Gafrarium* population.

While these trends work out perfectly for the broad chronological steps, a detailed spit to spit analysis of individual sites gave no clear results. The shell size varies, alternately increasing and decreasing throughout the sequence.

A more complicated pattern arises for *Anadara*. The sequence shows no trend, either in the broad chronological steps, or in the detailed spit to spit analysis, but alternately increases and decreases in shell size.

Comparison of the modern sample and the natural death assemblage beneath site TO–Pe–1 shows that the modern shells are much bigger. Since living conditions for *Anadara* on the reef islet have not changed, this observation reflects the fact that some environmental change had already taken place, affecting the natural *Anadara* population around the area of TO–Pe–1.

That the size of the *Anadara* sp. shells is dependent on the location of the site in the bight/lagoon can be documented by data from an as yet undated but definitely post-Lapita shell sample from Velitoa Hahake (TO–Vi–1), a small islet off Tongatapu. These shells are comparable in size to those from Makaha'a, and the environmental conditions at both sites are similar.

Shells originating from the early Lapita site TO–Nk–2, situated at the modern lagoon entrance, are very significantly larger than both the pooled early Lapita and the natural death assemblages, but are significantly smaller than the samples from Makaha'a and Velitoa.

The lower layers of site TO–Pe–1, also belonging to the early Lapita phase, are almost identical in size to the natural death assemblage derived from the underlying subsoil, but are very significantly smaller than those from site TO–Nk–2.

Another set of chronologically contemporary, but geographically distant, sites shows the same pattern: the late Lapita site TO–Mu–2, situated at the Fanga Kakau lagoon, shows larger shells than site TO–Pe–6, situated at the Fanga 'Utu lagoon. If the *Anadara* shell size is a highly sensitive tool for observing the changing environment, then the size pattern observable within three sites located near to each other allows for dating of the closure of the Nuku'alofa peninsula: the natural death assemblage, the early Lapita layers from TO–Pe–1, and the middle Lapita series from TO–Pe–3 show all similarly big shells. Those from the late Lapita site TO–Pe–6, however, are very significantly smaller. This implies that the Fanga 'Uta lagoon was completely cut off by the end of the middle Lapita period, i.e. about 700–500 BC.

Conclusions

Summarizing these findings, we can conclude that the Lapita population had to face a dwindling supply in large *Anadara* sp. shells, which was caused by both predation and changing environment. The other main species, *Gafrarium*, became continuously smaller in size.

This implies that the same number of shells collected contained less meat than before. If we assume that the Lapita population was growing in number – as can be documented to some extent by the high number of middle Lapita and especially late Lapita sites along the lagoon – then these resources must have become less and less sufficient. In this case, the meat supply had to be topped up by exploiting other shellfish species, in particular those given in Table 28.1. Some clue for this is given by comparing the percentages of *Gafrarium* and *Anadara* shells to all shells in the sample. The amount of other shells increases within the individual middens as well as between the chronological steps. In addition Poulsen (1987) observed an overall decrease of shells compared to the total midden debris (stones, earth, etc.).

The subsistence economy of the later Lapita times suffered from diminishing shellfish returns both in quality and quantity. It has to be asked whether this lack was made good by intensified exploitation of lagoon and reef fish, or whether the horticultural aspect was increased. The giving up of pottery and the introduction of a new food-preparation method, the *'umu*, as well as a significant increase of both pit-digging for presumed food storage, and pig bones as another indicator for horticulture, would seem to support the latter solution, which is further underlined by the relocation of settlements into the inland areas. However, it has to be mentioned that, to date, studies on the exploitation of fish are insufficient to give conclusive evidence.

Implications for further research

The data available to date are not geographically equally distributed throughout the lagoon, as can be seen in Figure 28.1. Further sampling and midden analysis is necessary to provide a broader basis of data. It is necessary to understand the regional component of the shellfish exploitation, as hinted by the ratio between *Gafrarium* and *Anadara*, as well as by the representation of the other species in the later sites.

In particular, data from sites on the Nuku'alofa-Ma'ofanga peninsula are needed for a better understanding of the environmental conditions at the time of the arrival of humans and the environmental change thereafter.

Acknowledgement

I am gratefully indebted to J. Poulsen (Aarhus) who readily allowed reworking of the data gathered by him, and for discussion of the matter.

References

Belz, L. H. 1984. *Nuku'alofa sanitation and reclamation.* Nuku'alofa: Ministry of Health.

Bloom, A. L. 1980. Late Quaternary sea level change on south Pacific coasts: the study in tectonic diversity. In: *Earth rheology, isostasy and eustasy,* N.–A. Moerner (ed.), 505–16. Chichester: J. Wiley.

Braley, R. D. 1979. Penaeid prawns in Fanga'uta Lagoon, Tongatapu. *Pacific Science* **33**, 315–21.

Crook, K. A. W. 1967. Appendix III. Analysis of soil samples from To.1. In *A contribution to the prehistory of the Tongan Islands.* Unpublished PhD thesis by J.I. Poulsen, Australian National University, Canberra.

Groube, L. M. 1971. Tonga, Lapita pottery and Polynesian origins. *Journal of the Polynesian Society* **81**, 278–316.

Jennings, J. D. 1980. Introduction. In *Archaeological excavations in western Samoa,* J. D. Jennings & R. N. Holmer (eds), 1–4. Pacific Anthropological Records 32. Honolulu: Bernice P. Bishop Museum.

Kirch, P. V. 1978. The Lapitoid period in West-Polynesia: excavations and survey in Niuatoputapu, Tonga. *Journal of Field Archaeology* **5**, 1–13.

Poulsen, J. I. 1967. *A contribution to the prehistory of the Tongan Islands.* Unpublished PhD thesis. Australian National University, Canberra.

Poulsen, J. I. 1983. The chronology of early Tongan prehistory and the Lapita-Ware. *Journal de la Société des Océanistes* **39**, 46–56.

Poulsen, J. I. 1984. Analysis of length measurements on *Gafrarium* shells. MS on file, Department of Prehistory, Research School of Pacific Studies, Australian National University.

Poulsen, J. I. 1987. Early Tongan prehistory. *Terra Australis* **12**. Canberra: Australian National University.

Spennemann, D. H. R. 1985a. *A comparative analysis on some Gafrarium sp. and Anadara sp. shells from shellmiddens on Tongatapu.* Osteological Report DRS 33 (1985). Ms on file, Department of Prehistory, Research School of Pacific Studies, Australian National University, Canberra.

Spennemann, D. H. R. 1985b. *Length measurements of a modern comparative sample of Gafrarium sp. shells from Tongatapu, Tonga.* MS on file, Department of Prehistory, Research School of Pacific Studies, Australian National University, Canberra.

Spennemann, D. H. R. 1985c. *A re-analysis of some shell-measurements on Gafrarium sp. and Anadara sp. shells from Tongatapu, Tonga.* Osteological Report DRS 37 (1985). MS on file, Department of Prehistory, Research School of Pacific Studies, Australian National University, Canberra.

Spennemann, D. H. R. 1986. Zum gegenwärtigen Stand der archäologischen Forschung auf den Tonga–Inseln. Ergebnisse und Perspektiven. *Anthropos* **81**. 469–95.

Taylor, F. W. & A. L. Bloom 1977. Coral reefs on tectonic blocks, Tonga Island arc. *Proceedings of the Third International Coral Reef Symposium.* Vol. 2: *Geology.* Miami: Rosentiel School of Marine and Athmospheric Science, University of Miami.

29 *Rocky Cape revisited – new light on prehistoric Tasmanian fishing*

SARAH M. COLLEY and RHYS JONES

The Tasmanian fish problem

Ethnographic accounts indicate that at the time of European contact the Tasmanians did not eat fish. Jones' excavations in two sites, North Cave and South Cave at Rocky Cape, northwest Tasmania, carried out in the 1960s (Fig. 29.1), revealed abundant fish remains in midden layer dated before *c*. 3800 BP (Jones 1971). More recent layers contained no fish bones. A similar pattern was subsequently discovered elsewhere in Tasmania (Lourandos 1970, Jones 1978, Bowdler 1984, Vanderwal & Horton 1984). It was clear that the Tasmanians had previously eaten fish but had stopped sometime between *c*. 3800 and 3500 BP.

After careful consideration of the evidence, Jones proposed a cultural explanation for this behaviour. In his view, the decision to drop fish from the diet did not make economic sense. He saw it as a cultural aberration which survived only because the Tasmanians were an isolated population cut off from competing societies on the Australian mainland.

Jones' views on prehistoric Tasmanian society prompted a storm of discussion which attracted world-wide archaeological attention (for example see Vanderwal 1978, Allen 1979, Horton 1979, Bowdler 1980, 1984, Walters 1981, Sutton 1982, White & O'Connell 1982, pp. 157–70). Fierce criticism was levelled at the political implications of some of Jones' conclusions, such as his contention that Tasmanian society was 'doomed to a slow strangulation of the mind' (1977, p. 203). Other critics questioned his interpretation of the archaeological data, in particular arguing that loss of fish from the Tasmanian diet did in fact make economic sense. Following an initial surge of interest, discussion of the Tasmanian fish problem subsided somewhat. However, reference to the question as an important archaeological issue in two recent publications (Mellars 1984, White 1984) confirms its continuing relevance.

Revived interest in an unanswered question

Currently the question 'Why did the Tasmanians stop eating fish?' remains unresolved. For several reasons we believe it now merits further attention. It impinges on a key area of archaeological interest – the question of

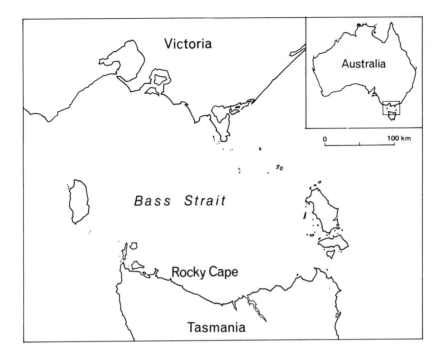

Figure 29.1   Map to show the location of Rocky Cape.

'cultural' versus 'ecological' explanations of human behaviour. The material itself is of great archaeological value due to the nature of the site, which according to Bowdler (1984, p. 2) is 'still unparalleled in Australian archaeology'.

All discussion of the Tasmanian fish question to date has been based on Jones' original analysis of faunal samples from the Rocky Cape cave sites. The work was carried out at a time when little archaeozoology had been done in Australia and when the zoological expertise and comparative collections necessary to identify and fully analyse the whole range of the Rocky Cape fish fauna were not available. Particular problems with the fish bones also arose because the soil was mostly sieved through a 5 mm mesh, now considered too large to retain small bones.

Almost all contributors to the Tasmanian fish debate have taken Jones' results at face value, ignoring any associated methodological problems. White even commented that: 'The facts of the Tasmanian case now seem beyond dispute . . .' (1984, p. 13). On the contrary, new analysis of fish remains from Rocky Cape shows that there is still much to learn.

Samples from Inner Cave, Rocky Cape South

Preliminary results suggest that further analysis of Jones' original faunal samples from the North and South Caves are likely to produce valuable new results. Of even greater potential interest is the material from an undisturbed living floor, with excellent preservation of environmental remains, which was sealed off soon after 6700 BP – the Inner Cave or Enclosed Chamber at Rocky Cape South (Jones 1971, pp. 558–84, Jones 1980). Recognizing the great importance of the site, and that contemporary archaeological methods would soon become obsolete, Jones conducted only limited excavations, preferring to leave the site for future investigation.

One hundred soil samples were taken from transects *c.* 0.3 m wide and 50 mm deep across the living floor (Fig. 29.2). Three-quarters of this soil was sieved through 10 mm, 5 mm, and 1 mm meshes, and preliminary sorting was carried out. One-quarter of excavated soil was retained as bulk soil samples, together with all sieved material and a sample of the residue, as an archive for future analysis. Jones' preliminary study of the site is presented in his thesis (1971), but detailed analysis of faunal remains is yet to be done.

The Inner Cave samples contain many types of fish remains which are absent in the samples recovered by Jones from the main midden sequences. Findings from such a unique site as the Inner Cave are interesting in themselves. They can also provide a check on what may have been lost elsewhere on the site due to worse preservation conditions, and sampling and excavation techniques.

Preliminary results of the fish-bone analysis

Seventeen families or types of fish have been identified so far (Colley & Jones 1987). At least 14 more are present but not yet identified owing to lack of comparative skeletons (Table 29.1). Commonest are bones from all areas of the skeleton of wrasses and leatherjackets. These occur in most samples from both the main midden sequence and the Inner Cave. Most of the porcupinefishes and boxfishes originate from the Inner Cave, represented almost exclusively by dermal spines or dermal plates. Bones of other fishes, represented by vertebrae and a few jaw bones, are found occasionally in both the main sequence and the Inner Cave.

Full discussion must await detailed quantitative data. However, preliminary results show that many more taxa are represented than were at first identified. Jones described fishing based almost exclusively on wrasses (1978, p. 27). Leatherjackets, which are common across the site, were not recognized at the time. Taking differential preservation into account, leatherjackets were probably as common in prehistoric Tasmanian fishing catches as wrasses. The relative numeric importance of other fishes is difficult to assess at this stage. Some types, such as Australian salmon and cartilaginous fishes, have very fragile skeletons and are almost certainly under-represented. The fish listed in Table 29.1 represent a variety of habitats and feeding behaviours, suggesting several different fishing strategies. Species not yet identified may well indicate yet further variety in fishing methods.

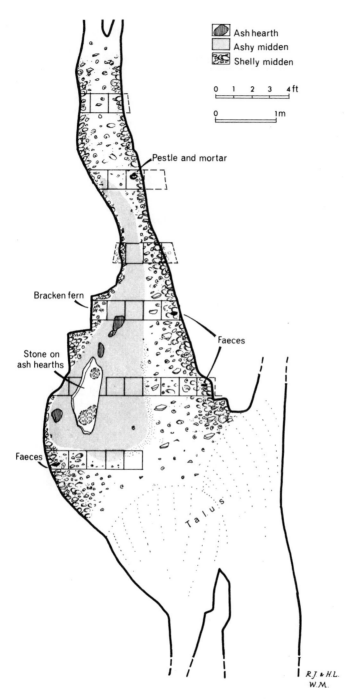

Figure 29.2 Plan of Inner Cave, Rocky Cape South, showing sampled transects.

Table 29.1 Fish identified from preliminary analysis of Rocky Cape samples.

	Ja	He	Sp	Sc	Ve	Frequency
Rocky reef fish						
marblefish (Aplodactylidae)	x	—	—	—	—	occasional
conger eel (Congridae)	x	—	—	—	x	occasional
porcupinefish (Diodontidae)	x	—	—	x	—	common
wrasse (Labridae)	x	x	x	x?	x	very common
leatherjacket (Monacanthidae)	x	x	x	x	x	very common
morid cod (Moridae)	x	—	—	—	x	occasional
ling (Ophidiidae)	—	x	—	—	—	occasional
boxfish (Ostraciontidae)	—	—	—	—	x	common
Bay and estuary fish						
freshwater eel (Anguillidae)	x	—	—	—	x	occasional
Australian salmon (Arripidae)	x	—	—	—	x	occasional
trevally (Carangidae)	—	—	—	x	—	occasional
barracouta (Gempylidae)	—	—	—	—	x	occasional
mullet (Mugilidae)	—	—	—	—	x	occasional
temperate icefish (Notothenea)	x	—	—	—	—	occasional
whiting (Silaginidae)	x	—	—	—	x	occasional
flatfish (Pleuronectiformes)	—	—	—	—	x	occasional
shark, skate, or ray	—	—	—	—	x	occasional

Plus at least 14 other types of fish, all represented occasionally, and all recognized solely by vertebrae. Ja = jawbones; He = other head bones; Sp = spines and fin rays; Sc = scales, scutes and dermal denticles; Ve = vertebrae.

The second conclusion involves differential recovery. Spines of the toothbrush leatherjacket, *Penicipelta vittiger*, are common in the Inner Cave 1 mm residues and bulk samples. They do not occur in the main midden sequence, although other leatherjacket bones are common. The spines, which are extremely small, were presumably missed during excavation and fell through the 5 mm sieves. A similar explanation may account for the differential spatial distribution of boxfish and porcupinefish dermal plates and spines.

Much of the Tasmanian fish debate centres on the relative importance of fish compared to other food resources. Obviously, answering this question will require further study of evidence for other subsistence activities. Likewise, the Tasmanian fish question is only part of a group of archaeologically visible phenomena (e.g. changes in stone-tool types, the disappearance of bone points) which may be related, and are currently receiving renewed study.

'Relative importance' can refer to the relative contribution of each type of food to the diet, which involves calculating relative dietary contribution from relative proportions of food remains in archaeological samples, taking into account factors of site-formation, differential survival of evidence, and excavation bias. Some of the difficulties are already apparent in the above preliminary discussion of the Rocky Cape fish remains.

Table 29.2 Fish taken in simple baited-box traps (based on information supplied by P. Last, C. Turner, and I. Whitehouse).

Fish type	Identified in Rocky Cape samples to date
leatherjackets (Monacanthidae)	Yes
wrasses (Labridae)	Yes
porcupinefishes (Diodontidae)	Yes
cowfishes (Ostraciontidae)*	Yes
morid cods (Moridae)	Yes
lings (Ophidiidae)	Yes
conger (Congridae)	Yes
marblefish (Aplodactylidae)	Yes
sergeant baker (Aulopodidae)	?+
small sharks	?‡
sweeps (Scorpidae)§	No
pufferfish (Tetraodontidea)	No¶

*Very small specimens would escape unless trap was closely woven.
+Could be present but not yet identified, due to lack of a comparative skeleton.
‡Vertebrae of cartilaginous fish have been identified but not yet separated into shark and skate or ray.
§An active swimmer which requires a well-built trap to retain it.
¶Highly poisonous.

A related definition of 'relative importance' involves trying to reconstruct how and when different foods were taken, to try to understand the relative costs, returns, risks, and social implications involved in their exploitation. The reconstruction of past fishing methods is important to this approach.

Fish traps

Table 29.2 lists fish most likely to be taken in a simple baited-box trap (Fig. 29.3, P. Last pers. comm.). Stockton suggested this as a likely method for catching the wrasses originally identified by Jones from the Rocky Cape excavations (1982, pp. 112–13). The suite of fishes most likely to be found together in simple baited-box traps (e.g. wrasses, leatherjackets, porcupinefishes, cods, etc.) are also fish which occur most often in the Rocky Cape samples. Cowfishes or boxfishes, also common in the archaeological material, can be taken in baited box traps provided the mesh size of the trap is small enough to retain them.

The absence of certain fish from the samples is also significant. Two very common species in the Rocky Cape area today – magpie perch (*Cheilodactylus nigripes*) and herring cale (*Odax cyanomelas*) – are never taken in such traps, but are very easily speared while diving (P. Last pers. comm.). The Rocky Cape archaeological fish remains have been checked for these fish, using modern comparative skeletons, but neither has been

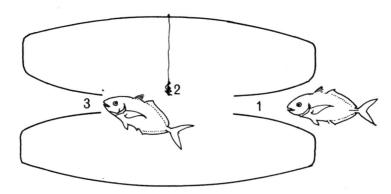

Figure 29.3 Diagram of a simple baited-box trap. Fish swim into the funnel-shaped entrance (1) attracted by the suspended bait (2) and are unable to escape through the narrow exit (3).

identified in the samples to date. Had spearing been an important fishing method these species would be highly likely to occur at the site. Sweep, *Scorpis aequipinnis*, a strong swimmer which can escape from a very simple trap, has also not yet been identified in the samples, despite careful checking against a modern skeleton. Current evidence therefore strongly favours the use of simply constructed baited-box traps for prehistoric fishing at Rocky Cape.

Some fish identified in the Rocky Cape samples, e.g. barracouta (Gempylidae), trevally (Carangidae), mullet (Mugilidae), and flatfish (Order Pleuronectiformes) could not be taken in baited-box traps (P. Last pers. comm.). It seems likely that some sort of tidal fish-trap was also used at Rocky Cape.

A stone-walled tidal fish-trap was initially noted near Rocky Cape by Jones (1971). Several more in the area have since been described by Stockton (1982). All have existed longer than any local resident can remember, and have been used in recent times, but they are otherwise undated. It is most unlikely that any stone fish-trap existing in the area today was used in prehistoric times by Tasmanian Aborigines. However, traps based on similar principles could have been used around Rocky Cape then.

Tidal traps in the area of Jacob's Boat Harbour and Sister's Beach (Figs 29.4 & 29.5) were used in recent times to catch a few specified fish: cocky salmon (young *Arripis trutta* up to about 0.5 kg); black-back salmon (adult *Arripis trutta*); warehou (*Seriollela brama*)[1] and black bream (*Acanthopagrus butcheri*) (I. Whitehouse & C. Turner pers. comm.).

Stockton (1982, p. 112) claimed that similar traps near Penguin, a few kilometres east along the coast, 'retained all the sizes and types of fish that come into the area', and his informant (G. Paine) named at least 19 types of fish caught in these traps, which are, unfortunately, not listed by Stockton. Stockton himself caught three types of fish in a tidal trap near

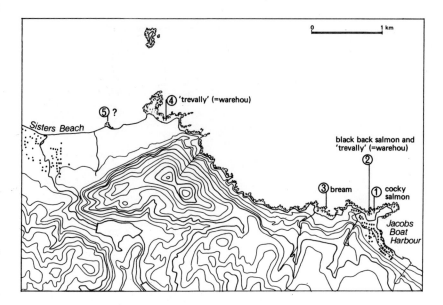

Figure 29.4 Sketch map showing the approximate location of five stone-walled fish traps in the Jacob's Boat Harbour and Sister's Beach area, NW Tasmania, and main catch taken in each.

Table 29.3 Fish taken in stone-walled tidal fish-traps (based on information supplied by P. Last, C. Turner, I. Whitehouse, and published in Stockton 1982).

Fish type	Identified in Rocky Cape samples to date
Australian salmons (Arripidae)	Yes
barracouta (Gempylidae)	Yes
flatfishes (Pleuronectiformes)	Yes
porcupinefish (Diodontidae)	Yes
leatherjackets (Monacanthidae)	Yes
mullets (Mugilidae)	Yes
whitings (Silaginidae)	Yes
wrasses (Labridae)	Yes
trevallys (Carangidae)	Yes
sting rays (Dasyatidae)	?★
warehou (Centrolophidae)	?+
black bream (Sparidae)	No
garfishes (Hemiramphidae)	No

★Vertebrae of cartilaginous fish have been identified but not yet separated into shark and skate or ray.
+Could be present but not yet identified, due to lack of a comparative skeleton.

Figure 29.5 A stone-walled fish-trap at Jacob's Boat Harbour, NW Tasmania (photo Richard Cosgrove).

Penguin: garfish (*Hemiramphus melanochir*), yellow-eyed or freshwater mullet (*Aldrichetta forsteri*), and yellowtail scad (*Trachurus mcchullochi*).

Table 29.3 lists fish which could be taken in tidal fish-traps. The bones of many of these fish occur in the Rocky Cape samples, albeit in small numbers. However, it should be remembered that since the bones of many of these fish are relatively fragile, they are unlikely to survive well in archaeological contexts, and are therefore probably grossly under-represented.

Discovery of a wide range of fish in the Rocky Cape samples has altered the view that prehistoric Tasmanian fishing was restricted to a few species. Ecological and historical information has proved to be useful for suggesting 'suites' of fish likely to be taken together by different fishing methods. The absence of some fish on the site may be as significant as the presence of others. Preliminary results suggest that at least two fishing methods (simple baited-box traps and some sort of tidal trap) may have been used by the Aboriginal inhabitants of Rocky Cape.

Acknowledgements

We wish to thank the following people: Ann Nicholson, Neil Sanderson, and Kindi Smith for sorting soil samples; Barry Blain, Richard Cosgrove, Peter Last, Betty Meehan, Don Ranson, Fran Pease, Eric Staedler, Charles Turner, and Ivor Whitehouse for help in Tasmania; Bill Orders, Alan Scrymgeor, and Nicholas Thorne for supplying fish; Andrew McNee for cleaning fish skeletons; Winifred Mumford for preparing the illustrations.

Note

1 Warehou, which belongs to the Family Centrolophidae, is confusingly called 'trevally' in Tasmania, although it is not a true trevally (Family Carangidae).

References

Allen, H. 1979. Left out in the cold: why the Tasmanians stopped eating fish. *The Artefact* **4**, 1–10.

Bowdler, S. 1980. Fish and culture: a Tasmanian polemic. *Mankind* **12**, 334–40.

Bowdler, S. 1984. Hunter Hill, Hunter Island. *Terra Australis* **8**. Canberra: Department of Prehistory, Research School of Pacific Studies, The Australian National University.

Colley, S. M. & R. Jones 1987. New fish bone data from Rocky Cape, north west Tasmania. *Archaeology in Oceania* **22**, 41–67.

Horton, D. R. 1979. Tasmanian adaptation. *Mankind* **12**, 28–34.

Jones, R. 1971. *Rocky Cape and the problem of the Tasmanians*. Unpublished PhD dissertation, Department of Anthropology, University of Sydney, Australia.

Jones, R. 1977. The Tasmanian paradox. In *Stone tools as cultural markers*, R. V. S. Wright (ed.), 189–204. Canberra: Australian Institute of Aboriginal Studies.

Jones, R. 1978. Why did the Tasmanians stop eating fish? In *Explorations in ethnoarchaeology*, R. Gould (ed.), 11–48. Albuquerque: University of New Mexico Press.

Jones, R. 1980. Different strokes for different folks: sites, scale and strategy. In *Holier than thou*, I. Johnson (ed.), 151–71. Canberra: Department of Prehistory, Research School of Pacific Studies, The Australian National University.

Lourandos, H. 1970. *Coast and hinterland: the archaeological sites of eastern Tasmania.* Unpublished MA dissertation, Department of Prehistory & Anthropology, Australian National University, Canberra.

Mellars, P. 1984. Review of 'J.Flood: archaeology of the dream-time: the story of aboriginal Australia and its people'. *Antiquity* **58**, 232–3.

Stockton, J. 1982. Stone wall fish-traps in Tasmania. *Australian Archaeology* **14**, 107–14.

Sutton, D. G. 1982. Towards the recognition of convergent cultural adaptation in the Subantarctic Zone. *Cultural Anthropology* **23**, 77–97.

Vanderwal, R. 1978. Adaptive technology in South West Tasmania. *Australian Archaeology* **8**, 107–27.

Vanderwal, R. & D. Horton 1984. Coastal southwest Tasmania. *Terra Australis* **9**. Canberra: Department of Prehistory, Research School of Pacific Studies, The Australian National University.

Walters, I. 1981. Why did the Tasmanians stop eating fish: a theoretical consideration. *Artefact* **6**, 71–7.

White, J. P. 1984. Absence of evidence? *The Quarterly Review of Archaeology* **5**, 13.

White, J. P. & J. F. O'Connell 1982. *A prehistory of Australia, New Guinea and Sahul.* Sydney: Academic Press.

30 *Mutualism between man and honeyguide*

ALEX HOOPER

While early man was certainly a hunter, fisher, and scavenger, the story of his relationship with animals in sub-Saharan Africa is by no means fully known. There is evidence that in Egypt successful attempts were made to tame and control gazelle, bubal and other antelopes, Barbary sheep, ibex, and hyenas, and, at Jebel Uweinat, giraffes and ostriches, but none of these species seems to have become fully domesticated in the way that the only known purely African domesticates – the ass, the cat, and the Guinea fowl – did (Shaw 1975, but see also p. 204, this volume). It has been suggested that because domesticates like cow, sheep, goat, and pig, introduced from southwestern Asia and perhaps elsewhere, adapted well to tropical conditions, there was little need to find local substitutes (Ajayi 1971). Also, an abundance of game and plant food in Africa might go towards explaining this lack of domestication of aboriginal animals. Further enlightenment on African approaches to animals might be found by looking at other forms of man–animal interaction, such as man's association with the honeyguide.

The honeyguides, the Indicatoridae, are a family unique among vertebrates in that its members have the ability to digest beeswax. The greater, or black-throated, honeyguide (*Indicator indicator*) has, among other specialities, the behavioural characteristic of attracting man's attention by its calling and movements – it chatters loudly and persistently and flutters a short distance – and, if followed, it will repeat this performance until it has led the man close to a bees' nest. When the man breaks into the nest to take the honey, the bird is rewarded by having access to the broken honeycomb. Apparently, the bird cannot easily break into the bees' nest by itself because it has only a small beak. Nevertheless, species of honeyguide that do not guide have been found to contain beeswax in the stomach. The main food of honeyguides is insects, especially bees and wasps, and it is not known exactly what contribution the beeswax makes to the birds' diet although it has been established that a bacterium (*Micrococcus cerolyticus*) helps the bird to digest the wax.

A mutual-assistance relationship, or mutualism, of the kind just outlined also obtains between the honeyguide and the ratel (sometimes known as the honey-badger or honey-weasel), *Mellivora capensis*. Friedmann reported that baboons also sometimes respond positively to the bird's guidance, but that monkeys and mongooses, although known to be solicited by the bird, have never been reliably observed to co-operate in the venture (Friedmann 1955). No instances of mutualism between honeyguide and chimpanzee or gorilla have been recorded, although it is reasonable to assume that the bird would approach these animals from time to time. Formerly, it was supposed that a discrepancy existed in the ratel–honeyguide association in that the former was known to be active by night while the latter is diurnal. However, Friedmann established that the ratel will operate diurnally and also that it will climb suitable trees in search of honey, although it normally prefers to proceed on the ground.

It is not known whether, in the beginning, man observed the honeyguide–ratel association and then imitated the ratel, whether the bird instigated the partnership with man by its soliciting activities, or whether some other course of events brought about the mutualism between them. Whatever the exact origin of the relationship may have been, some weight is lent to the man-emulating-ratel possibility by the fact that modern man on the honey quest may, in order to stimulate the honeyguide into action, mimic with his voice the ratel's guide grunts and, by knocking on trees, the noises the ratel makes breaking into bees' nests (see Friedmann 1955, whose Zulu guide performed this mimicry and described his intentions). The bird tends to be solitary in its daily habits, not often mixing with others of its own kind, and this together with the brood-parasitism of its reproduction process (i.e. the honeyguides do not construct their own nests or other incubation devices but leave their eggs in the nests of other selected species, and on hatching the infant honeyguide kills any other infants in the nest) suggests that its behaviour in its relationships with the other animals is instinctive rather than learned from its parents or other honeyguides. This instinct may be not specifically directed at man, ratel, baboon, or the other known recipients of the honeyguide's attentions, but rather at any animal the bird meets with at bees' nests, to be reinforced with positive responses by the collaborating species.

A further interesting phenomenon reported by Friedmann is that while the honeyguide may frequent areas of permanent human settlement, such as villages or the edges of towns, it does not try to apply its guidance routine to persons in such settlements. Normally, it approaches individuals or small groups in the open but will also make overtures to them in temporary camps. This suggests that the honeyguide's guidance of man probably was more frequent in times past, before permanent settlement became the normal mode of living for so many peoples in Africa.

Man's relationship with the honeyguide seems complete. There is no need for restraint (such as cages or leashes) or training and continual re-training of the bird, as in the case of man's association with the birds of falconry or with the cormorants involved with the well-known fishing method practised in China (Forde 1950).

It is likely that a very long time would be needed for a bird species to evolve such a complex pattern of behaviour (Ezealor 1981). Man's taste for honey is known to be ancient, being documented in rock art from early times in Europe and Africa, and it is more than probable that the predecessors of Homo sapiens also favoured honey. In view of the low number of millions of years that have elapsed since speciation from the Australopithecines, it could even be conjectured that the honeyguide's mutualism with hominids might pre-date the emergence of the genus *Homo*.

However, this remains conjectural. A counter-hypothesis, equally lacking in material basis, could be put forward that it might be a species of *Homo* that originally activated the relationship with the bird, in that it would not be until man's emergence from the Australopithecines, concomitant with the increase in the size of his brain and intelligence, that he would be able to recognize that the honeyguide (presumably, already in possession of the intestinal bacterium needed for the digestion of beeswax) was an excellent indicator of bees' nests. Thus, man would seek out the bird and, in some unknown way, the bird's current behavioural set became instinctive. Then, *mutatis mutandis*, the bird would seek partners (not necessarily man) for its raids on bees' nests.

Whatever the exact mechanisms may have been, it is clear that man's relationship with the honeyguide is a very ancient one. It is perhaps his oldest surviving partnership in predation.

References

Ajayi, S. S. 1971. Wild life as a source of protein in Nigeria. *Nigerian Field* **36**(3), 115–27.

Ezealor, A. U. 1981. The honeyguide: a partner in an early symbiotic relationship with man. *Zaria Archaeology Papers* **III**, 55–7.

Forde, C. D. 1950. *Habitat, economy and society*. London: Methuen.

Friedmann, H. 1955. *The honey-guides*. US National Museum Bulletin, no. 208, Smithsonian Institution, Washington.

Shaw, T. 1975. *Why 'darkest' Africa? Archaeological light on an old problem*. Ibadan: Ibadan University Press.

31 Cova Negra and Gorham's Cave: evidence of the place of birds in Mousterian communities

ANNE EASTHAM

Introduction

Cova Negra de Bellus at Jativa in Alicante province and Gorham's Cave, Gibraltar, are both sites which have produced a sequence of Mousterian industries. No other sites belonging to the first two phases of the Würm have been so extensively excavated in southern and southeastern Spain, and, although the locations of the two caves appear to be totally dissimilar, cultural comparisons have been made (Waechter 1964, Villaverde 1984) regarding the human occupation of the caves.

In this chapter I will attempt to make a comparison of the remains of birds recovered from the caves and draw some conclusions regarding indications about the environment, local habitat, and exploitation of the birds.

Cova Negra is situated on the Rio Albaida some 3 km south of Jativa and, at the present day, 30 km as the crow flies, or 40–45 km up valley, from the sea (Fig.31.1). At the maximum glacial periods of Würm I this distance may have been doubled as the sea-level was lowered. The cave faces east and is in the side of the gorge through which the Albaida flows at this stage. Fourteen levels of Mousterian occupation have been identified.

Gorham's Cave is situated at the base of the cliffs on the east side of Gibraltar (Fig.31.2) (Eastham 1968). The earliest Mousterian deposits rest on the Monastirian II beach. Above this are eight levels of Mousterian industry, interspersed with sterile layers (Waechter 1951, Zeuner 1953).

Birds in Mousterian communities

The specific identifications of the birds in Würm I show that at Cova Negra there was a variety of duck, all freshwater dabbling species, a number of small hawks and owls, a proportion of partridge, some quail, several thrushes, swifts, and chough. The commonest species throughout was the rock dove, whose bones numbered nearly 56 per cent of the total.

The other most frequent species were mallard, partridge, Alpine swift,

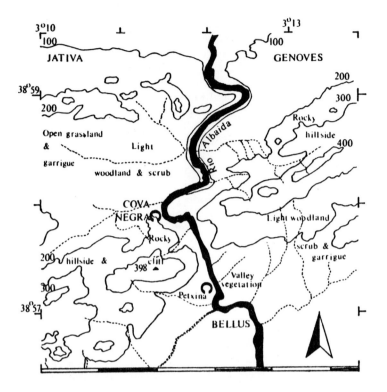

Figure 31.1 Map of Cova Negra.

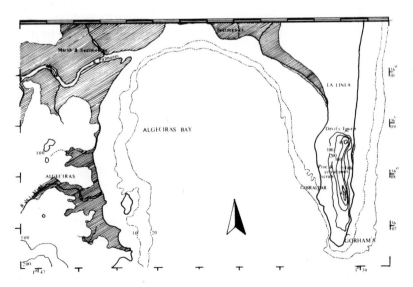

Figure 31.2 Map of Gorham's Cave, Gibraltar.

and chough. All the birds recovered from these levels may be described as 'temperate' rather than 'cold' species.

At Gorham's Cave, during Würm I, the species are different but the pattern remains the same. There is a group of aquatic birds, but these include cliff and coastal species such as the shag, gulls, great and little auks, red-throated diver, coot, and ducks, all of which are able to live in estuarine, lagoon, and coastal waters and some which are more common in temperate and boreal regions today. The predators include Bonelli's and white-tailed eagles, black kite, griffon vulture, and falcon, all of which are compatible with the different localities on the rock. Partridge, rock dove, and chough are the most common species throughout all levels, following the pattern of Cova Negra.

Plant remains from the cave include pine-cones, probably *Pinus pinea*, *Erica*, and box. There are no so-called 'cold' inland species, except a single femur of snowy owl (*Nyctea scandiaca*), which is also found on a number of sites during the Middle and Upper Palaeolithic of south and southwest France. There is also a humerus and ulna of an owl very much larger than the tawny owl (*Strix aluco*) and close in size to that of the Ural owl (*Strix uralensis*).

During the Würm I–II interstadial, because of the rise in sea-level, human occupation ceased at Gorham's Cave; the two birds, raven and chough, could be intrusive or have been sedimented naturally. But at Cova Negra the sequence continues and, in fact, there is no change in the bird population: duck, partridge, pigeon, swift, and chough are the predominant species and there is little to indicate any change of climate. The pattern is the same as during Würm I.

It is in the Würm II deposits of Cova Negra that a single coracoid of shag (*Phalocrocorax aristotelis*) was recovered. Since Cova Negra is quite a distance from the sea, and the shag, unlike the cormorant, is essentially marine, it is difficult to reconcile this species with the environment of the area. To do so demands acceptance of a hunting range of up to 70 or 80 km which implies an expedition of 2–4 days, depending on whether the party was male members only or included women and children. There is always the possibility that the shag at Cova Negra and Gibraltar were individuals kept in captivity and trained for fishing, a supposition that can only be speculative. Alternative possibilities are to question whether the present marine habitat of the species was always its only environment, or whether this specimen was transported back to the cave as an unusual find. For whichever reason it was deposited at Cova Negra, it is interesting to find a shag bone at such an inland location.

The onset of Würm II meant a lowering once more of the sea-level at Gibraltar. But the range of bird species in these deposits is reduced; fewer waterfowl, no waders, and a reduction in the number of predators. The predominant species are partridge, rock dove, and chough, with a few eagle, vulture, kite, and falcon. Even though there are fewer species, the pattern of Würm I is retained, a pattern which is repeated in similar levels at Devil's Tower (Garrod & Bate 1928).

The same may be said in general terms for the Würm II deposits at

Cova Negra, but here waterfowl remain a significant element. There are fewer partridge, in terms of numbers of bones, than in Würm I but a few snipe and woodcock appear as well. There are a number of owls, two bones of which are problematic. One is from level III where a single tibiotarsus of *Strix* sp. appears. Like the Gorham's Cave specimens, it is very much larger than *Strix aluco* and closely matches the dimensions given for *Strix intermedia* by Mourer Chauviré (1975). The other owl bone, a distal end of a humerus from level V, has been identified as eagle owl, *Bubo bubo*, but it is also larger than any recent example of this species. The measurements of this humerus match those given for *Bubo bubo davidii* from the site of L'Escale St. Esteve Janson, also identified by Mourer Chauviré (1975). However, all the material from this site is dated much earlier, to the Mindel glaciation, so on dating grounds the bones are difficult to equate, and the large size of these owls still poses an interesting problem. The other species in these levels present few difficulties. Rock doves, a number of species of thrush and corvines remain abundant and the passerines, swift, and shrike act as seasonal indicators.

At Cova Negra it is the swifts, mainly the Alpine swift (*Apus melba*), which give the clearest indication of a season when humans were certainly present at the cave, although they may have lived there at all seasons. Swifts are extremely useful in this respect, since their passage through an area is limited, within a few days, to dates of arrival during the last weeks of April and of departure during the third and fourth week in August.

Seasonal factors at Gorham's Cave point to a possible winter occupation on this site during the period of Würm I. Here there are no swifts or summer migrants. The best indicators are the wildfowl, among which pochards (Aythyini) and scoters (Melanittini), though not truly seasonal, have a tendency to move south by sea during the late autumn, returning to northerly inland waters to breed. The pine-cones and kernels recovered during excavation (Waechter 1964) also point to an autumn/winter date. There is no such firm evidence for Würm II. The isolated bone of a long-tailed duck (*Clangula hyemalis*) suggests 'cold' conditions but there are no other elements to support it.

The evidence from these two caves and from other sites in southern Spain gives some indication that certain species may have had a slightly more extensive range in Palaeolithic times than is normally ascribed to them at the present, or has been recorded from ornithological observations. For example, the great auk (*Alca impennis*) which was found in Layer K of Gorham's Cave, in Devil's Tower on Gibraltar, in later Magdalenian levels of the Cueva de Nerja (near Malaga), and at Urtiaga in Guipuzcoa, was never, in historic times, recorded in the Mediterranean. Brittany seems to have been its recent southernmost area of colonization in western Europe (Symington Grieve 1895).

The snowy owl (*Nyctea scandiaca*), although not common in Würm II, is found to be far more widespread in southerly latitudes during the middle Pleistocene. It occurs in the Mousterian of Gorham's Cave and on a number of sites in France, mainly in the Rhône valley, Pyrenees, and in Aquitaine. It is normally regarded as a 'cold' species, and in ecological terms should reflect lower summer temperatures. However, there is always the possibility that the frequency of the snowy owl in Palaeolithic deposits was due, in part, to larger populations of the species, which caused their range to be extended over a wider area.

Again, supposing that comparable migration patterns existed, there is some indication that summer temperatures were not significantly lower during the Pleniglacial phases of the Würm. In Spain, Alpine swifts, which were found in equally large numbers in all levels of Würm I, the I–II interstadial, and Würm II at Cova Negra, are also present in contemporary deposits at the Cueva de Zafarraya, near Malaga. In France, likewise, the Alpine swift is found on a number of sites belonging to different phases of Würm: in Würm II deposits at L'Hortus at Valflaunes, Herault, with the red-rumped swallow (*Hirundo daurica*). In the Würm IV deposits of La Balme des Grottes, Isère, on the left bank of the Rhône, the Alpine swift is found in association with large numbers of reindeer and a bird population dominated by 'cold' species. It also occurs as far up the Rhône valley as la Grotte des Romains à Pierre Chatel, Commune de Virignin, Ain, once again in association with 'cold' species (Mourer Chauviré 1975, p. 179). And yet today the northern limit for the Alpine swift is that of the July isotherm of 21.1°C.

Hence there is the possibility that the presence of migrants like the swift can be doubly significant: suggesting not only a seasonal occupation but also indicating a minimum summer temperature for the locality.

Ecological interpretations

The particular taxonomic groupings of birds found in each of these caves depends upon suitable habitat provided by the topography, vegetation, mammalian, and insect resources being available in the vicinity of the cave, each species having its preferred niche within the local pattern.

At Cova Negra (Fig.31.1) the habitat types near the cave at the present day fall into four general categories:

(a) Light woodland and scrub, with *Quercus*, *Pinus*, *Olea*, *Certonia*, *Ficus*, and *Myrtus*; suitable habitats for jay, magpie, shrike, owls, excluding barn owl and Tengmal's owl.
(b) Open land with grasses and herbs, *Dactylis*, *Thymus*, *Rosmaria*, *Jasonium*, and *Erica*; feeding-ground of partridge, quail, thrush, bunting, crow, and chough.
(c) Rocky hillside and cliff, providing a roosting habitat for accepiters, rock dove, thrush, and chough.
(d) River and river-bank at the bottom of the gorge with *Phragmites* and other reeds, and tree cover, in which all waterfowl and many of the woodland and grassland species would be found.

It would not appear to have been necessary to walk more than 5–8 km from the immediate vicinity of the cave in order to capture all or any of the species of bird r . .red in the Mousterian levels.

At Gorham's Cave (rig 31.2) there are essentially three kinds of habitat within 5–10 km of the cave:

(a) The pine and ericaceous scrub on the summit of the rock, which would provide a suitable habitat for the predators, partridge, snowy owl, thrush, and corvid.
(b) The cliff-face and rocks, descending to sea-level around the cave mouth, providing roosting and nesting facilities for shag, gulls, auks, starling, and chough.
(c) The waters of Algeciras bay surrounded by mud flats, marsh, and reed, and connected during the periods of low sea-level to the cave entrance by a broad stretch of beach. Here would be found the waterfowl, divers, heron, stork, oyster-catcher, and here the white-tailed eagle would catch its prey, and gulls would feed.

Hunting of birds in the Mousterian

It appears that the birds found on a number of Middle and Upper Palaeolithic sites were, in general, hunted at a distance from the cave within a maximum radius of 5–6 km. This is certainly true, as may be seen from Figures 31.1 and 31.2, of both Cova Negra and Gorham's Cave. It also holds good for the cave of El Salt at Alcoy, and for Upper Palaeolithic and later occupation sites – la Cueva de Nerja near Malaga, in eastern Spain at Volcan de Faro, and in the northwestern Cantabrian belt – where studies of caves like La Riera, Ekain, Eralla, Atrialda, Urtiaga, and Amalda showed that there were suitable habitats for most of the species found in the deposits within a relatively short walking distance of the cave, 5–6 km being a maximum radius.

The question remains whether all the bird material in the cave was brought in by humans. At Gorham's Cave there is clear evidence that at times when man was in residence, bird remains are relatively abundant and varied. In levels where there is no sign of human occupation, the only bird bones are those of chough. Since this cave was excavated many years ago (1951–54), the precise location of finds was not individually recorded and there is no indication of any area of concentration of microfauna which might demonstrate the activities of an avian predator. At Cova Negra, by contrast, bird bones were recovered from all levels, and the recording of finds was carried out sufficiently precisely to be able to locate the source of all bone material. However, this has revealed no concentrations to suggest an owl roost inside the cave; and only one of the owl species, the barn owl (*Tyto alba*) from the upper levels of Würm II, would have roosted inside the cave.

On balance, it seems that an argument favouring man as the principal hunter on these two Mousterian sites is a valid one. Some species – swift, swallows, choughs, and doves especially – are subject to accidental death, by collision with the cliff-face and other hazards, and small passerines die as a result of other circumstances, but despite these factors there are signs that humans exploited fully the bird resources within an easy distance of their habitation.

Comparative material from other sites on the Iberian peninsula is limited but similar. Devil's Tower, Gibraltar (Garrod & Bate 1928) has an avifauna very close to that of Gorham's Cave. The Cueva de Zafarraya, Malaga, with a Mousterian industry, has an eagle, kestrel (*Falco tinnunculus*), red-legged partridge (*Alectoris rufa*), Alpine swift (*Apus melba*), swallow (*Hirundo* sp.), blue rock-thrush (*Monticola solitarius*), and chough (*Pyrrhocorax* sp.). In the Cova Negra region there are a number of small Mousterian sites – Cochino, Petxina (Villaverde Bonilla 1984), and El Salt – in the cliff-face beside the steep waterfall of the River Serpis descending into the swampy valley below the town of Alcoy. The birds of these places would have had access to a variety of habitats, very like Cova Negra, and the very limited excavation by Pascual produced bones of mallard, partridge, dove, and chough.

In the north, in Cantabria, no bird studies have yet been done on Mousterian sites, except for Amalda Cave, and it is to the important work of Mourer Chauviré in France one must refer for the nearest comparisons. Studies of bird faunas from Würm I–II have been carried out at Grotte du Prince Grimaldi (Boule 1910), the Grotte de l'Observatoire, Monaco (Boule & Villeneuve 1927), in Herault L'Hortus and La Grotte du Salpêtre, and in Aude La Grotte Tournal. In Ariège there is the Grotte d'Aurensan, the Grotte de Balazuc in the Ardèche, Pech de l'Aze and Combe Grenal in the Dordogne.

Most of these sites contained species of mixed climate with some regarded as 'temperate' and some 'cold'; a mixture in which the influence of habitat, mountain, river valley, wetland, forest, or steppe would appear to have had a greater effect on the range of birds than strictly climatic factors.

The checklists from Pech de L'Aze and Combe Grenal show a particular similarity to that of Cova Negra, especially in respect of the small birds and passerines, unusual species such as the Calandra lark, the blue rock-thrush, and shrikes appearing at Combe Grenal and Cova Negra.

However, it may be even more important to note that many of the sites which show intensive occupation during Würm I and II, and are rich as regards cultural remains, are also rich in bird remains in terms of quantity and of variety of species and habitat. This is true of Cova Negra, of Gorham's Cave, and of Combe Grenal. And the rich bird fauna argues that the human inhabitants had a considerable knowledge of bird behaviour and the best means of exploiting it, even though material evidence for the methods of exploitation has not been found. Netting in various forms seems the most likely means of capturing many of the more difficult species such as the passerines, the game birds, and waterfowl, although missiles such as sling stones or bolas have been suggested; yet, above all, knowledge of the birds and expertise in pursuing them must have been the most effective weapon available to the Mousterian hunters of these caves.

There is certainly some suggestion at Gorham's Cave that the Mousterian hunters were making more use of avian resources and transporting bird carcasses from further away than their Upper Palaeolithic successors. The waterfowl, the crane, heron, stork, and other marshland species disappear from the scene altogether after Würm II. Levels B and D, with their Aurignacian- and Magdalenian-type industries, show a much more restricted range of species, and none of those is strictly marshland.

One is drawn, therefore, to the conclusion that despite the fact that birds formed only a minimal part of the diet of these people, the avian material of Cova Negra and Gorham's Cave, along with other sites of the period, does appear to show extensive exploitation of all the bird resources which they could find within easy reach of the homebase.

References

Boule, M. 1910. *Les grottes de Grimaldi: géologie et paléontologie.* Vol. I, Fasc. III: *Les oiseaux*, 299–331. Monaco: Institut Océanographie.

Boule, M. & L. de Villeneuve 1927. La grotte de l'observatoire à Monaco. Archives de l'institute de paléontologie. *Humaine Memoire* **I**.

Eastham, A. S. 1968. The avifauna of Gorham's Cave, Gibraltar. *Bulletin of the Institute of Archaeology* **7**, 37–42.

Garrod, A. E. & D. M. A. Bate 1928. Excavations of a Mousterians rock shelter at Devil's tower, Gibraltar. *Journal of the Royal Anthropological Institute* **58**, 33–113.

Mourer Chauviré, C. 1975. *Les Oiseaux du Pleistocene moyen et supérieur de France.* Unpublished PhD dissertation, University of Lyon.

Symington Grieve of Edinburgh 1895. *The great auk or gare fowl, its history, archaeology and remains.* London: Thomas Jack.

Villaverde, B. V. 1984. *La Cova Negra de Xativa y el Musteriense de la region central del Mediterraneo Espanol.* Obra Editada con la Colaboración del excelentisimo Ayuntamiento de Xativa.

Waechter, J. D'A. 1951. Excavations at Gorham's Cave, Gibraltar. *Proceedings of the Prehistoric Society* **17**(1), 83–92.

Waechter, J. D'A. 1964 Excavations at Gorham's Cave, Gibraltar. *Bulletin of the Institute of Archaeology* **4**, 189–221.

Zeuner, F. E. 1953. The chronology of Gorham's Cave, Gibraltar. *Proceedings of the Prehistoric Society* **19**(2), 180–8.

Index

Table and figure numbers are printed in italics after the number of the page on which they occur.